FUSION ENERGY

FUSION ENERGY

ROBERT A. GROSS

School of Engineering & Applied Science
Columbia University, New York

A Wiley-Interscience Publication

JOHN WILEY & SONS

New York   Chichester   Brisbane   Toronto   Singapore

Library of Congress Cataloging in Publication Data:

Gross, Robert A., 1927–
  Fusion energy.

  "A Wiley-Interscience publication."
  Includes bibliographical references and index.
  1. Nuclear fusion. 2. Fusion reactors. I. Title.

TK9204.G74 1984   621.48′4   84-2220
ISBN 0-471-88470-7

Printed in the United States of America

10 9 8 7 6 5 4 3 2 1

PREFACE

The worldwide effort to develop the fusion process as a major new source of energy has been going on for about 30 years. At the beginning, small groups of scientists, flush with the excitement of success in fission energy, set forth to bring the energy source of the stars—fusion—into practical use. Since then, there have been times of substantial progress and there have been periods of great frustration and despair. We are now close to scientific success.

It is my purpose in writing this book to provide an introduction to the physics and technology upon which fusion power reactors will be based. The contents have been distilled from courses given over the past 23 years by the author and his colleagues to engineering and science students at Columbia University. These students were interested in fusion. They had diverse backgrounds and interests: physics; astronomy; and electrical, nuclear, and mechanical engineering. Some of them had very little knowledge of plasma physics, the science on which fusion depends. Most of them were graduate students and some were professional engineers. Doctoral students specializing in plasma physics took this course to learn how their field of science would be incorporated into a new technology.

Fusion has proved to be a very difficult challenge. The early question was—Can fusion be done, and, if so, how? Great deficiencies in the understanding of the behavior of very-high-temperature gases had to be corrected by broad and sustained research in fundamental plasma physics. As fundamental understanding grew, the nature of the question had changed subtly and significantly. There is very little doubt that controlled fusion can be achieved. Now the challenge lies in whether fusion can be done in a reliable, an economical, and socially acceptable way so that eventually it will win a share of the energy marketplace. Consequently, the new science of plasma physics must now be taken up by the great engineering professions and applied in skillful and imaginative ways to answer this challenge successfully and positively.

Early books on fusion, written in the 1960s, usually set forth some basic plasma physics and described plasma devices and what experimental scientists observed from them. Often there was little direct relationship between theory and experiment. The situation is different now. There are excellent texts in plasma physics, some of which are very specialized and detailed. There is a broad and growing experimental data base. There is a large and growing literature. The role of large-scale computational simulation of plasmas has had a remarkable effect on fusion research. No large and expensive fusion experiment is now built unless extensive computational plasma simulation is used to predict its behavior and to optimize its performance. On the other hand, experiments that explore new plasma-confinement concepts or previously untested plasma conditions can still provide surprises.

With greatly increased plasma-physics understanding and some outstanding experimental successes in laboratories around the world, attention has turned toward engineering aspects of fusion energy. Preliminary designs of numerous types of fusion-power plants have been made, and these have highlighted some of the practical problems of fusion systems. Simplicity, reliability, and maintainability are new challenges for plasma scientists and fusion engineers to conquer.

This book consists of essentially two parts. The first five chapters cover a brief history of fusion, followed by the fundamentals of fusion reaction physics, plasma physics, heating, and confinement physics. The second part begins by describing and analyzing some of the important technology features of a fusion-power plant. This is followed by a detailed description of a commercial fusion-power plant, STARFIRE, based on the tokamak concept. The MARS tandem-mirror fusion-reactor power plant is then described. Finally, there are introductory descriptions of other fusion-reactor concepts. Emphasis has been given to the tokamak concept because, to date, it is the most successful and the most studied concept. Nonetheless, the eventual outcome from the competition among different fusion confinement concepts is not certain. Fusion is a field of research and development that is still rich in ideas.

The complex, interesting, and important interplay between science and engineering is nowhere better illustrated than in fusion. It will require the very best of the engineering professions to take up the concepts and the scientific foundations now unfolding and to forge out of them realistic and competitive power reactors. This book is written to help achieve that goal.

ROBERT A. GROSS

New York, New York
May 1984

ACKNOWLEDGMENTS

Many persons have helped in the development of this book. It is a pleasure to acknowledge contributions made by C. C. Baker, C. G. Bathke, R. Cherdack, C. K. Chu, R. W. Conn, S. O. Dean, T. K. Fowler, H. P. Furth, S. Gralnick, C. Henning, J. Hosea, R. A. Krakowski, E. E. Kintner, W. Kunkel, T. C. Marshall, D. M. Meade, R. W. Moir, G. A. Navratil, F. Ribe, E. Selcow, A. K. Sen, J. Sheffield, T. C. Simonen, W. M. Stacey, Jr., and E. Teller. The manuscript was expertly typed by S. K. Brown, L. Howes, and M. Pierce. Mollie Kauffmann deserves special recognition because she shared her home in Florida with Elee and me during a sabbatical leave when the first draft of this book was written.

This book is dedicated to my wife and best friend Elee, whose understanding, patience, and love made this book possible.

R.A.G.

CONTENTS

FUSION ENERGY

1 | INTRODUCTION AND HISTORY

1.1. ENERGY RESOURCES

Energy has been an essential ingredient in the development of civilization. Large amounts of energy will be necessary for sustaining industrial nations and improving the state of less-developed countries. An adult human requires a minimum of 10^7J/day, that is, 2500 food calories per day or on the average about 100W to sustain life. In the United States today, the per capita consumption of all forms of energy is about 100 times that amount. For the entire world population, energy consumption is about fifteen times that basic energy rate. A nation's energy consumption reflects industrial activity, food production, transportation, heating and cooling of buildings, and production of electricity. A nation requires large quantities of energy to develop sufficient food for its citizens, to provide material resources for industry, and to provide an environment for the well-being of its citizens.

Are there sufficient energy resources available? How long will they last? When will new sources be required? There are many ways to view these questions and a wide range of opinions, depending upon the length of time considered and how one imagines the future. Economics and politics play a substantial role in evaluating these questions. Some insight into the availability of large, reasonably priced energy resources can be obtained by examining the

1

data of Table 1.1, which were obtained from several energy studies [1−3]. Much of the uncertainty involved in projections of this kind is associated with attempting to predict future costs and economic growth patterns in the world. Besides the uncertainty in economic forecasting, it is very difficult to predict new discoveries in hydrocarbon resources. These large uncertainties make it a very speculative exercise to try to precisely forecast a future date when large new energy sources must be available.

To gain some perspective concerning the magnitudes of energies listed in Table 1.1, consider the following. The kinetic energy in the earth's atmosphere, together with the oceanic circulation, is about 10^{21}J. The earth's total recoverable fossil fuel is about 10^{23}J. Nuclear fission, employing breeding, can provide energy of the order of 10^{24}J. Fusion of deuterium is a resource whose potential is 10^{31}J. The solar energy intercepted by the earth is about 5.4×10^{24}J/yr. The heat flux from the earth's interior through the surface is about 10^{21}J/yr, and the energy dissipated in the slowing down of the earth's rotation due to tidal attraction is of the order of 10^{20}J/yr. The nuclear weapons arsenal in the world is estimated to be 10^4-10^5 megatons, the latter figure corresponding to 4×10^{20}J. These data are taken from the book by Sørensen [4].

Nonetheless, we can see from the data of Table 1.1 that in another 25 to perhaps 100 years we will need large new energy sources, because our present energy resources—hydrocarbon fossil fuels and U^{235} fission—will have been seriously depleted. Other available energy resources such as water and wind power, tidal power, geothermal energy, refuse-derived fuel, and biomass can make contributions to our energy supplies. Most forecasts doubt that these sources will make a large contribution. Conservation will help extend the time available for an orderly transition from our present fossil energy sources to new energy sources, but conservation does not change the ultimate conclusion that new energy sources will be needed. Considerable conservation efforts are already evident in the reduced energy growth rate in some industrialized

TABLE 1.1.   Energy from Depletable, Economically Recoverable Supplies[a]

	World Energy (J)	United States Energy (J)
Coal	$(200-400) \times 10^{20}$	$(50-75) \times 10^{20}$
Petroleum	$(30-70) \times 10^{20}$	$(6-12) \times 10^{20}$
Gas	$(20-60) \times 10^{20}$	$(6-12) \times 10^{20}$
^{235}U fission (light water reactor)	$\sim 10^{23}$	$\sim 10^{22}$

[a]The cumulative demand of world energy consumption for the period 1975−2000 is estimated to be about 67×10^{20} J and that for the United States, 23×10^{20} J. Sources: References 1−3.

countries, but this is offset by the substantial rate of growth of those non-industrial nations where rapid economic development is taking place.

Having reached the conclusion that very large new energy sources should, and perhaps must, be developed in the next few decades, we ask the question: What can these sources be? There are three very large and as yet relatively untapped energy sources, which are so large that they are sometimes described as inexhaustible. They are (1) solar energy, (2) fission breeders, and (3) fusion. So far as the energy resource problem is understood, these are the *only large-energy options* available for future generations; science knows of no others. None of these three inexhaustible energy sources is presently developed to the state where it can be considered an economic and reliable large energy source. We will in fact need all three of them, since a healthy, stable society needs multiple energy sources. No nation should be dependent on a single energy source.

It is estimated that to develop any one of these inexhaustible energy sources to the point where it can be considered for large-scale deployment will cost about 20—30 billion dollars. To develop all three to the point of commercialization will involve perhaps 100 billion dollars. This is not a large sum when compared to the trillion dollars predicted for U.S. investment in energy facilities during the period 1980—1990 [5]. Much greater sums will be needed for energy investment in the 21st century. The social cost of having no new energy option available when needed would be a catastrophe.

1.2. NUCLEAR ENERGY

The inexhaustible energy sources all rely fundamentally on energy derived from the nucleus of the atom. Consider the curve shown in Figure 1.1, which displays the binding energy per nucleon as a function of atomic mass number.

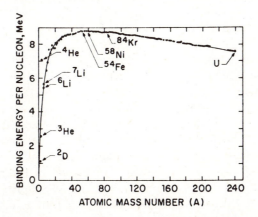

FIGURE 1.1.   The binding energy per nucleon as a function of atomic mass number A. The most stable atoms occur in the range $50 \leqslant A \leqslant 90$. Data taken from Reference 6.

The mass of a nucleon (a neutron or a proton) is found to be slightly different in nuclei of the various chemical elements. By convention, the carbon-12 nucleus is defined to have a mass of exactly 12 amu (atomic mass units) and other nuclei are then compared with it. It is found that free neutrons ($_0^1n$)* and free protons ($_1^1H$) have more mass than any neutrons and protons bound within nuclei. This means that when atoms and chemical elements formed, mass disappeared. It became binding energy in accordance with the formula $E = mc^2$. The nucleons in the middle atomic weight elements, for example, iron (Fe), nickel (Ni), and krypton (Kr), lost the most mass. Such elements are the most stable because any rearrangement of their nucleons requires energy. Breaking up the heaviest nuclei into lighter ones releases energy; this is nuclear *fission*. Combining the lightest nuclei into heavier ones also releases energy; this is *fusion*.

Fusion of the light elements offers the possibility of an essentially inexhaustible energy source that is considered to be relatively safe and environmentally acceptable. Fusion is emerging from its research phase and entering engineering development. The original question—can controlled fusion be achieved—is disappearing from conversations because most scientists are now convinced that it can be done by several techniques. It has not yet been experimentally completely demonstrated, but there are a number of large experiments now being built that are expected to achieve fusion "scientific breakeven," that is, where the power developed from fusion reactions is at least equal to the external power input to the plasma. This important scientific milestone for fusion is expected to be demonstrated within the next few years.

During the past decade, anticipating that scientific success was coming, teams of engineers and plasma scientists made design studies of fusion–electric powerplants. These studies called attention to the important need for fusion system reliability, safety, maintainability, and competitive economics, when compared with other major energy sources. As the science of fusion plasma physics matures, the technology needed for fusion is beginning to be developed.

1.3. FUSION HISTORY

Fusion physics has its origins in the fundamental studies of atomic structure and the desire to understand the energy source of the sun and the stars. Among the earliest experimental achievements that had a large effect on fusion-energy developments are the discovery of deuterium by H. C. Urey and colleagues in 1932 [7], the first pinch-effect experiment of W. H. Bennett in 1934 [8], and early observations of nuclear reactions from colliding deuterons by E. Rutherford and colleagues in 1934 [9]. These latter experiments are truly the beginning of the fusion era. In Rutherford's discussion of the transmutation of elements, he calls attention to the experiments of Oliphant, Harteck, and Rutherford, done at Cambridge, and he discussed for the first time two fusion

*The superscript indicates the number of nucleons and the subscript indicates the number of protons.

reactions that were observed to occur when deuterons of sufficient energy collide. These historic fusion experiments resulted in the very first observation of tritium, an isotope of hydrogen.

It was postulated in 1929 that the energy production in most stars is due to nuclear reactions involving light elements [10], and by 1939 the nuclear-fusion cycle for energy production in our sun had been analyzed in detail by H. Bethe [11]. Thus, by the start of World War II, it was a well-recognized fact that the fusion of light elements represented a very fundamental and very large energy source.

Prior to about 1950, there was very little effort devoted to developing fusion as a useful energy source. In 1942 E. Fermi, while having lunch at the Columbia Faculty Club, suggested to E. Teller the possibility of burning deuterium to develop a large source of energy. Following this suggestion, Teller and E. Konopinsky made a calculation that indicated that the fusion process was not possible! Somewhat later, attempting to rigorously prove that fusion was impossible, they realized that their prior calculation was wrong and that fusion, and in particular deuterium–tritium fusion, was an attractive possibility. This recollection was told by Teller when he revisited Columbia in 1980. Attention was directed toward the use of fusion reactions in nuclear weapons [12]. Uncontrolled fusion in nuclear weapons became a reality with the "George" shot in May, 1951, and a much bigger yield was demonstrated with "Mike", in November, 1952 [13]. Beginning about 1950, organized research efforts to achieve controlled fusion energy began in the United States, England, and the Soviet Union. This work was initially secret because of the concern that fusion reactions could be a large source of energetic neutrons, which, in turn, could be used to breed plutonium, a material used in the manufacture of atomic fission bombs. By about 1955 it was realized that controlled fusion development was going to be a very difficult task. Also, by then there had been developed much easier and cheaper sources of neutrons (from atomic piles), which were available to breed plutonium for military applications. The need for fusion secrecy disappeared. The need for scientific exchange of information also came into sharper focus because of a growing realization that an adequate understanding of some fundamental physics was lacking. Dr. H. J. Bhabha, Chairman of the Atomic Energy Commission of India, in his presidential address to the 1955 Geneva Conference [14], referred to the possibility of using fusion reactions for power production. He also ventured to predict "that a method of liberating fusion energy in a controlled manner will be found within the next two decades." His prediction may well prove correct, since by 1975 several methods, now expected to be successful, were known. Shortly after Bhabha's talk, speculations on the feasibility of thermonuclear power began to appear in journals [15].

In 1956 the Russian academician I. V. Kurchatov, while visiting the British Atomic Energy Research Laboratory at Harwell, presented a paper [16] describing Soviet work on controlled fusion. This research was being carried out under the direction of academicians Artsimovich and Leontovich. Kurchatov mentioned deuterium and tritium as particularly interesting fusion fuels, and he described experimental research on the linear pinch effect, including observations of neutron production. A description of early Soviet fusion research is given in a book written by Melnikova [17]. It was clearly recognized at that time that plasma instabilities were important phenomena. P. C. Thonemann, then deputy chief scientist at Harwell, reviewed [18] the Soviet experiments described by Kurchatov, and stated that "this far-sighted work will undoubtedly go down in scientific history as the first demonstration of nuclear reactions in a gas discharge plasma." But, he further remarked that the Russian workers were well aware that the nuclear reactions that were detected occurred at a rate too great to be accounted for by the estimated temperatures, and are attributed to some unexplained accelerating mechanism. He concluded his review by emphasizing that "an enormous gap lies between this demonstration and the building of a thermonuclear reactor which produces more electrical power than it consumes, if indeed, this is possible at all."

In 1957 J. D. Lawson of the British Atomic Energy Research Establishment at Harwell published a short and famous paper [19], "Some Criteria for a Power Producing Thermonuclear Reactor." Concentrating on only the most essential features, he calculated the power balance in fusion reactors operating under idealized conditions. He found that for a successful power-producing fusion reactor, not only must the temperature be sufficiently high (for deuterium−tritium fuel, $\geq 50 \times 10^6$ K), but the hot gas must be isolated from its surroundings for a sufficient time so that the fusion reactions can produce more energy than required to heat the plasma. Thus, for fusion power production, the simultaneous requirements of high temperature and energy confinement were quantified for the first time.

In the United States some of the problems of fusion power had been under study in AEC laboratories since before the end of World War II. Some basic theoretical aspects of fusion energy had been developed by E. Teller, E. Fermi, J. Tuck, and others. In early 1952, a small experimental program on the pinch effect was begun under Tuck at the Los Alamos Laboratory. Its purpose was to confirm and extend the pinch experiments previously done in England by A. A. Ware [20].

L. Spitzer, Jr. in 1951 had his attention drawn to the subject of controlled fusion by a press report that a German, R. Richter, working in Argentina, had achieved significant progress toward a fusion reactor. The report later proved to be incorrect. Nonetheless, stimulated by the news report, and unaware of the classified Los Alamos work, Spitzer conceived a different approach to fusion, the figure-eight stellarator concept. He submitted his ideas to the U. S. AEC, and this agency decided to support research on the stellarator concept. Thus began Project Matterhorn at Princeton University [21]. Shortly thereafter, H. York at the University of California Radiation Laboratory, having learned of the work of the Los Alamos and Princeton groups, suggested several other approaches. He formed a small experimental group under the direction of Richard F. Post, who began work on magnetic mirror confinement of plasma at the Livermore Laboratory.

In the spring of 1952, T. H. Johnson, Director of the Research Division of the U.S. AEC, convened a classified conference at Denver, Colorado, where these groups discussed their fusion research. These "Sherwood" conferences were repeated each year and they continue to this day. A. S. Bishop was given overall program direction for the fusion branch of the Division of Research within the U.S. AEC. In 1956 Post published a paper reviewing the status of controlled fusion research in the United States [22], and L. Spitzer, Jr. published an important small book describing some of the theoretical foundations of plasma physics [23].

By 1958 the difficulties of achieving controlled fusion were recognized and the benefits of open discussion were seen to far outweigh any perceived risks. Accordingly, controlled fusion research was declassified in the United States, England, and the USSR. Nearly a decade's worth of research performed in many countries was presented at the Second Geneva Conference on the Peaceful Uses of Atomic Energy [24]. Since then fusion has been an unclassified, worldwide, research effort—truly international in scope and style. The history of the U.S. fusion program from 1951 to 1958 is described in a book by Bishop [25].

The Atomic Energy Committee of Japan set up a division of fusion research in 1958 under the chairmanship of H. Yukawa, and in 1961 the Institute of Plasma Physics was established at Nagoya University under K. Husimi. Japan, an industrial nation with very little internal energy resources, has, since that time, maintained a broadly based and strong fusion development program.

Since 1958, the International Atomic Energy Agency (IAEA) has convened meetings every few years in which scientists and engineers from all countries report their fusion-research findings. Proceedings of all these conferences are published [26]. Fusion scientists and engineers now publish many of their findings in their archival journals, and there is a large secondary fusion literature composed of laboratory reports. Most of the major groups working on fusion research and development are listed in a publication of the IAEA [27]. The declassification of fusion research by the United Kingdom, United States, and the USSR in 1958, together with the publication of the Geneva Conference on the Peaceful Uses of Atomic Energy, stimulated interest in fusion among many scientists and engineers who were previously unaware or unable to contribute because of secrecy. Following the 2nd Geneva Conference several comprehensive books on fusion were published. Noteworthy are those by Glasstone and Lovberg [28], Rose and Clark [29], Artsimovich [30], Chandrasekhar [31], Longmire [32], and Simon [33].

In the early fusion research period there were a few pioneering attempts to forecast what a power plant might look like and what new technology may be needed. Perhaps the first fusion-power-plant study was that done by L. Spitzer and colleagues, based on the stellerator confinement concept. In Figure 1.2 is shown the shape and size of that 1954 fusion reactor. As you will see later in this text, fusion reactors, as perceived now, do not look like this.

During the 1960–1970 decade, design of fusion power plants was put aside

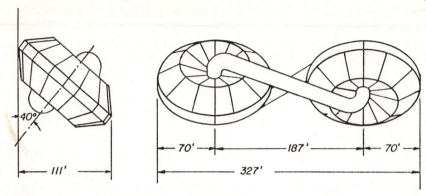

FIGURE 1.2.   Top and end views of a stellerator power-plant-reactor configuration as conceived by L. Spitzer and a few colleagues in 1954 [34]. The predicted saleable power is 600 MW when the magnetic field strength is 5 T.

because it was realized that a much better understanding of plasma confinement and plasma heating was needed. Very encouraging results from plasma experiments were obtained in the early 1970s, and fusion-power-plant design studies began again [35] and they continue to this day.

REFERENCES

1. *Resources and Man*, A Study and Recommendations by a Committee of the U.S. National Academy of Sciences, Freeman, San Francisco, 1969. See, in particular, Chapter 8, "Energy Resources," by M. King Hubbert.

2. *World Energy Outlook*, A Background Paper Prepared by Exxon Corp., Public Affairs Dept. (Dec. 1979). Also, Exxon Corp., *U.S.A's Energy Outlook*, 1980–2000 (Dec. 1979).

3. C. Starr, *The Energy Crisis; Long Term Solutions—A Scientist's View*, ASME paper, Winter meeting, Nov. 1972.

4. B. Sørensen, *Renewable Energy*, Academic Press, New York, 1974.

5. *U.S. Energy and Capital, A Forecast 1980–1990*, Banker's Trust Company, 280 Park Ave., New York, N.Y., 1980.

6. J. R. Lamarsh, *Introduction to Nuclear Engineering*, Addison-Wesley, Reading, MA, 1962.

7. H. C. Urey, F. G. Brickwedde, and G. N. Murphy, *Phys. Rev.* **39**, 164L (1932).

8. W. H. Bennett, *Phys. Rev* **45**, 890 (1934).

9. E. Rutherford, "The New Hydrogen," *Science* **80**, 221 (1934). See also the historic letter, M. L. Oliphant, P. Harteck, and E. Rutherford, "Transmutation Effects Observed with Heavy Hydrogen," *Nature* **133**, 413 (1934).

10. R. d'E. Atkinson and F. G. Houtermans, *Z. Phys.* **54**, 656 (1929).

11. H. A. Bethe, *Phys. Rev.* **55**, 434 (1939).

12. R. G. Hewlett and O. Anderson, Jr., *A History of the United States Atomic Energy Commission*, Vol. I, 1939–46, "The New World," WASH 1214; Vol. II, 1947–52, "Atomic Shield," WASH 1215.

13. H. York, *The Advisors*, Freeman, San Francisco, 1976, p. 77.

14. H. J. Bhabha, *Engineering* **180**, 234 (1955). Bhabha's address is reproduced in this journal in abridged form.

15. H. Thirring, *Nucleonics* **13**, 62 (1955).

16. I. V. Kurchatov, "On the Possibility of Producing Thermonuclear Reactions in a Gas Discharge," *Engineering* **181**, 322 (1956).

17. Z. Melnikova, *Tokamak: Towards Thermonuclear Energy*, Novosti Press Agency, Publishing House, Moscow, USSR, 1982.

18. P. C. Thonemann, *Nuclear Power* **1**, 169 (1956).

19. J. D. Lawson, *Proc. Phys. Soc.* **70**, part 1, no. 445 B, 6–10 (1 January 1957).

20. A. A. Ware, *Trans. Roy. Soc. London Ser.A* **243**, 863 (1951).

21. E. C. Tanner, *Project Matterhorn, 1951–1961; An Informal History*, Princeton Plasma Physics Laboratory Report (Sept. 1977), ERDA Contract EY-76-C-02-3073; *The Model C Decade, 1961–1969; An Informal History*, Princeton Plasma Physics Laboratory Report (Feb. 1980), DOE Contract EY-76-C-02-3073.

22. R. F. Post, *Rev. Mod. Phys.* **28**, 338 (1956).

23. L. Spitzer, Jr., *Physics of Fully Ionized Gases*, Interscience, New York, 1956.

24. *Proceedings of the 2nd International Conference on the Peaceful Uses of Atomic Energy*, Geneva, Sept. 1958: Vol. 31, "Theoretical and Experimental Aspects of Controlled Nuclear Fusion"; Vol. 32, "Controlled Fusion Devices." United Nations, 1958.

25. A. S. Bishop, *Project Sherwood, the U.S. Program in Controlled Fusion*, Addison-Wesley, Reading, MA, 1958.

26. *Plasma Physics and Controlled Nuclear Fusion Research*, IAEA Proceedings of International Conferences: 1st Conference, Salzburg, 1961; 2nd Conference, Culham, 1965; 3rd Conference, Novosibirsk, 1968; 4th Conference, Madison, WI, 1971; 5th Conference, Tokyo, 1974; 6th Conference, Berchtesgaden, 1976; 7th Conference, Innsbruch, 1978; 8th Conference, Brussels, 1980, 9th Conference, Baltimore, 1982.

27. "World Survey of Major Facilities in Controlled Fusion Research," 1976 Edition, Special Supplement, 1976, *Journal of Plasma Physics and Thermonuclear Fusion*, IAEA, Vienna.

28. S. Glasstone and R. H. Lovberg, *Controlled Thermonuclear Reactions*, Van Nostrand, New York, 1960.

29. D. J. Rose and M. Clark, Jr., *Plasmas and Controlled Fusion*, Wiley, New York, 1961.

30. L. A. Artsimovich, *Controlled Thermonuclear Reactions* (1961, Russian edition), Gordon and Breach, New York, 1964 (English edition).

31. S. Chandrasekhar, *Plasma Physics*, University of Chicago Press, Chicago, 1960.

32. C. L. Longmire, *Elementary Plasma Physics*, Interscience, New York, 1963. Developed from a series of lectures given at Los Alamos during 1956–1957.

33. A. Simon, *An Introduction to Thermonuclear Research*, Pergamon, New York, 1959.

34. L. Spitzer, Jr., D. J. Grove, W. E. Johnson, L. Tonks, and W. F. Westendorp, *Problems of the Stellerator as a Useful Power Source*, PM-S-14, NYO-6047 (1 Aug. 1954), Project Matterhorn, Princeton University.

35. See, for example, *A Fusion Power Plant*, R. G. Mills (ed.), MATT-1050, Aug. 1974, Princeton Plasma Physics Lab. Rept. or, UWMAK-I, A University of Wisconsin Toroidal Fusion Reactor Design Rept., Nov. 20, 1973.

2 | FUSION REACTIONS AND FUEL RESOURCES

2.1. NUCLEAR REACTIONS

A nuclear reaction is the process of two nuclear particles interacting and producing new nuclear particles and/or γ rays. If the initial nuclei (or reactants) are denoted by a and b, and the products of the reaction by c and d, then the reaction is symbolized as follows:

$$a + b \rightarrow c + d \qquad (2.1)$$

There may be more than two products. In experiments used to study nuclear reactions, one of the reactants, a, is usually at rest (the target) and the other reactant, b, is accelerated to high energy before impacting the target. Reaction (2.1) is often written as:

$$a(b,c)d \quad \text{or} \quad a(b,d)c$$

For modest particle energies (less than about 1 MeV) there are four fundamental conservation laws governing nuclear reactions. These are:

1. *Conservation of Nucleons.* The total number of nucleons before and after a reaction are the same.

2. *Conservation of Charge.*   The sum of the charges on all the particles before and after a reaction are the same.

3. *Conservation of Momentum.*   The total momentum of the interacting particles before and after a reaction are the same.

4. *Conservation of Energy.*   Energy, including rest mass energy, is conserved in nuclear reactions.

Conservation laws 2, 3, and 4 are always true, while relation 1 is valid for energies less than about 1 MeV.

The equation expressing conservation of energy is

$$E_a + E_b + [M_a + M_b]c^2 = E_c + E_d + [M_c + M_d]c^2 + E_\gamma \quad (2.2)$$

where E_i is the kinetic energy of the ith particle, $E_i = \frac{1}{2}M_i v_i^2$, M_i is the particle mass, v_i is the particle velocity, and c is the speed of light. E_γ is the energy of γ rays emitted from the reaction. The difference in the kinetic energy of the reactants and product particles is therefore

$$(E_c + E_d + E_\gamma) - (E_a + E_b) = [(M_a + M_b) - (M_c + M_d)]c^2 \quad (2.3)$$

The change in the kinetic energy of reactants and products is the difference in the rest mass of the corresponding particles. The right-hand side of Equation (2.3) is called the Q value of the reaction; that is,

$$
\begin{aligned}
Q &\equiv [(M_a + M_b) - (M_c + M_d)]c^2 \\
&= (E_c + E_d + E_\gamma) - (E_a + E_b)
\end{aligned}
\quad (2.4)
$$

If $Q > 0$, the product particles have a greater kinetic energy than the reactants, and the reactants have more mass than the products; $Q > 0$ reactions are called exothermic. They convert mass into kinetic energy of particles, or heat. For $Q < 0$, the reactions are said to be endothermic.

Fusion is defined as the exothermal process that occurs when light nuclei come together forming stable particles different from the original colliding nuclei.

To calculate the distribution of energy among the products of the reaction, both conservation of momentum and energy are used. Assuming the net momentum of the reactants is zero,

$$M_a \mathbf{v} + M_b \mathbf{v}_b = M_c \mathbf{v}_c + M_d \mathbf{v}_d = 0 \quad (2.5)$$

or

$$M_a \mathbf{v} = -M_b \mathbf{v}_b \quad (2.6)$$

and

$$M_c \mathbf{v}_c = -M_d \mathbf{v}_d \quad (2.7)$$

The momenta and kinetic energy of the product particles are, for exothermic reactions, about 10^3 times larger than the reactants, so their initial values are negligible. Thus, squaring Equation (2.7),

$$M_c E_c = M_d E_d \qquad (2.8)$$

together with

$$Q \approx E_c + E_d \qquad (2.9)$$

gives

$$E_c = \left(\frac{M_d}{M_c + M_d}\right)Q \quad \text{and} \quad E_d = \left(\frac{M_c}{M_c + M_d}\right)Q \qquad (2.10)$$

The energy of the nuclear reaction, which manifests itself as kinetic energy of the products, is shared between the two product particles in inverse proportion to their masses.

In fusion all the reactants are ionized atoms; that is, the atoms will have lost their electrons. Thus, the colliding particles that result in fusion reactions will be positively charged ions. Tables of the isotopes list the atomic mass of the neutral atom. It is, however, not necessary to subtract the mass of the electrons from the atomic mass of the reactants, since charge conservation causes the electron mass to cancel in the definition of Q. The computation of Q can be made using the mass values of the neutral atoms. The table in Appendix B lists the atomic masses of the light elements.

The masses of all nuclei are smaller than the sum of the masses of the neutrons and protons contained in them. The mass defect Δ for an arbitrary nucleus is:

$$\Delta = Z m_p + N m_n - M_A^+ \qquad (2.11)$$

where Z is the number of protons, m_p is the proton mass, N is the number of neutrons, m_n is the neutron mass, and M_A^+ is the mass of the nucleus. Equation (2.11) can be written

$$\Delta = Z(m_p + m_e) + N m_n - (M_A^+ + Z m_e) \qquad (2.12)$$

or

$$\Delta = Z M_{^1H} + N m_n - M_A$$

where m_e is the electron mass, M_A is the mass of the neutral atom and $M_{^1H}$ is the mass of neutral hydrogen, ^1H. Here, as before, we have neglected the energy of the bound electrons (a few electron volts), because this is usually quite negligible. The mass defect Δ, when expressed in units of energy, is called the binding energy of the system, since it represents the energy involved in the formation of the nuclei from free protons and neutrons.

Whenever it is possible to produce a more-stable configuration by combining two less-stable nuclei, energy is released in the process. Such reactions are possible for a great many ($\sim$ 80) pairs of nuclides. This is seen in Figure 1.1, where the binding energy per nucleon is plotted as a function of atomic mass number. The binding energy per nucleon has a maximum around iron, $A \sim 56$. A list of some of the most interesting nuclear fusion reactions, together with their associated energy release value Q, is given in the table in Appendix C.

Study of the many possible fusion reactions of the light elements, with reference to initial collision energy required to cause fusion events, substantially reduces the number of interesting, that is, practical reactions. For example, reaction 10 in the table in Appendix C, the overall carbon–nitrogen-catalyzed cycle that takes place in the sun, proceeds far too slowly to be of serious interest in a practical engineering device. The reactions most seriously considered for fusion energy are

1. *The Deuterium–Tritium Reaction.*

$$^2D + {}^3T \rightarrow {}^4He \ (3.5 \ MeV) + {}^1n(14.1 \ MeV) \quad (2.13)$$

As will be shown in Chapter 3, this reaction has the lowest ignition temperature and produces the highest fusion-power density. Because of this, it is the reaction chosen for first-generation fusion-power plants. Although deuterium is plentiful and inexpensive, tritium is extremely rare and radioactive (β^- decay). It will have to be manufactured. The end products or ashes of this reaction are α particles ($^4He^+$) and energetic neutrons. The neutrons will promptly leave the reacting plasma, but the 3.5-MeV α particles, because they are charged particles (ions), can be confined in the plasma and transfer their excess energy to the background plasma. Twenty percent (3.5/17.6) of the D–T reaction energy is in ions. There are no radioactive ashes, but the energetic fusion neutrons* present difficult technical problems, particularly for the first wall of the plasma-confinement chamber.

2. *The Deuterium–Deuterium Reactions.*

$$^2D + {}^2D \rightarrow {}^3He \ (0.82 \ MeV) + {}^1n \ (2.45 \ MeV) \quad (2.14)$$

$$^2D + {}^2D \rightarrow {}^3T \ (1.01 \ MeV) + {}^1H \ (3.02 \ MeV) \quad (2.15)$$

Each of these deuterium reactions takes place with nearly equal probability, and, therefore, the products of each reaction form at about the same rate. The

---

*These neutrons carry with them not only their kinetic energy but the capacity to undergo exothermal nuclear fission reactions. For example, if the neutron should be absorbed by ^6Li, to produce ^3T in a blanket surrounding the fusion chamber, then an additional 4.8 MeV per neutron is produced. If the neutron were absorbed by sodium, about 13 MeV would be produced. These neutrons have value; they can be used to breed tritium for the D–T fuel cycle, they can generate further thermal energy via fission events, and they can be used to breed fissile fuel. The neutrons also cause problems because they activate the structure surrounding the fusion chamber, as well as cause physical radiation damage.

tritium and the helium-3 so formed can react with deuterium in further fusion events. Thus,

$$
\begin{array}{ll}
{}^2D + {}^2D \rightarrow {}^3He + {}^1_0n & 3.27 \text{ MeV} \\
{}^2D + {}^2D \rightarrow {}^1T + {}^1H & 4.03 \text{ MeV} \\
{}^2D + {}^3T \rightarrow {}^4He + {}^1_0n & 17.6 \text{ MeV} \\
{}^2D + {}^3He \rightarrow {}^4He + {}^1H & 18.3 \text{ MeV} \\
\hline
6{}^2D \rightarrow 2{}^1n + 2{}^1H + 2{}^4He & 43.2 \text{ MeV}
\end{array} \tag{2.16}
$$

or

$$3{}^2D \rightarrow {}^4He + {}^1H + {}^1n + 21.6 \text{ MeV}$$

The possibility of proton–proton, proton–deuterium, and proton–tritium reactions are not considered because, at the energies relevant for contemplated fusion reactors, their probabilities for occurring are very small compared to the reactions listed in Equation (2.16). Because the overall reaction involves just deuterium, that is, the intermediaries ${}^3He$ and ${}^3T$ burn, Equation (2.16) is called the catalyzed deuterium reaction. If the rate at which ${}^3T$ and ${}^3He$ burn is exactly equal to the rate at which they are created, the reaction is called fully catalyzed. (see Chapter 3). The ashes of this reaction are two neutrons (2.45 and 14.1 MeV), two protons (3.02 and 14.6 MeV), and two α particles (3.5 and 3.7 MeV). The average energy yield per deuteron for the catalyzed reaction is 43.2/6 = 7.2 MeV. Fifty-seven percent (24.8/43.2) of the energy yield is contained in charged particles. At low energy (below 100 keV) the ${}^3He$ reaction proceeds more slowly that the D–T reaction. If the helium-3 does not react, then five deuterons react, yielding 24.9 MeV or about 5 MeV per deuteron. In this partially catalyzed deuterium reaction, one-third of the energy (8.3 out of 24.9 MeV) can reside in charged particles in the ashes. The pure-deuterium-fueled fusion power plant is perceived to be a second-generation device. It will be free of the problems created by manufacturing tritium, but it will have energetic neutrons as reaction products, and radioactive tritium is present in the plasma because of reaction (2.15). Hence, some tritium will be in the plasma exhaust.

3. *Some Neutron-Free Reactions.*

$$
{}^1H + {}^6Li \rightarrow {}^4He + {}^3He \tag{2.17}
$$

$$
{}^1H + {}^7Li \rightarrow 2{}^4He \tag{2.18}
$$

$$
{}^1H + {}^{11}B \rightarrow 3{}^4He \tag{2.19}
$$

$$
{}^2D + {}^3He \rightarrow {}^4He + {}^1H \tag{2.20}
$$

$$
{}^3He + {}^3He \rightarrow {}^4He + 2{}^1H \tag{2.21}
$$

Equation (2.17), the proton−lithium-6 reaction, usually involves further reactions because helium-3 can also react. Thus the catalyzed and fully chain reacted ^{1}H + ^{6}Li cycle involves at least the following reactions:

$$^1\text{H} + {}^6\text{Li} \rightarrow {}^3\text{He} + {}^4\text{He} \qquad 4.0 \text{ MeV}$$

$$^3\text{He} + {}^6\text{Li} \rightarrow {}^1\bar{\bar{\text{H}}} + 2{}^4\text{He} \qquad 16.8 \text{ MeV}$$

$$^3\text{He} + {}^3\text{He} \rightarrow 2{}^1\bar{\text{H}} + {}^4\text{He} \qquad 12.9 \text{ MeV} \qquad (2.22)$$

The fast protons $^1\bar{\bar{\text{H}}}$ and $^1\bar{\text{H}}$ have substantial probability of reacting with ^{6}Li, prior to thermalization when $T_e > 100$ keV.

The proton−boron-11 reaction [Equation (2.19)] involves the intermediary fast ($\sim 10^{-16}$ s) reactions:

$$^{11}\text{B} + {}^1\text{H} \rightarrow {}^{12}\text{C} \Big\langle \begin{array}{l} {}^4\text{He} \\ {}^8\text{Be} \rightarrow 2{}^4\text{He} \end{array} \qquad (2.23)$$

The overall reaction is ^{11}B(^{1}H,2α)^{4}He, where $\alpha \equiv {}^4$He.

The ^{2}D−^{3}He and ^{3}He−^{3}He reactions also yield ashes consisting of only protons and α particles. While protons, lithium, and boron-11 have reasonable abundances, helium-3 is very rare (about one part in 10^6 of all helium), so that it would have to be artificially manufactured. There are a number of nuclear reactions that can produce helium-3, but the need to manufacture the fuel is a significant problem for systems in which ^{3}He is to be a primary reactant.

One of these neutron-free fusion reactions [Equations (2.17)−(2.21)] is thought to be the fuel for a third-generation fusion device. There are a number of significant advantages for advanced-fueled fusion reactors and they have been discussed by J. Rand McNally and others [1,2]. These fuels are, however, much more difficult to burn because they require higher energies to react, and consequently, the plasma will have higher losses from bremsstrahlung and synchrotron radiation. Nonetheless, they are perceived as very desirable because they involve very little or no gaseous radioactivity, there is essentially no neutron damage to structural materials, and the safety and maintenance aspects may be similar to a chemical boiler system. The heat-transfer system would be much simpler than those that involve radioactive materials and intermediate heat-transfer loops. There would be relatively low environmental impact via air pollution, mining, and long-term solid-waste disposal.

Because the products of these latter fusion fuels are all charged particles, it is theoretically possible to employ direct energy conversion to electrical power, circumventing Carnot and other losses associated with a thermal cycle. Overall system energy efficiencies of the order of 80% might be achieved. Fusion reactions that have a substantial fraction of their energy in neutrons must use a thermal cycle, and, hence, are limited by high-temperature properties of material and are generally constrained by thermodynamics to an overall efficiency $\eta \sim 40\%$.

There seems little doubt that the first fusion power plants will employ the D−T reaction. It is hoped that once this is achieved, the pure D−D fusion-power plant will soon follow. Whether a reactor using neutron-free fusion fuels can eventually be developed is still a matter for basic research (see, e.g., Reference 3).

An entirely different approach to fusion is muon-catalyzed or "cold" fusion of deuterium and tritium. It was first suggested by Zel'dovich [4], but calculations indicated that the concept would not be practical. Recently [5], it was discovered that there is a weakly bound state of the mesic molecule DTμ, and it has been calculated that one μ^- meson in a mixture of DT can catalyze on the order of 100 fusion reactions releasing about 2GeV of energy. There has also been at least one experiment [6] that measured the rate of formation of these interesting DTμ molecules. Petrov [7] has calculated that a muon-catalyzed fusion reactor, together with a fissile blanket, would produce a net energy gain. Whether "cold fusion" can be made practical remains to be seen.

2.2.   FUSION-FUEL RESOURCES

The basic fuels for fusion are deuterium, tritium, helium, lithium, and boron. Their availability, together with some of their properties and their cost, are discussed in the following sections.

2.2.1.   Deuterium

Deuterium [8] is the designation for the hydrogen isotope of mass 2 (^{2}H or ^{2}D). It is a stable isotope and occurs in natural hydrogen, water, and other hydrogen-bearing compounds in an average abundance of 0.015 mole percent, that is, about one part D in 6670 hydrogen atoms. The term "light water" is applied to the oxide of the light isotope having a molecular weight of 18, and "heavy water" to deuterium oxide having a molecular weight of 20. As noted in Chapter 1, deuterium was first discovered by H. C. Urey and co-workers.* Nearly pure D_2O was first concentrated by G. N. Lewis in 1933 from water by electrolysis. Heavy water is used as a moderator in some types of nuclear reactors, and, consequently, its demand was sufficient to have several heavy-water production plants constructed in the United States and Canada. These include the Savannah River Plant in South Carolina, which is capable of producing 500 tons of heavy water annually.

The physical and chemical differences between the hydrogen isotopes are relatively much greater than those among the isotopes of all the other elements because of their large relative difference in mass. Physical, chemical, and

---

*Professor Urey discovered ^{2}D when he was working in the Pupin Building of Columbia University.

thermodynamic properties of hydrogen and deuterium and their respective oxides are given in Reference 8.

When small amounts (a few grams) of deuterium are required, electrolysis is the preferred production method since it gives a large separation in a single piece of equipment. For large, industrial-type production where economics is an important consideration, the H_2S/H_2O chemical exchange process [8] serves as the preenrichment process prior to electrolysis.

The oceans of the world have a volume of about $1.5 \times 10^{18} \, m^3$. The energy from deuterium reactions [Equations (2.14) and (2.15)], together with its natural abundance, leads to the conclusion that the fusion energy available from the deuterium in $1 m^3$ of water is about 8×10^{12} J. This quantity of energy is equivalent to that from about 1360 barrels of oil or 270 tons of coal. The quantity of deuterium in water on earth is estimated to be about 5×10^{13} tons. Thus the enormous energy content available from natural deuterium is sufficient to sustain an energy-affluent world's needs for more than 10^9 years. Hence, fusion is called an inexhaustible energy source.

In small quantities, chemically pure deuterium gas cost about \$1/liter (STP) in 1980, or about \$10/g of deuterium. It is much cheaper when large quantities are purchased. Recalling that the catalyzed $D-D$ reaction produces 7.2 MeV per deuteron, implies that deuterium energy has a cost of about 3.4×10^{10} J/dollar (about $\$0.03/10^6$ Btu). U.S. coal fuel energy costs were about 5×10^8 J/dollar in 1980 (about $\$2/10^6$ Btu). This does not imply, however, that fusion energy will be cheap. Fusion fuel is cheap. But to produce useful energy, fusion needs a very-high-technology system whose capital costs will be large. The capital and maintenance costs for a fusion electric power plant are expected to be at least comparable to those of modern coal or nuclear-fission power plants.

2.2.2. Tritium

Tritium [9] is the designation for the hydrogen isotope of mass 3 (^3H or ^3T). The molecular form is T_2. Tritium is energetically unstable and decays radioactively by the emission of a low-energy β^- particle (electron). The half-life is relatively short (12.36 y) and, therefore, tritium does not occur naturally, except in equilibrium with amounts produced by cosmic-rays or man-made nuclear devices. The cosmic-ray source of tritium results from the effect of high-energy protons or neutrons interacting with nitrogen-14. Typical reactions are,

$$^1H + {}^{14}N \rightarrow {}^3T + \text{fragments}$$

$$^1n + {}^{14}N \rightarrow {}^3T + {}^{12}C$$

Tritium is also created by neutron irradiation of deuterium and is therefore produced in heavy-water CANDU fission reactors.

A unit for measuring natural tritium, the TU, signifies the ratio of 1 atom of tritium to 10^{18} atoms of hydrogen, testifying to its natural scarcity. In 1958 the world's inventory of natural tritium was estimated to be 2×10^4 g in water and 2×10^2 g in atmospheric hydrogen.

Tritium is produced on a large scale by neutron irradiation of lithium. The two principal reactions are

$$^6\text{Li} + {}^1n \rightarrow {}^4\text{He} + {}^3\text{T} + 4.8 \text{ Mev} \qquad (2.24)$$

$$^7\text{Li} + {}^1n \rightarrow {}^4\text{He} + {}^3\text{T} + {}^1n' - 2.5 \text{MeV} \qquad (2.25)$$

The $^1n-{}^7\text{Li}$ reaction is endothermic. Cross sections for these reactions, as a function of neutron energy, are shown in Figure 2.1. The $^6\text{Li}(n, \alpha)$T reaction can use slow (thermal neutrons), while the $^7\text{Li}(n,n',\alpha)$T reaction has a threshold near 3 MeV and needs fast neutrons. This latter reaction produces another (slower) neutron, which provides the basis for breeding tritium in a lithium blanket surrounding a fusion plasma. Neutron multipliers, like beryllium, can be used to assist the neutron economics in a fusion breeding blanket. Tritium produced by neutron irradiation of lithium can be recovered by electrolysis of tritiated water, use of a thermal diffusion column, gas chromatography, or other well-established techniques.

Since it is planned that the neutrons will come from the $D-T$ fusion reactor

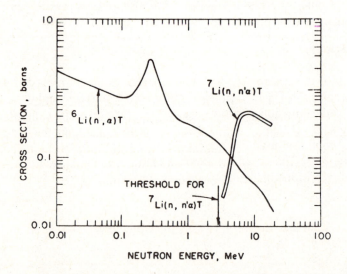

FIGURE 2.1.   Reaction cross sections versus neutron energy for the $^6\text{Li}(n, \alpha)$T and $^7\text{Li}(n, n', \alpha)$T reactions used to manufacture tritium. Data are taken from Reference 10. One barn $= 10^{-28}$ m^2.

itself, the amount of tritium that can be ultimately produced is dependent on the size of the lithium resource.

Estimates of the amount of lithium needed for a D−T fusion power plant range between 20 and 50 kg/MWe of fusion capacity [11]. This appears to be no problem. But, the initial fueling of tritium in a first generation D−T reactor, that is, about 10 kg of tritium, will be expensive. The cost of the start-up tritium inventory may be of the order of 50 million dollars, since the cost of tritium (1981) is of the order of $9000/g.* Once a successful fusion breeder is in operation, however, the cost of tritium should be greatly reduced. A 1 GW thermal D−T fusion plant will consume about 140 g tritium/day.

2.2.3. Lithium

Pure lithium is composed of 92.58% ^{7}Li and 6.42% ^{6}Li and is a soft, low-density metal whose physical appearance is similar to lead. It has a specific gravity of 0.53 (it floats on oil) and a melting point of 179°C. Its chemical, physical, and thermal properties are well known [13] and its chemical technology is well developed [11].

About 5×10^6 kg of lithium were produced in 1974, of which about 3.4×10^6 kg came from the United States [14]. Nearly all the latter came from the Kings mountain region in North Carolina. Known lithium resources are immense compared to present annual production. The average abundance of lithium in the earth's crust is 65 parts per million by weight (uranium's abundance is 4 ppm). The concentration of lithium in sea water is 0.17 g/m^3 or about 1/200th that of deuterium.

The presently known and inferred U.S. lithium reserves are equivalent to 3000 times the 1970 world energy consumption and 9000 times that of the United States. It is estimated [15] that this amount of lithium, used in D−T fusion cycles, could satisfy U.S. electrical demand for more than 600 years. Hubbert [16] has estimated that present lithium reserves are about equal to three times the world's initial total fossil fuels inventory. The adequacy of terrestial lithium reserves available at about $0.02/g seems assured on any time scale of present concern. When the energy potential from lithium extracted from sea water is also considered, the resource is adequate by any measure. Only if lithium batteries for automotive transportation become very widely adopted, could there be any problem with lithium as a fundamental resource in a fusion energy economy. Should that occur, one may anticipate a rapid evolution to the second-generation, pure-deuterium-fueled fusion systems.

2.2.4. Helium and Boron

Helium-3, which is a primary fuel for reactions (2.20) and (2.21), is a very scarce material. Its natural abundance is 0.00013%, or about one part per

---

*The price that the U.S. government charged for tritium in 1978 was $7200/g [12].

million in helium gas. It is found in slightly higher concentrations around active volcanoes. Helium-3 might be manufactured by the $^1\mathrm{H}(^6\mathrm{Li},^4\mathrm{He})^3\mathrm{He}$ reaction, the $^2\mathrm{D}(^2\mathrm{D},^1n)^3\mathrm{He}$ reaction, and the $^2\mathrm{D}(^6\mathrm{Li},^1n)^3\mathrm{He} + \alpha$ reaction (numbers 5, 12, and 18 in Appendix C). One could also produce $^3\mathrm{T}$ and allow it to decay (12-year half-life), that is, $^3\mathrm{T} \rightarrow {}^3\mathrm{He} + e^-$. It appears very doubtful that some $^3\mathrm{He}$ manufacturing cycle can be developed which is compatible with a fusion power system and still meet economic constraints and supply requirements.

Boron-11 is another neutron-free fusion fuel that has attracted some interest via the $^1\mathrm{H}(^{11}\mathrm{B},2\alpha)\alpha$ reaction. Boron is widely found in nature, usually as alkali or alkaline earth borates or as boric acid. It has two stable isotopes, of which $^{11}\mathrm{B}$ constitutes 80.2% of the natural abundance. It exists in an amorphous form and in at least three crystalline forms. It has a melting point near 2100°C and a boiling point near 2600°C. Further details concerning its technology can be found in Reference 17.

REFERENCES

1. J. R. McNally, Jr. *Advanced Fuels for Nuclear Fusion Reactors*, NBS SP 425. Proceedings of the Conference on Nuclear Cross Sections and Technology, R. A. Schrack and C. D. Bowman, (eds.), NBS SP 425, 1975, p. 683.

2. C. C. Baker et al., "The Impact of Alternate Fusion Fuels on Fusion Reactor Technology," Argonne National Laboratory Report, ANL/FPP/TM-128, Nov. 1979.

3. R. W. Conn et al., IAEA-CN-38/V-5 paper, *8th International Conference on Plasma Physics and Controlled Nuclear Fusion Research*, Brussels 1980; also, UCLA report on *Alternate Fusion Fuel Cycle Research*, PPG-492-July 1980.

4. Ya. B. Zel'dovich, *Dokl. Akad. Nauk. USSR* **95**, (1954); also, Ya. B. Zel'dovich and S. S. Gerstein, *Usp. Fiz. Nauk.* **71**, 581 (1960).

5. S. S. Gerstein and L. P. Ponomarev, *Phys. Lett.* **72B**, 80 (1977).

6. V. M. Bystritsky et al., *Phys. Lett.* **94B**, 476 (1980).

7. Ya. V. Pertrov, *Nature* **285**, 466 (1980).

8. J. F. Proctor, *Encyclopedia of Chemical Technology*, 2nd ed., Kirk-Othmer (ed.), Wiley, New York, 1967, vol. 6, p. 895.

9. L. H. Meyer, *Encyclopedia of Chemical Technology,* 2nd ed., Kirk-Othmer (ed.), Wiley, New York; 1967, vol. 6, p. 910; see also D. G. Jacobs, *Sources of Tritium and Its Behavior Upon Release to the Environment*, AEC Critical Review Series. TID-24635, 1968.

10. J. R. Stehn, M. D. Goldberg, B. A. Magurno, and R. Weiner-Chasman, *Neutron Cross Section*, US AEC Report BNL-325, 1964, Vol. I.

11. R. O. Bach, C. W. Kamienski, and R. B. Ellestad, *Encyclopedia of Chemical Technology*, 2nd ed. Kirk-Othmer (ed.), Wiley, New York, 1967 Vol. 12, pp. 529–556.

12. T. B. Rinehammer and L. J. Wittenberg, *An Evaluation of Fuel Resources and Requirements for the Magnetic Fusion Energy Program*, Mound Facility Report, MLM-2419, Oct., 1978.

13. V. A. Maroni, E. J. Cairns, and R. A. Cafasso, *A Review of the Chemical, Physical, and Thermal Properties of Lithium that Are Related to Its Use in Fusion Reactors*, Argonne National Laboratory Report ANL-8001 (1973).

14. A. L. Hammond, *Science* **191**, 1037 (1976)

15. J. P. Holdren, *Adequacy of Lithium Supplies as a Fusion Energy Source*, Lawrence Livermore Report UCID-15953, Dec. 1971.

16. M. K. Hubbert, "Energy Resources," in *Resources and Man*, Freeman, San Francisco, 1969.

17. J. G. Bower, *Encyclopedia of Chemical Technology*, 2nd ed., Kirk-Othmer (ed,), Wiley, New York, 1967, Vol. 3, pp. 602−605.

3 | REACTION RATES, IGNITION, AND CONFINEMENT

3.1. KINETIC THEORY FUNDAMENTALS

Consider a uniform number density N of free target particles and let a beam of test particles of density n, with uniform velocity v, pass through N. The loss of particles from the test beam dn, due to any physical process such as scattering or nuclear reactions, is given by the following expression:

$$dn = -\sigma n N \, dx \qquad (3.1)$$

where σ, the constant of proportionality, is called the cross section for the physical process, and dx is the differential distance along the beam path. Since n and N, the number densities of test and field particles, have the dimension m^{-3} (i.e., particles per cubic meter), σ has the dimensions of area (m^2). If all the particles were simply hard spheres of radius r_1 and r_2, respectively, the cross section for mechanical collisions is given from the geometry as simply:

$$\sigma_{col} = \pi(r_1^2 + r_2^2) \qquad (3.2)$$

If the cross section σ is independent of x, and $n << N$, then integration of Equation (3.1) yields the result that the test-beam number density decays exponentially in distance:

$$n = n_0 e^{-\sigma Nx} \qquad (3.3)$$

where n_0 is the test-particle number density at $x = 0$. The probability that a particle goes a distance x without an event of the kind characterized by the cross section σ is $e^{-\sigma Nx}$.

For example, the mean free path λ of a particle is defined as

$$\lambda = \frac{\text{total distance traveled by all test particles}}{\text{total number of test particles}}$$

or

$$\lambda = \frac{n_0 \int_0^\infty x e^{-\sigma Nx}\, dx}{n_0 \int_0^\infty e^{-\sigma Nx}\, dx} \qquad (3.4)$$

where in the integrand of the numerator, x is the distance a test particle has traveled and $e^{-\sigma Nx}$ is the probability of undergoing an event (characterized by σ) in the distance between x and $x + dx$. The denominator is simply the total number of test particles. Both integrals are well known, and yield the simple result:

$$\lambda = \frac{1}{N\sigma} \qquad (3.5)$$

The frequency ν of occurrence of the event is given by

$$\nu = \bar{v}\sigma N \qquad (3.6)$$

$$\nu = \frac{\bar{v}}{\lambda} \qquad (3.6a)$$

where $\bar{v}$ is the mean velocity.

Study of different kinds of particle interactions is a major activity in physical research. The book by McDanniel [1] is an excellent source of information on cross sections, and there are several compilations of cross sections particularly useful in fusion research [2−4]. Some of the most important types of cross sections, together with their order of magnitude, are given in Table 3.1.

Elastic scattering between charged particles in a plasma involves simultaneous many-body interactions. A change in the direction of the momentum of a

TABLE 3.1.   Some Common Cross Sections

Physical Process	σ (m^2)[a]	Remarks
Elastic scattering	10^{-20}	Neutral atoms at room temperature
Elastic scattering	10^{-19}	Electron−neutral atom
Atomic excitation	10^{-20}	Electron−hydrogen
Ionization	10^{-20}	Electron−hydrogen
Charge exchange	10^{-19}	$H^{+(fast)} + H(1,s) \rightarrow H^{fast}(1,s) + H^{+(slow)}$
Coulomb scattering	10^{-24}	Hydrogen ions at ~ 10 keV
Nuclear fusion	10^{-29}	D−T at ~ 25 keV

[a]The size of most cross sections is a function of the relative energy between interacting particles; the magnitudes listed here are typical values for laboratory plasmas. For more exact values and further detail, see References 1−4. Some cross sections, for example, excitation and ionization, have energy thresholds, below which the cross section is zero. For molecular and atomic processes, a common unit to express cross sections is the area of the Bohr atom $\pi a_0^2 = 0.88 \times 10^{-20}$ m^2. For nuclear processes the common unit is the barn = 10^{-28} m^2.

particle by 90° is used to define the Coulomb elastic 90° scattering cross section. It has the value,*

$$\sigma_{90°} = \frac{Z^4 e^4 \ln\Lambda}{25\pi\varepsilon_0^2(kT)^2} = 5.62 \times 10^{-10} \frac{Z^4 \ln\Lambda}{T^2} \tag{3.7}$$

For a derivation and discussion of this formula see Reference 5 and Section 4.3.4. In Equation (3.7), e is the proton charge, Ze is the charge of the interacting particles, ε_0 is the vacuum permittivity, k is Boltzmann's constant, and $\ln\Lambda$ is the Coulomb logarithm (see Section 4.3.4) given by

$$\Lambda = \frac{12\pi}{n_e^{1/2}} \left(\frac{\varepsilon_0 kT}{e^2}\right)^{3/2} = 1.24 \times 10^7 \frac{T^{3/2}}{n_e^{1/2}} \tag{3.8}$$

In a hydrogen fusion plasma with $T = 10^8$ K, and $n_e = 0.5 \times 10^{20}$ m^{-3}, we find $\ln\Lambda = 21.3$. The Coulomb logarithm is a slowly varying function whose value lies between 10 and 25 for most laboratory plasmas. The elastic scattering cross section at these fusion conditions is $\sigma = 9.4 \times 10^{-25}$ m^2. The mean free path for electrons and hydrogen ions at these fusion conditions is $\lambda_e = \lambda_i = 2.1 \times 10^4$ m.

The D−T fusion cross section at 10^8 K is about 10 millibarns or 10^{-30} m^2. The mean free path for a D−T fusion event in this plasma is

$$\lambda_f = \frac{1}{n_i \sigma_f} \approx \frac{1}{0.5 \times 10^{20} \times 10^{-30}} = 2 \times 10^{10} \text{ m}$$

*All equations are is SI units.

or about a million times the 90° elastic scattering mean free path.

If a gas is in thermodynamic equilibrium, its particles, each of mass m, have a velocity distribution given by the Maxwell−Boltzmann function,

$$f_0(v) = n\left(\frac{m}{2\pi kT}\right)^{3/2} e^{-mv^2/2kT}$$

(3.9)

The speed distribution function $f(v)$ is obtained by integrating the velocity distribution over all directions to obtain the result

$$f(v)dv = 4\pi n\left(\frac{m}{2\pi kT}\right)^{3/2} v^2 e^{-mv^2/2kT} dv$$

(3.10)

which is the number of particles per unit volume with speeds between v and $v + dv$. The most probable speed, sometimes called the thermal speed, v_{th}, corresponds to the maximum of the speed distribution function and is obtained by setting $df/dv = 0$. This gives

$$v_{th} = \left(\frac{2kT}{m}\right)^{1/2}$$

(3.11)

The mean speed $\bar{v}$ is, by definition,

$$\bar{v} = \frac{1}{n}\int_0^\infty vf(v)dv$$

(3.12)

or, for a Maxwellian distribution,

$$\bar{v} = \left(\frac{8kT}{\pi m}\right)^{1/2}$$

(3.12a)

Thus,

$$\bar{v} = 1.13v_{th}$$

(3.13)

If the gas consists of two unlike particles in a binary gas mixture, then it can be shown that the mean value of their relative velocities is

$$\bar{v}_r = (\bar{v}_1^2 + \bar{v}_2^2)^{1/2}$$

(3.14)

and if all the particles are alike, it follows that

$$\bar{v}_r = \sqrt{2}\,\bar{v}$$

(3.15)

In a similar way, if $n(E)dE$ is the number of particles per unit volume having kinetic energies between E and $E + dE$, then

$$n(E)dE = \frac{2\pi n}{(\pi kT)^{3/2}}E^{1/2}e^{-E/kT}dE$$

(3.16)

The most probable energy is defined as the maximum of the function $n(E)$ and is, for a Maxwellian distribution, given by

$$E_{mp} = \tfrac{1}{2}kT$$

(3.17)

The mean energy $\bar{E}$ is defined as

$$\bar{E} = \frac{1}{n}\int_0^\infty n(E)E\,dE$$

(3.18)

where, for a Maxwellian,

$$\bar{E} = \tfrac{3}{2}kT$$

(3.19)

The mean energy is three times the most probable energy. If a gas has a velocity distribution different than Equation (3.9), it is useful to define a temperature as:

$$\tfrac{1}{2}m\bar{v}_E^2 = \tfrac{3}{2}kT$$

(3.20)

where the mean speed is defined by Equation (3.12), but the distribution function $f(v)$ is not Maxwellian. This is a useful definition of temperature, but it should be noted that it coincides with the thermodynamic definition of temperature only for a Maxwell–Boltzmann velocity distribution. It is known from kinetic theory that a random distribution of particle velocities will approach closely that of a Maxwellian distribution within a few collision times.

It is common practice in fusion research to refer to the plasma temperature in units of electron volts, eV. The relationship between degrees Kelvin and eV is

$$kT = 1.0\ eV = 1.6030 \times 10^{-19}\ J$$

Since the Boltzmann constant $k = 1.38054 \times 10^{-23}$ J/K,

$$\frac{e}{k} = 11,600\quad K/V$$

or

$$T\ (eV) = 8.61 \times 10^{-5}T\ (K)$$

Since fusion involves gases whose temperatures may be tens of keV or larger, it is appropriate to enquire when relativistic effects become significant. The total energy of a particle, that is, its rest mass plus its kinetic energy, is given by

$$E_{\text{tot}} = mc^2 \tag{3.21}$$

The kinetic energy E is the difference between the total energy and the rest mass energy, namely,

$$E = (m - m_0)c^2 \tag{3.22}$$

The mass is related to the velocity, relative to an observer, according to

$$m = \frac{m_0}{[1 - (v/c)^2]^{\frac{1}{2}}} \tag{3.23}$$

Combining Equations (3.23) and (3.22) and expanding in a power series, the kinetic energy E is found to be, when terms of the order of $(v/c)^2$ are negligible compared to one,

$$E = \frac{1}{2} m_0 v^2 \tag{3.24}$$

If, for example, 5% is considered a significant fraction, then relativistic corrections to the kinetic energy can be ignored when

$$\frac{E}{m_0 c^2} \leq 0.05$$

Since the rest mass of an electron is 511 keV, relativistic effects begin to be significant for an electron gas whose energy is of the order of 25 keV. The rest mass of a proton is about 1000 MeV, so that relativistic corrections to a proton gas become significant at 50 MeV and higher.

3.2.   THERMONUCLEAR REACTION CROSS SECTIONS AND REACTION RATES

Values of cross sections for many fusion reactions have been experimentally measured and tabulated by several investigators [6–12]. For energies much less than the classical value necessary to overcome the Coulomb repulsion between two nuclei ($\sim 0.3Z_1Z_2$ MeV), Gamow [13] studied the fusion reaction cross section resulting from the quantum-mechanical tunneling effect, or Coulomb barrier penetration. He found that the total nuclear reaction cross section at low energies can be written in the form

$$\sigma_{\text{fusion}} = \left(\frac{S}{E_{\text{CM}}}\right) \exp\left[-\left(\frac{G}{E_{\text{CM}}^{1/2}}\right)\right] \tag{3.25}$$

where E_{CM} is the energy in the center of momentum coordinate system. The constants S and G are in practice evaluated from experimental data, and are usually in reasonable agreement with their theoretical predictions [11,14]. Values of S and G are listed for some of the more important fusion reactions in Table 3.2. The largest cross section at low energy (< 200 keV) is the D–T reaction. At higher energy the D–^{3}He reaction approaches that of D–T. The D–T reaction cross section, as a function of relative energy, is shown in Figure 3.1. The reaction cross section has a maximum at about 100 keV. A detailed review of $T(D,n)^4$He cross-section data, with particular attention to values at low energy, is given in Reference 12.

The rate R_{ij} at which a reaction proceeds in a uniform gas mixture of species i and j may be expressed in terms of the reaction cross section σ and the number densities of the reactants [15]. If every particle were moving relative to each other at just one speed v, then

$$R_{ij} = a_{ij}n_i n_j \sigma v \tag{3.26}$$

where $a_{ij} = 1$ if there are two distinct species (like D–T), that is, $i \neq j$;
$a_{ij} = \frac{1}{2}$ if there is only one reactant (like D–D), that is, $i = j$,
so that particle reactions are not counted twice.

But, the more common situation in fusion is one where the particles have a distribution of velocities. Then

$$R_{ij} = a_{ij}n_i n_j \langle \sigma v \rangle \tag{3.27}$$

TABLE 3.2.   Some Fusion Reaction Cross-Section Parameters[a]

Reaction	S [barn-keV]	G [(keV)$^{1/2}$]
D + T → ^{4}He + n	11,000	34.4
D + D → T + H	53	31.4
D + D → ^{3}He + n	53	31.4
D + ^{3}He → ^{4}He + H	7,000	88.8
H + ^{6}Li → ^{3}He + ^{4}He	5,500	87.2
H + ^{7}Li → 3^{4}He	125	88.1
D + ^{6}Li → 2^{4}He	3,060	104.0
H + ^{11}B → 3^{4}He	100,000	150.3

[a]Data from Reference 11.

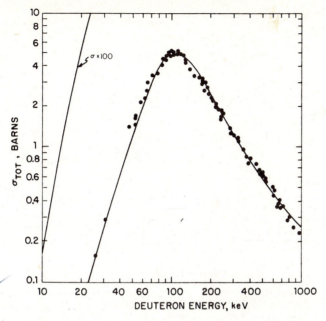

FIGURE 3.1. Fusion cross section for the reaction D + T → ^1n + ^{4}He. Data from Reference 12; 1 barn = 10^{-28} m^2.

where $\langle \sigma v \rangle$, the reaction rate probability or reactivity, has been averaged over the particle velocity distribution. If the reactants are in thermal equilibrium so that the velocity distributions are Maxwellian, then

$$\langle \sigma v \rangle = \frac{\int \sigma v \, dn}{\int dn} \qquad (3.28)$$

where

$$dn = n \left(\frac{\mu}{2 \pi k T} \right)^{3/2} v^2 e^{-\mu v^2 / 2kT} dv \qquad (3.29)$$

and

$$\mu = \frac{m_1 m_2}{m_1 + m_2}$$

Thus

$$\langle \sigma v \rangle = \frac{\int_0^\infty \sigma v^3 e^{-\mu v^2 / 2kT} dv}{\int_0^\infty v^2 e^{-\mu v^2 / 2kT} dv} \qquad (3.30)$$

The denominator of Equation (3.30) can be integrated, giving

$$\langle \sigma v \rangle = \frac{4}{\pi^{1/2}} \left(\frac{\mu}{2kT} \right)^{3/2} \int_0^\infty \sigma(v) v^3 e^{-\mu v^2/2kT} dv \tag{3.31}$$

Or, in terms of the relative interaction energy $E_r \equiv \frac{1}{2}\mu v_r^2$, where v_r is the relative velocity,

$$\langle \sigma v \rangle = \frac{4}{(2\pi\mu)^{1/2} (kT)^{3/2}} \int_0^\infty \sigma(E_r) E_r e^{-E_r/kT} dE_r \tag{3.31a}$$

The reactivity $\langle \sigma v \rangle$ has been numerically evaluated for many of the important fusion reactions; see, for example, References 16–18. Graphs of $\langle \sigma v \rangle$ for some of the most important fusion reactions are shown in Figures 3.2 and 3.3.

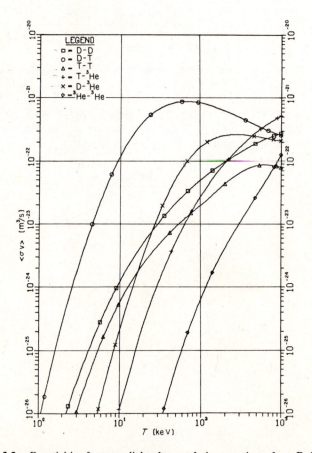

FIGURE 3.2.   Reactivities for some light-element fusion reactions; from Reference 17.

$y = mx + b$

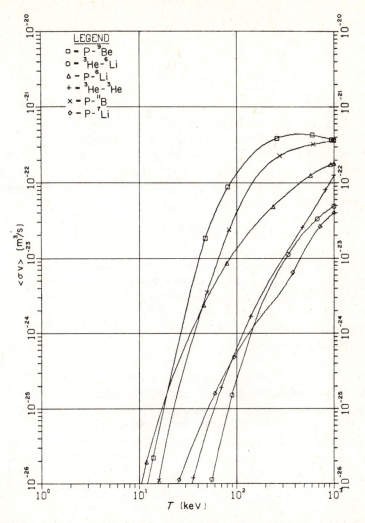

FIGURE 3.3. Some reactivities for advanced fusion reactions, which are essentially neutron free; from Reference 17.

How long does it take for the fusion fuel to burn? A characteristic reaction time τ_R for fusion can be estimated from Equation (3.27) as follows:

$$\frac{dn_i}{dt} = - a_{ij}n_i n_j \langle \sigma v \rangle \tag{3.32}$$

whose solution for constant $n_j \langle \sigma v \rangle$ is

$$n_i = n_{i0} e^{-t/\tau_R} \tag{3.33}$$

where

$$\tau_R = \frac{1}{a_{ij}n_j\langle\sigma v\rangle}$$

(3.34)

That is, fuel particles of kind i will decrease to e^{-1} of their initial value n_{i0} in a time τ_R. This assumes that the plasma temperature and n_j remain constant in time. For further discussion see Section 4.4.6. Although these quantities often change, τ_R is a useful measure of the order of magnitude of time needed to burn a significant fraction of the fuel. For example, a 10-keV D−T plasma whose initial fuel density is $n_D = n_T = 0.25 \times 10^{20}$ m^{-3} has (see Figure 3.3 and find, at 10 keV, $\langle\sigma v\rangle_{DT} \approx 10^{-22}$ m^3/s)

$$\tau_R = (0.25 \times 10^{20} \times 1 \times 10^{-22})^{-1} = 400 \text{ s}$$

It is interesting and important to note that for relatively low-temperature plasmas ($<10^8$ K) the particles in the high-energy tail of the Maxwell−Boltzmann distribution make the greatest contribution to the reaction. This is illustrated in Figure 3.4. For example, in a 1-keV ($\sim 10^7$ K) deuterium plasma, the largest contribution to the rate of the thermonuclear reaction comes from nuclear collisions between deuterons with a relative energy of about 6 keV. These very important particles constitute a very small fraction of the total number of particles. As the particles in the Maxwellian tail react, the depleted part of the velocity distribution gets repopulated by collisions, permitting the reaction to continue. To signify this important thermal repopulating process, fusion often carries with it the adjective "thermonuclear." Since the collision cross section for charged particles is small at high energies [see Equation (3.7); $\sigma \sim T^{-2}$], a truly Maxwellian distribution is established relatively slowly and if

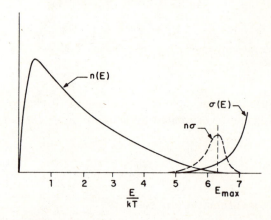

FIGURE 3.4. For low-temperature fusion plasmas (i.e., $<10^8$ K), particles in the high-energy tail of the Maxwell distribution contribute most to fusion.

the high-energy particles should become lost, the calculated reaction rates could not be achieved except at high temperatures.

Nuclear reaction rates can be larger than those indicated in Figure 3.3 if the nuclei are polarized. In 1982 it was suggested by Kulsrud et al. [19,20] that new techniques to produce bulk polarization might be used to fuel a fusion reactor with polarized hydrogenic atoms. Theoretical calculations indicate that, once the nuclei of a plasma are polarized in some preferred state, they may be able to maintain this state for a period of time that is long compared to the characteristic fusion reaction time. It is predicted that the D−T reaction rate can be enhanced by as much as 50% with the added intriguing consequence that the reaction products will be emitted perpendicular to the magnetic field. The D−D reaction rate enhancement is somewhat less certain, but might be changed by factors of 1.5−2.5. Fluctuations of the magnetic field, particle collisions, and interaction with the walls are among some of the phenomena that can depolarize nuclei. Whether polarized fusion is possible and practical is a topic of current research.

3.3. FUSION-POWER DENSITY

The rate at which fusion energy is produced per unit volume, P, is the product of the densities of the fuels, the reactivity, and the energy released per fusion event. Thus,

$$P_{ij} = a_{ij}n_in_j\langle\sigma v\rangle Q \qquad (3.35)$$

where $a_{ij} = \frac{1}{2}$ when $i = j$ and $a_{ij} = 1$ when $i \neq j$. For deuterium−tritium, $Q = 17.6$ MeV $= 2.82 \times 10^{-12}$ J, /mol

$$P_{DT} = 2.82 \times 10^{-12} n_D n_T \langle\sigma v\rangle_{DT} \qquad (3.36)$$

For D−T fuel with $n_D = n_T = 0.25 \times 10^{20}$ m^{-3} and $T = 10$ keV ($\sim 10^8$ K) the reactivity, from Figure 3.3, is $\langle\sigma v\rangle_{DT} \approx 1.2 \times 10^{-22}$ m^3/s, and the total D−T fusion power is 2.1×10^5 W/m^3.

The neutrons formed by the D−T reaction carry with them 80% of the energy, and they promptly leave the plasma. Their energy can be recovered in an absorbing blanket surrounding the reacting plasma. The alpha particles (^{4}He), carrying 20% of the D−T fusion energy, can remain within the plasma since they are ions, and their energy will be given in due time to the surrounding plasma particles. The maximum power remaining within the plasma from the ion ash is   20% of Q above in 3.36

$$P_{DT}(\text{ions}) = 5.64 \times 10^{-13} n_D n_T \langle\sigma v\rangle_{DT} \qquad (3.37)$$

or, for the plasma conditions listed above, $P_{DT}(\text{ions}) \approx 4.2 \times 10^4$ W/m^3.

See p 36. for Brem. radiation loss.

Power density vs, stress on reactor + heat-transfer + confinement pressure.

number density dualities

34   REACTION RATES, IGNITION, AND CONFINEMENT

Average acceptable mechanical stress and heat-transfer conditions on a fusion reactor wall usually limit the reaction power density to values less than 100 MW/m^3, and more typically, for steady-state, long-life conditions, to values between 1 and 10 MW/m^3. Therefore, at 10 keV, a steady-state D−T fusion reactor cannot sustain a plasma number density above about 10^{21} m^{-3}. The power density is proportional to the square of the reactant number density, and, therefore, power density is quite sensitive to changes in n. Fusion-reactor designers seek to have the reactant number density as large as is physically possible so as to keep the system size relatively small and, hence, economical. Reactor first-wall material strength and lifetime limitations together with plasma-confinement stability criteria usually set an upper limit on reactant density.

It is useful to note that a deuterium−tritium fusion reactor operating at average plasma conditions of 10 keV with $n_D = n_T = 1 \times 10^{21} \text{ m}^{-3}$ has a confined fusion power density $P_{DT} \sim 56 \text{ MW/m}^3$. The plasma pressure is given by

$$p = \Sigma n_j kT \qquad (3.38)$$

where n_j is the number density of the jth constitutent (electrons and all ions) which, for the conditions stated gives $p \simeq 30$ atm (or about 450 psi, or 3×10^6 Pa). The specific energy density e of a plasma is, from kinetic theory, given by

$$e = \tfrac{3}{2} \Sigma n_j kT \qquad (3.39)$$

which, for the stated plasma conditions, yields $e = 8 \times 10^6 \text{ J/m}^3$. Eight joules is an amount of energy that could raise the temperature of 1 g of water about 2 K. A 1000 MWe fusion-power plant operating at a power density of 50 MW/m^3 will have a plasma whose total mass is about 0.1 g, and whose total internal energy is about 5×10^8 J. This quantity of energy can only vaporize about 200 liters of water; that is, it is a relatively small quantity of energy. This is the basis for stating that the explosion hazard of the plasma in a fusion reactor is negligible. Also, should confinement of the plasma be lost, since the reaction rate is proportional to n^2, burning plasma would be quickly extinguished by rapid expansion.

An important plasma physics parameter, β, is defined as the dimensionless ratio of the plasma pressure to the magnetic field pressure; that is

$$\beta = \frac{p}{B^2/2\mu_0} = \frac{2\mu_0 nkT}{B^2} = 3.47 \times 10^{-29}\frac{nT}{B^2} \qquad (3.40)$$

where B is the magnetic induction and μ_0 is the vacuum magnetic permeability ($\mu_0 = 4 \pi \times 10^{-7}$ H/m). β is also proportional to the ratio of the plasma specific internal energy to the magnetic field energy density, and usually is a function of position; that is, $\beta (r)$ has a value at each point within a plasma that contains a magnetic field. Enclosed by brackets, $\langle \beta \rangle$, usually implies a volume average; β can have any positive value, but for magnetically confined equilibrium plasmas, $\beta \leqslant 1$.

Fusion power density P (see Section 3.3) has an important relation to the economics of magnetically confined fusion. The commercial output of a fusion-power plant is proportional to the fusion-power density, and, therefore, this quantity should be as large as the physics and technology will permit. Power density is related to β in the following way:

$$P \approx n^2 \langle \sigma v \rangle Q$$

and using Equations (3.38) and (3.40), we obtain

$$P = C_1 \beta^2 B^4 \frac{\langle \sigma v \rangle}{T^2} Q \tag{3.41}$$

where C_1 is a constant. Therefore, to maximize P one should have a fusion reactor that has as large values of β, B, and $\langle \sigma v \rangle Q/T^2$ as plasma physics and technology permit. The maximum value of β is determined by plasma-physics limits. The maximum value of B is determined mainly by strength of materials used to construct the magnetic coils, and the maximum value of $\langle \sigma v \rangle Q/T^2$ is determined by the choice of fusion fuel. In a real reactor the situation is more complex because of spatial gradients of all the quantities involved in Equation (3.41), but, nonetheless, fusion-reactor designers do seek to maximize β, B, and $\langle \sigma v \rangle Q/T^2$, so as to achieve economic power generation. Deuterium–tritium has, at relatively low temperatures, the largest values of $\langle \sigma v \rangle Q/T^2$ of any of the fusion fuels (by about a factor of 100), and that is one of the primary reasons it is considered as the fuel choice of early fusion devices.

3.4. IDEAL FUSION IGNITION

Consider a very large stationary volume of plasma whose temperature T and particle number density n are constant everywhere. Because there are no gradients, conduction and diffusion of energy are zero. By what mechanism, if any, will such an idealized plasma lose energy? It will lose energy by radiation. It is well known that when a free electron experiences acceleration, say from a collision with an ion, there is a probability of that electron emitting a light quantum and correspondingly decreasing its kinetic energy. If the electron remains free, such transitions are called free–free transitions, and the radiation emitted is called bremsstrahlung (braking radiation). There is also radiation associated with free–bound and bound–bound transitions, which are significant at low plasma temperatures ($E_i/kT \geqslant 1$, where E_i is the ionization energy of the atom). Since fusion concerns high-temperature plasma, for the ideal conditions considered in this section, we only need consider free-free radiation. If the electrons have a Maxwellian velocity distribution and are nonrelativistic, then the bremsstrahlung power radiated from a unit volume of plasma is given by

$$P_{ff} = \frac{32\pi e^6 n_e n_i Z^2}{3(4\pi\epsilon_0)^3 c^3 m_e h} \left(\frac{2\pi k T_e}{3m_e}\right)^{1/2} g_{ff} \tag{3.42}$$

where T_e is the electron gas temperature, Z is the ionic charge, h is Planck's constant, e is the electronic charge, c is the speed of light, and g_{ff} is the Gaunt factor, a quantum-mechanical correction whose magnitude at high temperature is $g_{ff} \approx 1.11$. An elementary derivation of Equation (3.42) can be found in Reference 5, and a more rigorous derivation can be found in the text by Bekefi [21]. Evaluating the physical constants, using the approximation $g_{ff} = 1.11$, and summing over different ionic species gives

$$P_{ff} = 1.59 \times 10^{-40} n_e \Sigma n_i Z_i^2 T_e^{1/2} \tag{3.43}$$

The mean free path of nearly all bremsstrahlung photons exceeds by many orders of magnitude the size of any magnetically confined fusion plasma (see Section 4.3). It is primarily for this reason that fusion plasmas do not radiate like a black body (power $\sim T^4$) but rather lose energy at a rate proportional to $T^{1/2}$. This lower radiation loss rate is what makes fusion ignition possible in a laboratory-sized plasma. But, the bremsstrahlung photons do provide an unavoidable energy loss from a fusion plasma. There are other mechanisms for energy loss, such as conduction, diffusion of particles, or cyclotron (or synchrotron) radiation. In a real system these losses must be taken into account. But, it is a convenient and useful concept to assume a finite but uniform plasma (sometimes referred to as zero dimensional), where the only energy loss from the plasma results from free–free radiation.

Ideal fusion ignition is defined as the lowest temperature at which the bremsstrahlung radiation power loss is exactly equal to the fusion power absorbed by the plasma. Neutrons, created by fusion reactions, have such long mean free paths that they, too, like bremsstrahlung photons, all leave the plasma. Only the ions that are created in the fusion events are considered to remain within a magnetically confined fusion plasma, where they slow down, depositing their energy to the surrounding plasma. Fusion-ignition temperature is obtained by solving the energy equation where there is a balance between bremsstrahlung power and fusion power that remains within the plasma. Thus,

$$a_{ij} n_i n_j \langle \sigma v \rangle Q_+ = 1.59 \times 10^{-40} n_e T_e^{1/2} \Sigma n_i Z_i^2 \tag{3.44}$$

where Q_+ is the fusion energy residing in charged (ions) fusion ash particles. Plots of both terms of Equation (3.44) are shown in Figure 3.5 for the D–T and D–D reactions. The ideal ignition temperature for D–T is found to be 4.4

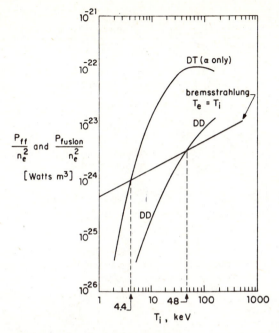

FIGURE 3.5. The reactant charged-particle fusion power for reactions D−T and D−D and the bremsstrahlung power are shown as a function of temperature. The intercepts indicate the ideal fusion-ignition temperatures. From Reference 22.

ignition temp.

keV and that for D−D is about 48 keV. Ignition in a real finite-sized plasma will require higher temperatures than the ideal values in order to account for energy losses from other than simply pure bremsstrahlung.

It is interesting to note that there are usually two roots of Equation (3.44). This is indicated schematically in Figure 3.6. The ignition root, shown as point 1, is unstable. That is, if the plasma temperature slightly exceeds T_{ig}, the charged-particle fusion power exceeds the bremsstrahlung loss, and, therefore, the plasma temperature will continue to increase. This process will, ideally, continue until these rates again come to equilibrium. This does happen because the fusion reaction power decreases at very high temperature. This is illustrated in Figure 3.7. The second root is stable, that is, small changes in T are such that the system will return to T_2. Although a temperature excursion will take place when ignition occurs in a real plasma, losses from increased impurity radiation, increased diffusion, or plasma instabilities will limit the temperature rise to less than the ideal maximum excursion value of about 400 keV for D−T.

High-Z impurities in the plasma can prevent ignition, because the bremsstrahlung loss rate increases very substantially with small concentrations of high-Z impurities in a fusion plasma. These impurities may come, for example, from sputtering of the vacuum wall, caused by impact of energetic plasma particles. Because free−free radiation power is proportional to $\Sigma n_i Z_i^2$, a very small percentage of a high-Z impurities can, as illustrated in Figure 3.6, raise

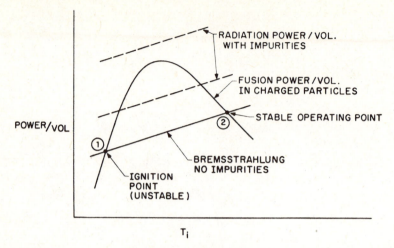

FIGURE 3.6. Schematic illustration of the two equilibrium points for fusion operation. Point 1, the ideal ignition condition, is unstable. Point 2, the higher-temperature point, is stable. A dashed line illustrates a case where the high-Z impurity concentration is sufficient to prevent ignition.

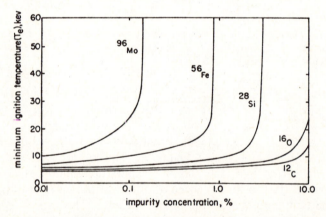

FIGURE 3.7. Effect of various impurities on ideal ignition temperature for a D—T plasma. Data from Reference 23.

the radiation loss rate above any fusion rate, so that no ignition may be physically possible. The percentages of some common impurities, which can prevent ideal fusion ignition, are shown in Figure 3.7.

3.5. IDEAL PLASMA-CONFINEMENT ($n\tau$) CRITERIA

What fundamental physical conditions must be satisfied in an ideal (zero-dimensional) fusion reactor so that interesting amounts of fusion power are produced? As we will now show, this question leads to a number of $n\tau$ criteria, whose value depends on what is considered "interesting" fusion power.

The simplest fusion-power objective to analyze is that obtained when the total fusion-power output is equal to the rate of power input to the plasma, that is, *power breakeven*. Thus,

$$\frac{a_{ij}n_i n_j \langle \sigma v \rangle Q}{\frac{3}{2}(kT/\tau_E)\Sigma n_j} = 1 \tag{3.45}$$

where τ_E is the energy-confinement time for the plasma, which includes all mechanisms by which energy is lost from the plasma. For a D−T reaction, $a_{ij} = 1$ and $n_e = 2n_D = 2n_T$, so that Equation (3.45) becomes

$$\frac{n_D \langle \sigma v \rangle Q \tau_E}{6kT} = 1$$

or

$$n_e \tau_E = \frac{12kT}{\langle \sigma v \rangle_{DT} Q} \tag{3.45a}$$

The minimum power-breakeven value of $n_e\tau_E$ for D−T ($Q = 17.6$ MeV = 2.82×10^{-12} J) is about 7×10^{19} m^{-3} s at $T \approx 10$ keV. Sometimes, an ignition-power breakeven is referred to and that implies Equation (3.45a), with Q replaced by Q_+. For D−T, ignition-power breakeven $n\tau$ is at least five times larger than the minimum $n_e\tau_E$ (based upon Q) and at D−T ignition, (at $T \approx 5$ keV, $n_e\tau_E = 5.7 \times 10^{20}$m^{-3} s.

Another interesting $n\tau$ value results from equating the energy produced by fusion to the energy input to the plasma, that is, *energy breakeven*. Thus, for D−T (50% each),

$$\frac{n_D n_T \langle \sigma v \rangle Q \tau_P}{P_{ff}\tau_p + \frac{3}{2}(n_e + n_D + n_T)kT} = 1 \tag{3.46}$$

where τ_p is the pulse length of the confinement system, during which all plasma quantities are ideally assumed to be constant. Again, solving for $n_e\tau_p$,

$$n_e\tau_p = \frac{12kT}{\langle \sigma v \rangle Q - 6.28 \times 10^{-40}T^{\frac{1}{2}}} \tag{3.46a}$$

which has a minimum $n_e\tau_p$ value at $T \approx 28$ keV where $n_e\tau_p \approx 3.2 \times 10^{19}$m^{-3} s.

J. D. Lawson [24] was the first person to recognize the importance of the quantity $n\tau$. He analyzed this general (zero-dimension) situation but asked for what conditions a fusion plant would have energy left over (for sale) after a pulse length τ_p, where some of the energy from the first pulse must be used to heat the next cold fuel input to a temperature T. He assumed an efficiency η for the conversion of thermal energy into electrical energy (or stored energy). The Lawson analysis began with

$$\eta(E_{\text{fusion}} + E_{\text{supplied}}) \geq E_{\text{supplied}} \tag{3.47}$$

where E_{fusion} is the fusion energy yield obtained during a pulse time τ_p and E_{supplied} is the energy input to heat the gas to temperature T and to make up for radiation loss. Dividing by E_{supplied}, one obtains,

$$\eta\left(\frac{a_{ij}n_i n_j \langle \sigma v \rangle Q\tau_p}{1.59 \times 10^{-40} n_e T^{1/2} \Sigma n_i Z_i^2 \tau_p + \frac{3}{2}(n_e + n_i)kT} + 1\right) \geq 1$$

For D–T with $n_e = 2n_D = 2n_T$, and solving for $n_e\tau_p$ ($= n_i\tau_p$) yields the Lawson $n\tau$ criterion,

$$n_e\tau_p = \frac{(\eta^{-1} - 1)12kT}{\langle \sigma v \rangle Q - (\eta^{-1} - 1)6.28 \times 10^{-40} T^{1/2}} \tag{3.47a}$$

Lawson originally found, for $\eta = \frac{1}{3}$, that for D–T, $n_i\tau_p \geq 10^{20}$ m^{-3} s and for D–D, $n_i\tau_p \geq 10^{22}$ m^{-3} s. The $n_e\tau_p$ confinement criterion for D–D is about two orders of magnitude larger than for D–T.

These $n\tau$ values are indicative of the minimum length of time a plasma of density n must be confined to yield a breakeven or net-energy-yield condition. It should be noted that the energy-confinement time, τ_E, is not the same as the pulse length τ_P. These simple expressions for $n\tau$, whether for power breakeven or Lawson net-energy breakeven, etc., are just indicative of the character of the confinement problem. In real systems, which have spatial gradients, impurities, etc., a much more detailed analysis is required to determine breakeven or net-energy-gain conditions. The effect of high-Z impurities on $n\tau$ and ignition conditions have been studied by a number of investigators, see, for example, References 25–27.

It is interesting to note that when the denominator of Equation (3.46a) uses Q_+ and is set equal to zero, the ignition temperature is recovered (i.e., when $n_e\tau_p \to \infty$). The same situation occurs for the Lawson criterion when $\eta = 0.5$. A plot of the Lawson criterion for D–T, $\eta = \frac{1}{2}$, and $Q_+ = 3.5$ MeV is shown in Figure 3.8, and $n_i\tau_p = 1.5 \times 10^{20}$ m^{-3} s is obtained as a minimum at $T \sim 25$ keV. As $n\tau \to \infty$, the ignition temperature is recovered.

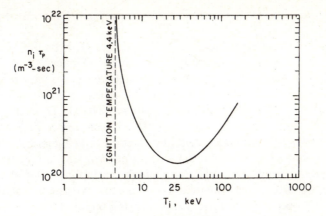

FIGURE 3.8. The Lawson criteria $n_i \tau_p$ is shown for a D−T fusion system energy breakeven, with $\eta = \frac{1}{2}$ and $Q_+ = 3.5$ MeV. For these conditions, minimum $n_i \tau_p \simeq 1.5 \times 10^{20}$ m^{-3} s. Data from Reference 22.

3.6. FUSION-FUEL-CYCLE CHARACTERISTICS

It is now possible to draw some conclusions about the potential advantages and disadvantages of different fusion fuels. Among the 50 or so fusion reactions listed in Appendix C, the following are the most important.

3.6.1. Deuterium−Tritium (Lithium)

The D−T reaction has, among all fusion fuels, the lowest ideal ignition temperature (~5 keV) and the highest power-density factor $\langle \sigma v \rangle Q / T^2$. It has the highest reactivity $\langle \sigma v \rangle$ below 100 keV. Eighty percent of the fusion energy is deposited away from the first wall surface, and it has the potential to burn steady state. Among the disadvantages are the facts that only 20% of Q is available to heat the plasma, and tritium is radioactive (12.3 year half life; β decay; $E_{max} = 18$ keV). The 14-MeV neutrons produced by the fusion reaction activate the structure and damage the first wall (atom displacements and helium swelling). The need to use lithium (a few hundred tons in a large power reactor) to breed tritium complicates the heat-transfer design, and in some cases may be a fire hazard. The need to separate the tritium from the lithium, and from the fusion exhaust, further complicates the reactor system. Nonetheless, the advantages of D−T are sufficient that nearly all research and development plans contemplate using this fuel in first-generation fusion reactors.

3.6.2.   Deuterium Only (Pure; Partially Catalyzed; Fully Catalyzed)

The advantages of a deuterium fusion reactor are that the fuel is available at very low cost and in very large quantity. A D−D reactor will have a very modest tritium inventory (∼ 1 g), and no fuel breeding is required. Although the total neutron flux to the first wall is comparable to that from a D−T reactor, only about 45% of these are 14-MeV neutrons. The D−D reactor has significantly higher ignition temperature (∼ 50 keV) and $n\tau$ requirements, and for reasonable reactor economics, requires a significantly higher β (about three to four times) than a D−T system. Because the neutrons are not required to breed fusion fuel, they could be used to breed nuclear weapons material, and there is thus a "safeguard" problem with a D−D reactor. A partially catalyzed D−D reaction will occur because the tritium created by one D−D reaction branch will rapidly burn. The fully catalyzed D−D cycle requires fuel concentrations of D, T, and ^{3}He such that all their burn rates are the same. Thus

$$\frac{n_T}{n_D} = \frac{\langle \sigma v \rangle_{DD}}{2\langle \sigma v \rangle_{DT}}$$

and

$$\frac{n_{^3He}}{n_D} = \frac{\langle \sigma v \rangle_{DD}}{2\langle \sigma v \rangle_{D^3He}}$$

The tritium concentration is small (typically 1%) because of the large reactivity of D−T, but at low temperature the ^{3}He concentration becomes large and the overall reaction rate is low because of the small reactivity of D−^{3}He. Studies have been made of the catalyzed deuterium fusion reactor by Mills [28] and McNally and Rothe [29].

3.6.3.   Advanced Fuels (D−^{3}He; ^{3}He−^{3}He; Proton-Based Fuels)

D−^{3}He is, second only to D−T, the most reactive fusion fuel. Neutron production (from D−D reactions) is low, and it has the potential for steady-state operation. However, ^{3}He is very rare, and this fuel would have to be bred. It has a higher ignition temperature (∼ 30 keV) and requires higher $n\tau$ and β values than D−T. The ^{3}He−^{3}He has a still lower reactivity and suffers from the serious defect that essentially no fuel is naturally available.

Proton-based fuel possibilities (p−^{6}Li; p−^{7}Li; p−^{9}Be; p−^{11}B) are considered "advanced" because they produce very little radioactivity and neutrons. They require very high temperatures (300−900 keV) and very high $n\tau$ values. It is uncertain whether they can ignite. Because of the very high temperatures required, there exists the prospect of chain reactions, but synchrotron radiation losses also become important. Basic studies of these advanced fusion reactions are being made by several groups [29−31]. Whether any of these advanced fuels can be a practical source of fusion energy is uncertain and is therefore a subject of current research.

REFERENCES

1. E. W. McDanniel, *Collision Phenomena in Ionized Gases,* Wiley, New York, 1964.

2. C. F. Barnett, J. A. Ray, and J. C. Thompson, *Atomic and Molecular Collision Cross Sections of Interest in CTR,* ORNL-Report 3113, revised, Aug. 1964.

3. R. L. Freeman and E. M. Jones, *Atomic Collision Processes in Plasma Physics Experiments,* Culham Report CLM-R-137, 1974.

4. M. R. C. McDowell and A. M. Ferendeci (eds.), *Atomic and Molecular Processes in Controlled Thermonuclear Fusion,* Plenum Press, New York, 1980.

5. K. Miyamoto, *Plasma Physics for Nuclear Fusion,* MIT Press, Cambridge, MA, 1980.

6. W. A. Fowler, *Phys. Rev.* **81**, 655 (1951).

7. H. V. Argo, R. Taschek, H. Agnew, A. Hemmendinger, and W. Leland, *Phys. Rev.* **87**, 612 (1952).

8. W. R. Arnold, J. A. Phillips, G. A. Sawyer, E. J. Stovall, Jr., and J. L. Tuck, *Phys. Rev.* **93**, 483 (1954).

9. R. F. Post, *Rev. Mod. Phys.* **28**, 338 (1956).

10. J. L. Tuck, *Nuclear Fusion* **1**, 201 (1961).

11. W. A. Fowler, G. R. Caughlan, and B. A. Zimmerman, *Ann. Rev. Astron. Astrophys.* **5**, 525 (1967).

12. L. Stewart and G. M. Hale, Los Alamos Scientific Laboratory Report LA-5828-MS, Jan. 1975.

13. G. Gamow, *Phys. Rev.* **53**, 59 (1938); see also, *Structure of Atomic Nuclei and Nuclear Transformations,* Oxford University Press, England, 1937.

14. L. I. Schiff. *Quantum Mechanics,* McGraw-Hill, New York, 1968, p. 278.

15. W. B. Thompson, *Proc. Phys. Soc.* **79**, B, 1 (1957).

16. S. L. Greene, Jr., *Maxwell Averaged Cross Sections for Some Thermonuclear Reactions on Light Isotopes,* Lawrence Radiation Laboratory Report UCRL-70522, May 1967.

17. J. R. McNally, Jr., K. E. Rothe, and R. D. Sharp, *Fusion Reactivity Graphs and Tables for Charged Particle Reactions,* ORNL/TM-6914, Aug. 1979.

18. R. J. Howerton, *Maxwell-Averaged Reaction Rates ($\overline{\sigma v}$) for Selected Reactions Between Ions with Atomic Mass ≤ 11,* UCRL-50400, Vol. 21, Part A, Feb. 1979.

19. R. M. Kulsrud, H. P. Furth, E. J. Valeo, and M. Goldhaber, *Phys. Rev. Lett.* **49**, 1248 (1982).

20. R. M. Kulsrud et al., *Fusion Reactor Plasmas with Polarized Nuclei,* IAEA-CN-41/P2, Baltimore, September 1982.

21. G. Bekefi, *Radiation Processes in Plasmas,* Wiley, New York, 1966.

22. S. S. Penner and A. Hockstim (eds.) *Nuclear Energy and Energy Policies,* Vol. 3. Addison-Wesley, Reading, MA., 1976, p. 175.

23. W. M. Stacey et al., *Tokamak Experimental Power Reactor Studies,* Argonne National Laboratory Report. ANL/CTR-75-2, June, 1975, p. 1−17.

24. J. D. Lawson, "Some Criteria for a Power Producing Thermonuclear Reactor," *Proc. Phys. Soc.* **70**, pt. 1, no. 445, B, 6−10 (1957).

25. R. V. Jensen, D. E. Post, W. H. Grasberger, C. B. Tarter, and W. A. Lokke, *Calculations of Impurity Radiation and its Effects on Tokamak Experiments*, Princeton Report PPPL-1334, March 1977.

26. D. M. Meade, "Effect of High Z Impurities on the Ignition and Lawson Conditions for Thermonuclear Reactor," *Nuclear Fusion* **14**, 289 (1974).

27. R. Carruthers, P. A. Davenport, and J. T. D. Mitchell, *The Economic Generation of Power from Thermonuclear Fusion*, Culham Report CLM-R-85, 1967.

28. R. G. Mills, *Catalyzed Deuterium Fusion Reactor*, Princeton Technical Memo, TM-259, April 1971.

29. J. R. McNally, Jr. and K. E. Rothe, "Advanced Fusion Fuels—A Review," Preprint, Second International Conference on Emerging Nuclear Energy Systems, Lausanne, Switzerland, April 23–25, 1980.

30. J. M. Dawson, *CTR Using the $p-{}^{11}B$ Reaction,* UCLA Report PPG-273, Aug. 1976. Also, Invited Lecture, APS New York meeting, Jan. 1979.

31. R. W. Conn et al., Paper IAEA-CN-38/V-5. 8th International Conference on Plasma Physics and Controlled Nuclear Fusion Research. Brussels, 1980. Also, UCLA Report on Alternate Fusion Fuel Cycle Research, PPG-492, July 1980.

32. J. R. McNally, Jr. and R. D. Sharp. "Cross Section Parameters Used in Advanced Fusion Fuels Code." ORNL Theory Section Memo #76/68 (Aug. 12, 1976).

4 | BASIC PLASMA DIRECTORY

4.1. INTRODUCTION

When undergoing thermonuclear reactions, fusion fuel is an ionized gas called a plasma. In physics, the plasma state usually means a high-temperature gas consisting of free electrons and ions exhibiting collective behavior. Such a collection of charged particles will display characteristics of a fluid that can interact with electromagnetic fields. Collective behavior, such as waves, is a manifestation of a true plasma. A plasma must have many particles within a sphere whose radius is the Debye length (see Section 4.3.1). Because of simultaneous electromagnetic interactions and particle–particle interactions, there are many time scales for a plasma. In contrast, an ordinary simple gas possesses only one basic time scale, the particle–particle collision time.

The physical characteristics and the behavior of plasmas constitute the field of study of plasma physics. Since much of the matter in our universe, for example, stars and interstellar gas clouds, is in the plasma state, astrophysicists have had a long-standing interest in this subject. There is also a long history of research by scientists and engineers studying gaseous discharges, electric arcs, and ionospheric phenomena. Fusion energy has attracted great attention to plasma physics and in particular to the behavior of magnetically confined, high-temperature laboratory plasmas. There exists a large literature and many good books on plasma physics. For example, there is the introductory text by Chen [1], the concise and pioneering book by Spitzer [2], and the newer and comprehensive text on fusion plasmas by Miyamoto [3].

It is important to have an understanding of plasma physics to work in developing fusion energy. Yet many who contribute to this field need not be plasma physics specialists. The development of fusion energy requires very important and substantial contributions from specialists in the fields of mechanical, electrical, civil, and nuclear engineering as well as from the fields of metallurgy, high vacuum, and cryogenics. In this chapter, some of the basic concepts from plasma physics are summarized and appropriate formulas are given. Either an original reference or a particularly clear exposition is cited so that the details of the derivations and more complete discussion can be sought. The reader is urged to study these sources. This chapter is a short handbook of basic plasma physics, and the usual caveats associated with collections of formulas must be borne in mind. This chapter is not an attempt to summarize all of plasma physics. It is a collection of some of the most important plasma length and time scales, and other characteristic phenomena relevant to fusion plasmas. An important part of plasma physics concerns different methods of analysis, and different analytical points of view such as magnetohydrodynamics or plasma kinetic theory. The study of waves and instabilities also constitutes a substantial part of plasma physics. These are worthy of study in their own right and are not included in this summary. The particular topics of plasma equilibrium, stability, heating, and transport are, for the most part, discussed in later chapters, where specific fusion concepts are analyzed.

In all formulas in this text, SI (metric) units [4] have been used. In particular, T is in K, n is in m^{-3}, B is in T, etc. For clarity the Boltzmann constant $k = 1.380 \times 10^{-23}$ J/K is always included in analytical formulas.

4.2. IONIZATION EQUILIBRIUM

What percentage of atoms are ionized in a gas in thermodynamic equilibrium at temperature T and pressure p? This question was first studied by Saha in 1920 [5] in his investigation of the solar chromosphere. Boltzmann had previously shown that a gas in thermal equilibrium has the relative number of atoms n_A and n_B in two separate energy levels A and B, in the ratio

$$\frac{n_B}{n_A} = \frac{g_B}{g_A} e^{-E_{AB}/kT} \tag{4.1}$$

where g_j is the statistical weight of state j (A or B), which is equal to $2J_i + 1$; J_i is the quantum number for angular momentum. E_{AB} is the energy difference between states A and B. By introducing the partition function $u(T)$ defined by

$$u(T) = \Sigma g_j e^{-E_j/kT} \tag{4.2}$$

together with the Boltzmann equation (4.1), Saha obtained the equilibrium number density of ionized atoms:

$$\frac{n_i n_e}{n_a} = \left(\frac{2\pi m_e kT}{h^2}\right)^{3/2} \frac{2u_i(T)}{u_a(T)} e^{-E_i/kT} \tag{4.3}$$

where subscripts i,e,a refer to ions, electrons, and atoms, respectively, h is Planck's constant $(6.62 \times 10^{-34}$ J/s$)$, and E_i is the ionization energy. A somewhat clearer derivation of Equation (4.3) can be found in the astrophysics text of Aller [6]. It is assumed that only ground-state or singly ionized atoms of one species are present in the gas. Tabulated values of partition functions for a variety of atomic species can be found in the book by Aller [6], or in Allen, *Astrophysical Quantities* [7].

Another useful form of Equation (4.3), with the electron gas pressure $p_e = n_e kT$, is

$$\frac{n_i p_e}{n_a} = \left(\frac{2\pi m_e}{h^2}\right)^{3/2} (kT)^{5/2} \frac{2u_i(T)}{u_a(T)} e^{-E_i/kT} \tag{4.4}$$

or, with the numerical constants evaluated,

$$\frac{n_i p_e}{n_a} = 6.67 \times 10^{-2} T^{5/2} \frac{u_i(T)}{u_a(T)} e^{-E_i/kT} \tag{4.4a}$$

where p_e is in pascals (1 standard atmosphere $= 1.013 \times 10^5$ Pa).

For higher stages of ionization,

$$\frac{n_{q+1} n_e}{n_q} = \left(\frac{2\pi m_e kT}{h^2}\right)^{3/2} \frac{2u_{q+1}(T)}{u_q(T)} e^{-E_{\Delta q}/kT} \tag{4.5}$$

where n_{q+1} is the number density of atoms ionized to the $q + 1$ stage, and $E_{\Delta q}$ is the energy needed to change the ionization from the q to the $q + 1$ state.

The partition function ratio u_i/u_a is a slowly varying function of temperature, which is $\leqslant 1$. An examination of, for example, Equation (4.4a) reveals that the exponential term dominates the equation as the percentage of ionization increases. When $kT \geqslant E_i$, the gas is highly ionized. For fusion conditions $E_i/kT \approx 10^{-3}$, so that the gas is completely ionized, although statistically there are always a few neutral atoms. However, the edge region of a plasma, near the walls, will be at much lower temperature ($kT \sim 1-100$ eV), and in such regions the gas will be only partially ionized and many neutral atoms can be present.

It takes some time, on the order of at least several collision times, for a gas to achieve thermodynamic equilibrium. There are physical situations, such as the start of preionization, shock waves, and some astrophysical phenomena, where the time scales are shorter than that required to achieve chemical equilibrium. In such circumstances the Saha equation should not be used.

4.3.   CHARACTERISTIC PLASMA LENGTH SCALES

4.3.1.   Debye Length

One of the most important and fundamental length scales in plasma physics is the Debye length λ_D defined as

$$\lambda_D = \left(\frac{\varepsilon_0 k T_e}{n_e e^2}\right)^{1/2}$$

(4.6)

or

$$\lambda_D = 6.90 \times 10^1 \left(\frac{T_e}{n_e}\right)^{1/2}$$

(4.6a)

The Debye length or shielding distance is a measure of the distance that the electric field of a particle will penetrate into the surrounding sea of charged particles. If the number of particles within a sphere of radius λ_D is much greater than one, then these charged particles will exhibit many-body or collective effects. Langmuir defined a plasma as an ionized gas in which all other length scales of interest are larger than λ_D and $\frac{4}{3} n\pi \lambda_D^3 \gg 1$.

The derivation of Equation (4.6) can be found in every textbook of plasma physics, for example, References 1–3. Plasmas are essentially neutral, that is, $n_e \approx n_i$ when averaged over volumes greater than a Debye sphere. Electric sheaths, where $n_e \neq n_i$, such as those found at the boundary of plasmas in contact with solid surfaces, have a thickness of the order of λ_D. For a fusion plasma with $T_e = 10^8$ K and $n_e = 10^{20}$ m^{-3}, $\lambda_D = 6.5 \times 10^{-5}$ m and within a Debye sphere there are about 10^8 electrons and a comparable number of ions.

4.3.2.   Impact Parameter for 90° Binary Coulomb Scattering

Kinetic theory is frequently used to predict the properties of plasmas. An important length scale that arises in kinetic theory of plasmas is the impact parameter b_0 for 90° binary Coulomb scattering. Two particles, each with charge Ze, will be scattered by 90° from their original trajectories if, prior to their interaction, their undeflected trajectories would have brought them to within a distance b_0 of each other. An analysis of binary interaction gives

$$b_0 = \frac{(Ze)^2}{4\pi\varepsilon_0 \mu v_r^2}$$

(4.7)

where μ is the reduced mass equal to $m_1 m_2 / (m_1 + m_2)$ and v_r is their relative velocity. An excellent esposition of binary scattering can be found, for example, in the text *Classical Mechanics* by Goldstein [8]. If the gas consists of particles that have a Maxwellian velocity distribution and the approximation $\mu v_r^2 = 3kT$ is made, then the average impact parameter is

$$\bar{b}_0 = \frac{(Ze)^2}{12\pi\varepsilon_0 kT} \tag{4.7a}$$

or

$$\bar{b}_0 = 5.57 \times 10^6 \frac{Z^2}{T} \tag{4.7b}$$

In plasmas, multiple small-angle scattering has a much larger effect on particles than the occasional binary large-angle scattering event. Also, electric field fluctuations and microturbulence can significantly alter plasma properties from those calculated from basic kinetic theory. For $Z = 1$ and $T = 10^8$ K, Equation (4.7b) gives $\bar{b}_0 \approx 10^{-14}$ m, that is, about 60 times the classical size of the nucleus. Classical mechanics description of the interaction of particles over distances smaller than the de Broglie wavelength (see Section 4.3.4) is not valid, and quantum-mechanical treatment is needed.

4.3.3. Gyro Radii

Charged particles spiral about magnetic lines of induction. A balance between the Lorentz and centrifugal forces results in a gyro, or cyclotron, radius r_c, given by

$$r_c = \frac{mv_\perp}{ZeB} \tag{4.8}$$

where $v_\perp$ ($v_\parallel$) is the particle's velocity component perpendicular (parallel) to **B**, and $v^2 = v_\perp^2 + v_\parallel^2$. For equipartition of energy, two-thirds of the particle's energy is $\frac{1}{2} mv_\perp^2$ and one-third is $\frac{1}{2} mv_\parallel^2$. Taking $mv^2 = 3kT$, gives

$$r_c = \frac{(2mkT)^{1/2}}{ZeB} \tag{4.8a}$$

The gyro radius is sometimes referred to as the cyclotron or Larmor radius. The direction of the gyration is always such that the magnetic field induced by the charged particle motion is opposite to the externally imposed magnetic field; that is, plasmas are a diamagnetic medium. For a derivation and further discussion see Chen [1].

For electrons and ions, respectively,

$$r_{ce} = 3.13 \times 10^{-8} \frac{T_e^{\frac{1}{2}}}{B} \tag{4.8b}$$

and

$$r_{ci} = 1.34 \times 10^{-6} \frac{A T_i^{\frac{1}{2}}}{ZB} \tag{4.8c}$$

For hydrogen ($A = Z = 1$) at $T = 10^8$ K, in a magnetic field whose strength is $B = 1$ T, $r_{ce} = 3.13 \times 10^{-4}$ m and $r_{ci} = 1.35 \times 10^{-2}$ m.

A D–T fusion reaction generates alpha particles ($A = 4$, $Z = 2$) which are born with an initial energy of 3.5 MeV. Noting that $m_\alpha = 6.64 \times 10^{-27}$ kg, and on the average these newly born particles have $v_\perp = 1.06 \times 10^7$ m/s and $v_\parallel = 7.5 \times 10^6$ m/s, the average new-born alpha particle has a gyro radius given by

$$r_{c\alpha} = \frac{2.20 \times 10^{-1}}{B} \tag{4.8d}$$

For a fusion reactor with $B = 5$ T, we find $r_{c\alpha} = 4.4 \times 10^{-2}$ m or about 4 cm.

4.3.4. Mean Free Path

The mean free path of a particle in a plasma is defined by the distance required to deflect the particle's momentum 90° from its initial trajectory. By multiple small-angle scattering the mean free path is

$$\lambda = \frac{25\pi\varepsilon_0^2(kT)^2}{Z^4 n e^4 \ln\Lambda} \tag{4.9}$$

Derivation of this analytical expression for the mean free path of a plasma particle can be found in the text by Miyamoto [3]. This result is within a few percent of the more-detailed numerical calculation of Spitzer and Harm [9] and is sufficient for nearly all plasma estimates. The quantity Λ is the dimensionless ratio of the Debye length to the 90° impact parameter, that is,

$$\Lambda = \frac{\lambda_D}{b_0} = (12\pi n \lambda_D^3) = \frac{12\pi}{Z^2 n_e^{\frac{1}{2}}} \left(\frac{\varepsilon_0 kT}{e^2} \right)^{\frac{3}{2}} \tag{4.10}$$

This parameter, Λ, in plasma physics always occurs as $\ln \Lambda$, which is a slowly varying function of T and n_e; it is called the Coulomb logarithm. Evaluating the physical constants gives

$$\lambda = 1.78 \times 10^9 \frac{T^2}{Z^4 n \ln \Lambda} \tag{4.9a}$$

and

$$\Lambda = 1.24 \times 10^7 \frac{T^{3/2}}{Z^2 n_e^{1/2}} \tag{4.10a}$$

For a fusion plasma with $T_i = T_e = 10^8$ K, $Z = 1$, $n_e = n_i$, $n_e + n_i = 10^{20}$ m^{-3}, we find $\Lambda = 1.75 \times 10^9$, $\ln \Lambda = 21.3$, and $\lambda_i = \lambda_e = 1.6 \times 10^4$ m (16 km!). The 90° Coulomb scattering cross section $\sigma_{90°}$ is obtained from Equation (4.9), from the relationship

$$\lambda_{90°} = \frac{1}{n \sigma_{90°}} \tag{4.11}$$

giving

$$\sigma_{90°} = \frac{Z^4 e^4 \ln \Lambda}{25 \pi \varepsilon_0^2 (kT)^2} \tag{4.12}$$

or

$$\sigma_{90°} = 5.62 \times 10^{-10} \frac{Z^4 \ln \Lambda}{T^2} \tag{4.12a}$$

The Coulomb elastic 90° scattering cross section at fusion conditions ($T = 10^8$ K, $Z = 1$, $n_e = 0.5 \times 10^{20}$ m^{-3}) is $\sigma_{90°} = 9.4 \times 10^{-25}$ m^2. This can be compared with the cross section for D−T fusion reaction (see Figure 3.2) which at 10^8 K(~ 10 keV) is $\sigma_{DT} \sim 2 \times 10^{-31}$ m^2. The reaction mean free path for D−T at 10^8 K is nearly 5×10^6 times longer than the 90° elastic-scattering mean free path.

Neutrons born in fusion events will have a mean free path in the plasma, assuming only nuclear force interactions with plasma ions ($\sigma_n \sim 1$ barn = 10^{-25} m^2), given by

$$\lambda_n \approx \frac{10^{28}}{n_i} \tag{4.13}$$

In a magnetically confined plasma with $n_i \approx 0.5 \times 10^{20}$ m^{-3}, the neutron mean free path $\lambda_n \sim 10^8$ m. Thus all fusion neutrons leave the fusion-reaction chamber.

The mean free path for bremsstrahlung photons of frequency v in a plasma is

$$\lambda_p^v = 2.05 \times 10^2 \frac{T_e^{\frac{1}{2}} v^3}{Z^3 n_i^2} \tag{4.14}$$

The power spectrum of bremsstrahlung photons has a sharp maximum. For a fusion plasma with $T_e = 10^8$ K, this maximum occurs at about 1Å. These hard x-rays in a hydrogen plasma with $n_i = 0.5 \times 10^{20}$ m^{-3} have a mean free path $\lambda_p = 6 \times 10^{22}$ m. Nearly all of the bremsstrahlung photons (except those with long wavelength, and therefore low energy) have mean free paths much greater than the size of a fusion plasma in a reactor and, hence, they too leave the plasma and will be absorbed in the surface of the first wall.

It should be recalled that classical physics is valid only when the physical length scales are larger than the de Broglie wavelength λ_{DB}, which for electrons is

$$\lambda_{DB} = \frac{h}{m_e v_e} \approx \frac{h}{(3 m_e kT)^{\frac{1}{2}}} \tag{4.15}$$

where h is Planck's constant $= 6.625 \times 10^{-34}$ J s. Or,

$$\lambda_{DB} \approx \frac{1.08 \times 10^{-7}}{T^{\frac{1}{2}}} \tag{4.15a}$$

For electrons at $T = 10^8$ K, $\lambda_{DB} = 10^{-11}$ m. Note that the Coulomb 90° impact parameter $\bar{b}_0$ is, for these conditions, about 10^{-14} m and hence is much smaller than λ_{DB}. Therefore, a correct analysis of close-scattering events and some radiation phenomena at fusion conditions requires quantum-mechanical treatment.

4.3.5. Skin Depth

The amplitude of an electromagnetic wave of frequency ω will attenuate to e^{-1} of its value at the surface of a media of conductivity σ in a distance δ, called the skin depth, given by

$$\delta = \left(\frac{2}{\mu_0 \omega \sigma}\right)^{\frac{1}{2}} \tag{4.16}$$

Using the classical expression for the plasma conductivity [see Equation (4.81a)], the skin depth for a uniform plasma slab is

$$\delta = 1.05 \times 10^4 \left(\frac{Z \ln\Lambda}{\omega T_e^{3/2}} \right)^{1/2} \qquad (4.16a)$$

where it is assumed that $\omega < \omega_p$.

For magnetically confined fusion plasmas these characteristic lengths nearly always have the following ordering:

$$\bar{b}_0 \ll \lambda_{\mathrm{DB}} \ll \lambda_D \lesssim r_{ce} \ll r_{ci} \ll r_\alpha \lesssim L \ll \lambda_i \approx \lambda_e \ll \lambda_n \ll \lambda_p^v$$

where L is the characteristic size of a plasma in a magnetic fusion reactor.

4.4.   CHARACTERISTIC PLASMA TIMES AND FREQUENCIES

4.4.1.   Collision Times

The mean time between collisions τ for a charged particle in a plasma is defined in terms of its mean thermal speed and the distance required to change the particle's momentum 90° from its initial trajectory, that is,

$$\tau = \frac{\lambda}{v} \qquad (4.17)$$

Using $\bar{v} = (3kT/m)^{1/2}$, the speed associated with the mean energy, and the mean free path given in Section 4.3.4, we obtain

$$\tau = \frac{25\pi m^{1/2} \varepsilon_0^2 (kT)^{3/2}}{3^{1/2} Z^4 n e^4 \ln\Lambda} \qquad (4.18)$$

The self-collision times for electrons τ_{ee} and ions τ_{ii} respectively, are:

$$\tau_{ee} = 2.64 \times 10^5 \frac{T_e^{3/2}}{n_e \ln\Lambda} \qquad (4.18a)$$

$$\tau_{ii} = 1.13 \times 10^7 \frac{A^{1/2} T_i^{3/2}}{Z^4 n_i \ln\Lambda} \qquad (4.18b)$$

When an ion is interacting with an electron plasma, the time to deflect the ion 90° from its initial momentum is called the ion−electron collision time τ_{ie} whose value is calculated [3] to be

$$\tau_{ie} = \frac{(2\pi)^{\frac{1}{2}}3\pi\varepsilon_0^2 m_i (kT_e)^{\frac{3}{2}}}{Z^2 n_e m_e^{\frac{1}{2}} e^4 \ln\Lambda} \tag{4.19}$$

or

$$\tau_{ie} = 2.51 \times 10^8 \frac{A T_e^{\frac{3}{2}}}{Z^2 n_e \ln\Lambda} \tag{4.19a}$$

For a D−T fusion plasma at $T = 10^8 \, K$, with $n_e = n_i$ and $n_e + n_i = 10^{20} \, \text{m}^{-3}$, $Z = 1$, $A = 2.5$ (averaged for 50% D−T), we find $\tau_{ee} = 2.5 \times 10^{-4}$ s, $\tau_{ii} = 1.7 \times 10^{-2}$ s, and $\tau_{ie} = 0.59$ s.

The collision times are seen to be in the ratios

$$\tau_{ee} : \tau_{ii} : \tau_{ie} = 1 : \frac{1}{Z^3}\left(\frac{m_i}{m_e}\right)^{\frac{1}{2}} \left(\frac{T_i}{T_e}\right)^{\frac{3}{2}} : \frac{m_i}{Z^2 m_e}$$

When there are more than one ionic species present in a plasma, it is convenient to define an "effective" Z. This Z_{eff} is used as a weighted average of the ionic charge in calculations of momentum transfer. It is defined as

$$Z_{\text{eff}} = \frac{\Sigma n_k Z_k^2}{n_e} = \frac{\Sigma n_k Z_k^2}{\Sigma n_k Z_k}$$

where the sum is taken over all the ionic species.

4.4.2. Energy Relaxation Times

The rate at which equipartition of energy is established between two groups of particles, each characterized initially by their own separate temperature, has been studied by Spitzer [2] and others. Following Spitzer, we define an energy equilibration time τ_{eq} by:

$$\frac{dT}{dT} = \frac{T^* - T}{\tau_{\text{eq}}} \tag{4.20}$$

where T^* and T are the different temperatures of each group of particles. The energy equilibration times is [3]

$$\tau_{eq} = \frac{(2\pi)^{1/2}3\pi\varepsilon_0^2 m^* m}{n^* Z^{*2}Z^2 e^4 \ln\Lambda} \left(\frac{kT}{m} + \frac{kT^*}{m^*}\right)^{3/2} \tag{4.21}$$

This equilibration time is called the temperature or energy relaxation time. If we denote the electron–electron, ion–ion, and electron–ion relaxation times by τ_{eq}^{ee}, τ_{eq}^{ii}, and τ_{eq}^{ei}, respectively, then to a good approximation of Equation (4.21), we find

$$\tau_{eq}^{ee} = \frac{(2\pi)^{1/2}6\pi\varepsilon_0^2 m_e^{1/2}(kT_e)^{3/2}}{n_e e^4 \ln\Lambda} \tag{4.22}$$

$$\tau_{eq}^{ii} = \frac{(2\pi)^{1/2}6\pi\varepsilon_0^2 m_i^{1/2}(kT_i)^{3/2}}{n_i Z^4 e^4 \ln\Lambda} \tag{4.23}$$

$$\tau_{eq}^{ei} = \frac{(2\pi)^{1/2}3\pi\varepsilon_0^2 m_i(kT_e)^{3/2}}{n_i Z^2 e^4 m_e^{1/2}\ln\Lambda} \tag{4.24}$$

or

$$\tau_{eq}^{ee} = 2.75 \times 10^5 \frac{T_e^{3/2}}{n_e \ln\Lambda} \tag{4.22a}$$

$$\tau_{eq}^{ii} = 1.17 \times 10^7 \frac{A^{1/2}T_i^{3/2}}{Z^4 n_i \ln\Lambda} \tag{4.23a}$$

$$\tau_{eq}^{ei} = 2.51 \times 10^8 \frac{AT_e^{3/2}}{Z^2 n_e \ln\Lambda} \tag{4.24a}$$

These energy relaxation times are functionally similar to the momentum relaxation times.

The collision frequencies ν_{ee}, ν_{ii}, ν_{ie}, and ν_{ei} are the inverse of the corresponding collision times, that is,

$$\nu_{ee} = 3.79 \times 10^{-6} \frac{n_e \ln\Lambda}{T_e^{3/2}} \tag{4.18c}$$

$$\nu_{ii} = 8.85 \times 10^{-8} \frac{Z^4 n_i \ln\Lambda}{A^{1/2}T_i^{3/2}} \tag{4.18d}$$

$$\nu_{ie} = 3.98 \times 10^{-9} \; \frac{Z^2 n_e \ln\Lambda}{A T_e^{3/2}} \tag{4.19b}$$

$$\nu_{ei} = 3.88 \times 10^{-6} \; \frac{Z n_e \ln\Lambda}{T_e^{3/2}} \tag{4.19c}$$

The time that it takes for an energetic ion to slow down to the energy of the background plasma has been studied by Stix and others [3, 10]. Energetic ions occur in fusion, for example, by injection of ion beams into a plasma for supplementary heating, or fast alpha particles born in fusion events. It is related to τ_{eq}^{ei} in the following way. Let the slowing down time be τ_s, where

$$\tau_s = -\int_0^E \frac{dE_{bi}}{dE_{bi}/dt} \tag{4.25}$$

where E_{bi} is the energy of a fast beam ion. It is found to be approximately

$$\tau_s \approx \frac{\tau_{eq}^{ei}}{1.5} \ln\left[1 + \left(\frac{E_{bi}}{E_{cr}}\right)^{3/2}\right] \tag{4.26}$$

where E_{cr} is that energy where the energetic ion is transferring its energy at an equal rate to both the electron and ion gases. At energy greater than E_{cr}, the particle primarily heats the electron gas because the plasma electrons are moving faster than the plasma ions and in subsequent collisions, the fast ion preferentially heats the electrons. This critical energy is

$$E_{cr} \approx 15 k T_e \left[A_b^{3/2} \frac{\Sigma n_i Z_i^2}{n_e A_i}\right]^{2/3} \tag{4.27}$$

where A_b and A_i are the atomic numbers of the beam ion and the plasma ion, respectively. As an interesting example, consider an alpha particle, created with energy 3.5 MeV, slowing down in a D−T fusion plasma whose energy is $kT_e = 10$ keV. For $A = 4$, $n_e = n_i Z_i^2$, and $A_i = 2.5$, we find $E_{cr} = 48$ keV. Using expression (4.24a) with $A = 4$, $n_e = 0.5 \times 10^{20}$ m^{-3}, find $\tau_{eq}^{ei} = 1.2$ s, and $\tau_s = 5.1$ s. Since an alpha particle from a D−T reaction is created with a velocity of 1.3×10^7 m/s, in the time τ_s it will have traveled a distance $l \sim 3 \times 10^7$ m.

4.4.3. Gyro Periods

Charged particles spiraling about magnetic lines of induction do so with a gyro or cyclotron period

$$\tau_c = \frac{2\pi m}{ZeB}$$

(4.28)

For electrons, using the absolute value of $|e|$,

$$\tau_{ce} = \frac{3.57 \times 10^{-11}}{B}$$

(4.28a)

and for ions

$$\tau_{ci} = 6.49 \times 10^{-8} \frac{A}{ZB}$$

(4.28b)

The gyro frequency $\nu_c = \tau_c^{-1}$, or, in radians per second, $\omega_c = 2\pi\nu_c$. Derivation of these classical gyro periods can be found in all plasma physics texts, for example, Chen [1]. Gyro periods are sometimes called cyclotron or Larmor periods. For a fusion reactor with $B = 5$ T, $\tau_{ce} = 7.1 \times 10^{-12}$ s, and for deuteron ions, $\tau_{ci} = 2.6 \times 10^{-8}$ s. The corresponding frequencies are $\nu_{ce} = 1.4 \times 10^{11}$ Hz and $\nu_{ci} = 3.85 \times 10^{7}$ Hz.

4.4.4. Plasma Periods

A natural or resonant period of a plasma is associated with the rate at which an excess electron charge density electrostatically oscillates about the heavy ions. In the limit of negligible thermal motion, this plasma period is

$$\tau_{pe} = 2\pi \left(\frac{m_e \varepsilon_0}{n_e e^2} \right)^{1/2}$$

(4.29)

or

$$\tau_{pe} = \frac{1.11 \times 10^{-1}}{n_e^{1/2}}$$

(4.29a)

An ion plasma period is similarly defined as

$$\tau_{pi} = 2\pi \left(\frac{m_i \varepsilon_0}{n_i Z^2 e^2} \right)^{1/2}$$

(4.30)

or

$$\tau_{pi} = 4.76 \frac{A^{1/2}}{Z n_i^{1/2}}$$

(4.30a)

There is no oscillation at the ion plasma period, but this expression appears frequently and is given here for convenience. Derivation of the plasma period can be found in every basic text on plasma physics. For hydrogen ($Z = A = 1$) with $n_e = n_i = 0.5 \times 10^{20}$ m^{-3}, $\tau_{pe} = 1.6 \times 10^{-11}$ s, and $\tau_{pi} = 6.7 \times 10^{-10}$ s. The corresponding frequencies are $\nu_{pe} = 6.2 \times 10^{10}$ Hz and $\nu_{pi} = 1.5 \times 10^{9}$ Hz. When the plasma temperature is not negligible, plasma oscillations become electrostatic waves.

4.4.5. Upper and Lower Hybrid Frequencies

The study of wave propagation in a plasma has revealed that there are many and diverse types of waves, some of which are unique to the plasma medium. A comprehensive treatment of this subject can be found in the text by Stix [11]. Two relatively simple characteristic frequencies of a plasma are composed of combinations of the gyro and plasma frequencies. The first is called the upper hybrid frequency ν_{UH}, which is

$$\nu_{UH}^2 = \nu_{ce}^2 + \nu_{pe}^2 \tag{4.31}$$

Note that when $B \to 0$, the upper hybrid frequency reduces to the electron plasma frequency.

The second is called the lower hybrid frequency ν_{LH}, which is

$$\frac{1}{\nu_{LH}^2} = \frac{1}{\nu_{ci}^2 + \nu_{pi}^2} + \frac{1}{\nu_{ci}\nu_{ce}} \tag{4.32}$$

When the plasma density is large such that $\nu_{pi}^2 >> \nu_{ce}\nu_{ci}$, then

$$\nu_{LH}^2 \approx \nu_{ci}\nu_{ce} \tag{4.32a}$$

When $\nu_{pi}^2 << \nu_{ei}\nu_{ce}$, then

$$\nu_{LH}^2 \approx \nu_{pi}^2 + \nu_{ci}^2 \tag{4.32b}$$

The upper and lower hybrid frequencies arise in the solution of the dispersion equation (the relationship between the wave progagation vector **k** and wave frequency ω) for waves with $\mathbf{k} \perp \mathbf{B}_0$ and $\mathbf{E} \perp \mathbf{B}_0$ in a cold plasma. These frequencies at plasma resonance occur when the index of refraction becomes infinite, and the wave energy is absorbed by the plasma. The electron and ion gyro frequencies also are plasma resonances for $\mathbf{k} \parallel \mathbf{B}_0$ and $\mathbf{E} \perp \mathbf{B}_0$. The importance of such resonances is that waves of resonant frequency can be used to heat a plasma. When a hot plasma is studied, the analysis becomes more complex and the resonant frequencies are slightly altered. Even when particle collisions do not occur and, consequently, classical dissipation is zero, Landau damping does occur. These interesting effects are analyzed in texts such as References 3, 11, and 12.

For a hydrogen plasma ($A = Z = 1$) with $n_e = n_i = 0.5 \times 10^{20}$ m^{-3}, and $B = 5$ T, we find $\nu_{UH} = 1.5 \times 10^{11}$ Hz and $\nu_{LH} = 1.3 \times 10^9$ Hz. The corresponding times are $\tau_{UH} = 6.5 \times 10^{-12}$ s and $\tau_{LH} = 7.9 \times 10^{-10}$ s. At these conditions, $\nu_{UH} \sim \nu_{ce} < \nu_{pe}, \nu_{LH} \lesssim \nu_{pi}, \ll \nu_{ce}$, and $\nu_{LH} \gg \nu_{ci}$.

4.4.6. Fusion Reaction Time

Calculation of the rate at which fusion proceeds involves the simultaneous solution of a system of nonlinear differential equations. For example, consider a D−T system where only the D−D, D−T and D−^{3}He reactions are important. At a given temperature, the rate at which each fuel is consumed is

$$\frac{dn_D}{dt} = S_D - n_D^2 \langle \sigma v \rangle_{DD} - n_D n_T \langle \sigma v \rangle_{DT} - n_D n_{^3He} \langle \sigma v \rangle_{D^3He} \tag{4.33}$$

$$\frac{dn_T}{dt} = S_T + \frac{n_D^2}{4} \langle \sigma v \rangle_{DD} - n_D n_T \langle \sigma v \rangle_{DT} \tag{4.34}$$

$$\frac{dn_{^3He}}{dt} = S_{^3He} + \frac{n_D^2}{4} \langle \sigma v \rangle_{DD} - n_D n_{^3He} \langle \sigma v \rangle_{D^3He} \tag{4.35}$$

where S_j is the source rate term for the jth fuel. Note that the numerical coefficient for the D−D reactivity in Equation (4.33) is 2/2 = 1, since two deuterons are consumed each time this reaction takes place. For simplicity, consider that the major term in Equation (4.33) is the D−T reactivity and $S_D = 0$. Then

$$\frac{dn_D}{dt} \approx - n_D n_T \langle \sigma v \rangle_{DT} \tag{4.36}$$

whose solution at constant temperature is

$$n_D = n_D(0) e^{-t/\tau_R} \tag{4.37}$$

where τ_R, defined as the reaction time, is

$$\tau_R = \frac{1}{n_T \langle \sigma v \rangle_{DT}} \tag{4.37a}$$

That is, the mean lifetime of each fusion fuel particle is $(n \langle \sigma v \rangle)^{-1}$. This reaction time is a measure of how long it will take, under ideal circumstances, for a given fuel to have e^{-1} of its initial concentration react. More accurate estimates require solution of the system of equations like (4.33)−(4.35). The value of τ_R for D−T at 10 keV with $n_D = n_T = 0.25 \times 10^{20}$ m^{-3} is about 400 s.

The fraction of fuel burned f_b is defined, for the D−T reaction, as

$$f_b = \frac{n_\alpha}{n_{D,T}(0)} = 1 - \frac{n_{D,T}(t)}{n_{D,T}(0)} \tag{4.37b}$$

where n_α is the alpha particle number density.
   Using Equation (4.37) for a 50−50 mixture, we obtain

$$f_b = 1 - e^{-t/\tau_R} \tag{4.37c}$$

4.4.7. Plasma Cooling Time

Plasma at temperature T, cooled by radiation power loss at the rate P_R, will, if there are no other sources of heating or cooling, be reduced to about $e^{-1}T$ in a time τ_B, given by

$$\tau_B = \frac{\tfrac{3}{2}(n_e + n_i)kT}{P_R} \tag{4.38}$$

As discussed in Section 4.6, the bremsstrahlung radiation power per unit volume for pure free−free radiation is

$$P_{ff} = 1.59 \times 10^{-40} n_e n_i Z^2 T_e^{\frac{1}{2}} \tag{4.39}$$

where, if there is a mixture of different ionic species, the $n_i Z^2$ is replaced by $\Sigma\, n_i Z^2$. Using expression (4.39) in (4.38) and choosing $n_e = n_i$ ($Z = 1$), we obtain for the bremsstrahlung cooling time

$$\tau_B = 2.60 \times 10^{17} \frac{T_e^{\frac{1}{2}}}{n_e} \tag{4.40}$$

For a pure hydrogen plasma at $T_e = 10^8$ K, $n_e = 0.5 \times 10^{20}$ m^{-3}, we find $\tau_B = 52$ s and $P_{ff} \approx 4$ kW/m^3.
   Radiation cooling of a plasma, particularly if higher -Z impurities are present, can be very significant. Radiation losses including free−bound and bound−bound transitions are more complicated to evaluate than free−free brems-

strahlung. Radiation rates have been numerically evaluated for hydrogen plasmas containing a wide class of impurities, and tables and graphs for these data are referenced in section 4.6. Synchrotron radiation may also be an important energy loss, but much of that radiation may be absorbed within the plasma.

For a fusion plasma with $T = 10^8$ K, $n_e = n_i = 0.5 \times 10^{20}$ m^{-3}, $B = 5$ T, $A = Z = 1$, these times are ordered as follows:

$$\tau_{UH} \approx \tau_{ce} < \tau_{pe} < \tau_{pi} \approx \tau_{LH} < \tau_{ci} \ll \tau_{ee} \approx$$
$$\tau_{ei} \ll \tau_{ii} \ll \tau_{ie} < \tau_{s\alpha} \ll \tau_B < \tau_R$$

4.4.8. Resistive Diffusion Time

The time it takes a magnetic field to diffuse through a uniform plasma to a depth l is called the resistive magnetic diffusion time τ_Ω. For a stationary plasma this magnetic diffusion time is defined as

$$\tau_\Omega \equiv \mu_0 \sigma l^2$$

and using the classical expression for the electrical conductivity [see Equation (4.81)], obtain

$$\tau_\Omega = 1.82 \times 10^{-8} \frac{l^2 T_e^{3/2}}{Z \ln\Lambda}$$

For $T_e = 10^8$ K, $\ln \Lambda \approx 20$, $l = 10^{-2}$ m, and $Z = 1$, $\tau_\Omega \approx 0.1$ s.

4.5. CHARACTERISTIC SPEEDS

4.5.1. Mean Thermal Particle Speed

A plasma that has a Maxwellian velocity distribution $f(v)$ has a particle mean thermal speed

$$\bar{v} = \left(\frac{8kT}{\pi m}\right)^{1/2} \tag{4.41}$$

or

$$\bar{v}_e = 6.21 \times 10^3 T_e^{1/2} \tag{4.41a}$$

$$\bar{v}_i = 1.45 \times 10^2 \left(\frac{T_i}{A}\right)^{1/2} \tag{4.41b}$$

For a D−T fusion plasma ($A = 2.5$) at $T = 10^8$ K, $\bar{v}_e = 6.2 \times 10^7$ m/s and $\bar{v}_i = 9.2 \times 10^5$ m/s. See Chapter 3, Section 1 for details and references.

There are two other characteristic particle speeds. They are the most probable speed, sometimes called the thermal speed $\bar{v}_{th}$, which is the speed corresponding to the maximum of the speed distribution function,

$$\bar{v}_{th} = \left(\frac{2kT}{m}\right)^{1/2} = 1.28 \times 10^2 \left(\frac{T}{A}\right)^{1/2} \tag{4.42}$$

and the speed related to the mean energy of the particles:

$$\bar{v}_E = \left(\frac{3kT}{m}\right)^{1/2} = 1.58 \times 10^2 \left(\frac{T}{A}\right)^{1/2} \tag{4.43}$$

The difference between these three speeds is relatively small, namely,

$$\bar{v} = 0.92\bar{v}_E = 1.13\bar{v}_{th}$$

In plasma physics, traditionally, $\bar{v}_E$ is, for simplicity, the usual reference speed.

4.5.2.   Acoustic Speed

The speed of a small-amplitude pressure wave in a gas is the sound speed, which is

$$a = \left(\frac{\gamma kT}{m}\right)^{1/2} \tag{4.44}$$

where γ is the ratio of specific heats, which for plasma particles with three degrees of freedom is $5/3$. Thus

$$a = 1.17 \times 10^2 \left(\frac{T}{A}\right)^{1/2} \tag{4.44a}$$

It should be noted that this macroscopic sound speed a is related directly to the microscopic mean thermal speed $\bar{v}$ by $\bar{v}/a = (24/\pi)^{1/2} = 1.24$. An analysis of ion acoustic waves in a two-component plasma in which $T_e \neq T_i$ leads to the following ion acoustic wave speed a_i:

$$a_i = \left[\frac{k(T_e + \gamma T_i)}{m_i}\right]^{1/2} \tag{4.45}$$

where it has been assumed that the electron gas is isothermal and therefore $\gamma_e = 1$. A derivation of this ion acoustic wave speed can be found in the text by Chen [1].

4.5.3. Alfvén Speed

If the plasma contains a magnetic field, then a low-frequency ($\omega < \omega_{ci}$) wave is possible, which was first analyzed by Alfvén [13]. The speed of a small-amplitude Alfvén wave b is

$$
b = \left(\frac{B^2}{\mu_0 n_i m_i} \right)^{1/2}
$$

(4.46)

or

$$
b = 2.19 \times 10^{16} \frac{B}{(n_i A)^{1/2}}
$$

(4.46a)

In a D−T plasma with $A = 2.5$, $B = 5$ T, and $n_i = 0.5 \times 10^{20}$ m^{-3}, $b = 9.8 \times 10^6$ m/s. It should be noted that for $T = 10^8$ K, the acoustic speed $a = 7.4 \times 10^5$ m/s, so that for these conditions $b > a$. For many macroscopic magnetohydrodynamic plasma instabilities the characteristic speed is the Alfvén speed. Transverse small-amplitude hydrodynamic waves propagate along the magnetic field lines at the Alfvén speed. An incompressible conducting fluid (like mercury) can have large-amplitude waves propagating along magnetic field lines at the speed b.

Small-amplitude magnetohydrodynamic signals can and will propagate throughout a plasma in any arbitrary direction relative to the magnetic field **B**, but the signal speed depends on the angle between the direction of propagation **k** and **B**. A plasma with a magnetic field is an anisotropic compressible media. The study of small-amplitude wave propagation in such media was first done by Friedrichs, and a thorough discussion can be found in the text by Jeffrey and Taniuti [14]. In the direction of **B** there are the signal speeds, a and b. Perpendicular to **B** there is only one signal speed, and it is $(a^2 + b^2)^{1/2}$. At an arbitrary angle with **B**, in a stationary plasma, there are generally three speeds, which depend on the angle between the wave vector **k** and **B**. These speeds are called the fast, intermediate (or Alfvén), and slow magnetohydrodynamic wave speeds.

4.5.4. Speed of Electromagnetic Waves

The theory of an electromagnetic wave propagating through a plasma containing a magnetic field is naturally part of any general study of plasma waves and can be found in the text by Stix [11]. The speed of an electromagnetic wave c is:

$$c = (\varepsilon\mu)^{-\frac{1}{2}} \tag{4.47}$$

In fusion plasmas the permeability μ is essentially its vacuum value, $\mu_0 = 4\pi \times 10^{-7}$ H/m. However, the dielectric constant ε is a function of the plasma state, and when a strong magnetic field ($\omega_{ci}\tau_i >> 1$) is present, it is necessary to represent ε by a second-order tensor. In the low-frequency limit ($\omega << \omega_{pe}$) the dielectric constant for wave motion transverse to **B** is particularly simple [1]. It is

$$\frac{\varepsilon}{\varepsilon_0} = \left(1 + \frac{c^2}{b^2}\right) \tag{4.48}$$

where b is the Alfvén speed and ε_0 is the vacuum value of the dielectric constant, $\varepsilon_0 = 8.854 \times 10^{-12}$ F/m. For plasmas where $b << c$, the dielectric constant can become much larger than its vacuum value, and the corresponding electromagnetic speed is much less than the vacuum value of the speed of light, $c = 2.9979 \times 10^8$ m/s.

4.5.5. Particle Drift Speeds

Charged particles subject to external forces, such as an electric field or gravity, or moving in a magnetic field that has gradients or curvature do not simply spiral along the field lines. They will drift orthogonal to the magnetic field as well as along it. Single, (i.e., collisionless) charged-particle adiabatic motion has been analyzed in great detail and excellent treatments, at different degrees of detail, can be found in References 1–3 and in the texts by Longmire [15] and Northrup [16]. Particularly useful in analyzing plasma-confinement properties of magnetic fields are the following three drift speeds.

If an electric field **E** has a component normal to **B**, then a particle will experience an electric field drift, $\mathbf{v}_{DE}$, given by

$$\mathbf{v}_{DE} = \frac{\mathbf{E}_\perp \times \mathbf{B}}{B^2} \tag{4.49}$$

The drift is in the direction orthogonal to both **E** and **B** and has the magnitude $|E/B|$. The motion of both electrons and ions consists of gyrations about **B** superimposed upon a drift velocity $\mathbf{v}_{DE}$. Since both electrons and ions have the same electric field drift velocity, there is no current associated with $\mathbf{v}_{DE}$.

If a magnetic field has curvature, where **R** is the radius of curvature of **B**, then charged particles experience a centrifugal force due to **R**, which results in a drift $\mathbf{v}_{DR}$ given by

$$\mathbf{v}_{DR} = \frac{2w_\parallel}{eR^2 B^2}(\mathbf{R} \times \mathbf{B}) \tag{4.50}$$

where $w_\parallel$ is the particle's kinetic energy along $\mathbf{B}$, and e is the proton $(+)$ charge. Since $w_\parallel = kT/2$,

$$\mathbf{v}_{DR} = \frac{kT}{eR^2 B^2}(\mathbf{R} \times \mathbf{B}) \tag{4.50a}$$

For example, in the van Allen belt of the earth's magnetic field, electrons drift westward and ions drift eastward as these particles oscillate between the magnetic poles. Evaluating the constants in Equation (4.50a),

$$\mathbf{v}_{DR} = \pm 8.62 \times 10^{-5} \ \frac{T(\mathbf{R} \times \mathbf{B})}{R^2 B^2} \tag{4.50b}$$

where the $+$ sign is for ions and $-$ for electrons. These expressions are valid only for $v_D \ll c$. Because electrons and ions drift in opposite directions, there is a current that results from radius-of-curvature drifts.

If a magnetic field has spatial gradients, then charged particles experience a drift that is caused by the particle moving through a varying magnitude of magnetic field during its spiraling motion. The magnetic gradient drift velocity $\mathbf{v}_{DG}$ is

$$\mathbf{v}_{DG} = \frac{w_\perp}{eB^2}(\hat{\mathbf{b}} \times \nabla B) \tag{4.51}$$

where $w_\perp$ is the particle's energy orthogonal to $\mathbf{B}$, and $\hat{\mathbf{b}}$ is the unit vector $\hat{\mathbf{b}} = \mathbf{B}/|\mathbf{B}|$. Since $w_\perp = kT$,

$$\mathbf{v}_{DG} = \frac{kT}{eB^2}(\hat{\mathbf{b}} \times \nabla B) \tag{4.51a}$$

or

$$\mathbf{v}_{DG} = \pm 8.62 \times 10^{-5} \ \frac{T}{B^2}(\hat{\mathbf{b}} \times \nabla B) \tag{4.51b}$$

where the $+$ and $-$ signs correspond to ions and electrons. Again, because the electrons and ions drift in opposite directions, there is a current which results from the ∇B drift.

4.6. PLASMA RADIATION

4.6.1. Bremsstrahlung

Plasma electrons are accelerated by collisions with ions, and this results in radiation called bremsstrahlung, which is a German word meaning "braking radiation." Since the electron is free or unbound, both before and after interacting with an ion, this type of radiation is sometimes referred to as free−free radiation. It is analyzed in detail in the text by Bekefi [17]. The total power per unit volume P_{ff} radiated by bremsstrahlung from a Maxwellian plasma is

$$P_{ff} = \frac{32\pi e^6 n_e n_i Z^2}{3(4\pi\varepsilon_0)^3 c^3 m_e h}\left(\frac{2\pi k T_e}{3m_e}\right)^{\frac{1}{2}} g_{ff} \tag{4.52}$$

A brief discussion of bremsstrahlung was given in Section 3.4. Taking the Gaunt factor $g_{ff} = 1.11$ and evaluating the physics constants gives

$$P_{ff} = 1.59 \times 10^{-40} n_e n_i Z^2 T_e^{\frac{1}{2}} \tag{4.52a}$$

At $T_e = 10^8$ K, $n_e = n_i = 0.5 \times 10^{20}$ m^{-3}, $Z = 1$, the bremsstrahlung power is $P_{ff} = 4$ kW/m^3, which is an order of magnitude less than the D−T alpha power at these same conditions (see Section 3.3).   *See p 262,3*

If there is a mixture of electrons and different ionic species, then

$$P_{ff} = 1.59 \times 10^{-40} n_e (\Sigma n_i Z_i^2) T_e^{\frac{1}{2}} \tag{4.52b}$$

Comparing Equations (4.52a) for a single ionic species and (4.52b) for a mixture of different ionic species leads to a convenient definition of "effective Z for radiation," namely,

$$\langle Z_{\text{eff}}\rangle_{\text{rad}} \equiv \frac{\Sigma n_i Z_i^2}{n_i Z_i} \tag{4.52c}$$

The bremsstrahlung power spectrum or the power radiated per unit volume between the frequencies v and $v + dv$ is

$$P_v dv = \frac{32\pi e^6 Z^2}{3(4\pi\varepsilon_0)^3 c^3 m_e}\left(\frac{2\pi}{3m_e k T_e}\right)^{\frac{1}{2}} n_e n_i e^{-hv/kT_e} dv \tag{4.53}$$

or

$$P_v dv = 1.18 \times 10^{-50} \frac{n_e n_i Z^2 e^{-hv/kT_e}}{T_e^{\frac{1}{2}}} \tag{4.53a}$$

The low-energy portion of the spectrum ($h\nu \leqslant kT_e$) is nearly constant, but the spectrum decreases exponentially when $h\nu > kT_e$. The wavelength corresponding to $h\nu = kT$ is

$$\lambda = \frac{1.439 \times 10^{-2}}{T_e} \tag{4.54}$$

which, for $T = 10^8$ K, is $\lambda = 1.4 \times 10^{-10}$ m or about 1.4 Å. The mean free path for all but the longest-wavelength (low-energy) photons is

$$\lambda_p^\nu \simeq 2.05 \times 10^2 \frac{T_e^{\frac{1}{2}}\nu^3}{Z^3 n_i^2} \tag{4.55}$$

Relativistic corrections to these simple bremsstrahlung formulas have been calculated by Maxon [18], and they begin to have an effect for $kT_e > 20$ keV.

4.6.2. Synchrotron Radiation

Plasma electrons in a magnetic field gyrate about the lines of magnetic induction, and the corresponding electron acceleration produces synchrotron (sometimes called cyclotron) radiation. Derivation of basic formulas concerning synchrotron radiation in a plasma can be found in the text by Bekefi [17].

Synchrotron radiation power from a single electron is

$$P_{sy} = \frac{e^2 \omega_{ce}^2}{6\pi\varepsilon_0 c^3} \frac{v_\perp^2}{1 - \beta^2} \tag{4.56}$$

where $\beta^2 = v^2/c^2$. For a mildly relativistic electron

$$P_{sy} = \frac{e^4 B^2}{6\pi\varepsilon_0 m_e^2 c} \left(\frac{kT_e}{m_e c^2} \right) \left[1 + \left(\frac{3kT_e}{mc^2} \right) + \cdots \right] \tag{4.56a}$$

The spectrum of this single particle radiation consists of a great many harmonics of ω_{ce}. At 50 keV approximately 94% of the emitted energy is in harmonics higher than the first. The angular distribution of the radiation is also quite complicated, and as the harmonic number is increased, the emission is concentrated more nearly in a cone whose axis is at right angles to the applied magnetic field. Although the first harmonics will be strongly absorbed within a distance of the size of a fusion plasma, higher harmonics will escape the plasma. However, much of this radiation can be redirected back into the plasma by using reflectors at the first wall seen by the plasma. By so increasing the effective path length for synchrotron radiation, much of it can be absorbed within the plasma.

The total plasma synchrotron radiation power was first calculated properly by Trubnikov and his results are summarized, for example, in Reference 3, Section 15.3. The total synchrotron power radiated per unit volume of plasma is

$$P_{sy} = \frac{e^2 \omega_{ce}^2 n_e}{3\pi\varepsilon_0 c}\left(\frac{kT_e}{mc^2}\right)\left[1 + \frac{5}{2}\frac{kT_e}{mc^2} + \cdots\right]$$

$$P_{sy} = 6.2 \times 10^{-17} B^2 n_e T_e \left[1 + \frac{Te}{204} + \cdots\right]$$

(4.57)

where, in the latter formula, T_e is in units of keV. Because of reabsorption, synchrotron radiation is not expected to be a serious energy loss for D-T fusion but it will be important for D-D, which ignites at higher temperature.

4.6.3. Free−Bound and Bound−Bound Radiation

For a clean hydrogen-fusion plasma, free−bound and bound−bound (or line) radiation losses are not expected to be important except near the edge (lower temperature) regions of the plasma. Impurity ions and atoms do, however, have a serious effect. By assuming coronal equilibrium, radiation cooling rates including free−bound and bound−bound radiation losses have been numerically evaluated [19] for hydrogen plasmas with impurities from 47 different elements ($2 \leqslant Z \leqslant 92$). These calculations are valid for low-density ($n_e \leqslant 10^{21}$ m^{-3}) and high-temperature ($2 \leqslant T_e \leqslant 10^5$ eV) plasmas. Functions are graphed and tabulated that permit calculation of the radiation rates for hydrogen plasma with impurities, including line and free−bound effects.

4.7. CLASSICAL PLASMA−TRANSPORT PROPERTIES

The flux rate of mass, momentum, energy, and electrical charge often can be expressed in terms of spatial gradients of a relevant physical property multiplied by the appropriate transport coefficient. Kinetic theory has been used to develop analytical expressions for these transport coefficients. A great deal of experimental effort has been devoted to measuring actual transport rates in physical situations similar to those anticipated in magnetic fusion reactors. Often there has been observed large differences between classical predictions of transport and that observed in experiments. This is not surprising because turbulent fluctuations and instabilities are common effects in plasma, and they are not taken into account in classical plasma kinetic theory.

A fluid dynamic analog to this situation is laminar and turbulent flow. It is well known in hydrodynamics when a fluid is expected to be laminar or turbulent by considering the magnitude of the appropriate Reynolds number.

When the flow is laminar, its behavior can be accurately predicted using hydrodynamic theory employing classical fluid transport properties. On the other hand, when the turbulent state is present, semiempirical formulations of the transport properties, based on experimental data, must be used to predict the flow characteristics.

In plasmas a similar situation exists, but there is as yet no simple criteria such as the Reynolds number that can be used to ascertain whether classical transport should be expected. Classical transport has been observed under special laboratory situations, but its occurrence seems to be rare. With fusion–like plasmas, with the large number of instabilities possible, observation of some form of "anomolous" or turbulent transport is the common experience. Since turbulence is an outstanding unsolved problem of classical physics, it is not surprising that plasmas, with their large number of fundamental time scales, should have complex transport properties. The values and scaling laws of transport properties for plasmas must be verified or ascertained from actual experiments.

Nonetheless, it is exceedingly useful and important to know the theoretical predictions of classical plasma transport properties. They often set a limit on what can be expected in a real situation, and sometimes they indicate what scaling laws can be expected. In turbulent situations the flux is usually much larger than that predicted by kinetic theory. Turbulent eddies and vorticies transport matter at a rate much in excess of particle flux calculations.

The methods of kinetic theory of ordinary gases are developed in detail in the classical text by Chapman and Cowling [20]. A comprehensive treatment of classical transport processes in plasmas is that given in the review by Braginskii [21]. Plasma physics texts, for example, References 1–3, devote varying degrees of effort to developing plasma transport theory. In this chapter, we present a set of plasma transport coefficients based on an elementary and basic kinetic theory model. More complex analyses result in the same functional form with slightly different numerical coefficients. Magnetic field geometry can have a fundamental effect on classical transport. Theoretical results for closed, toroidal field configurations called "neoclassical transport" have been reviewed in great detail by Hinton and Hazeltine [22], and are discussed in Chapter 5.

4.7.1. Diffusion

Consider a quiet and stable plasma in which the particle mean free path is sufficiently short that the flux of particles at a point in the plasma depends only on local properties. If the particle velocity distribution is nearly Maxwellian, then elementary kinetic theory predicts that the diffusion coefficient is

$$D_0 = \frac{\lambda^2}{\tau} = \bar{v}\lambda = \bar{v}^2\tau \tag{4.58}$$

When a magnetic field is present in the plasma, so that the space is aniso-tropic, then a second-order tensor is needed to describe the flux of particles [3]. The flux of particles in direction α is

$$n v_\alpha = - \frac{\partial}{\partial x_\beta} D_{\alpha\beta} n + \mu_{\alpha\beta} n E_\beta$$

(4.59)

where $D_{\alpha\beta}$ is the diffusion tensor, $\mu_{\alpha\beta}$ is the mobility tensor, and repeated subscripts in Equation (4.59) imply the summation $\Sigma_{\beta=1}^3$. The diffusion tensor consists of

$$D_{\alpha\beta}] = \begin{bmatrix} D_\perp & -D_H & 0 \\ D_H & D_\perp & 0 \\ 0 & 0 & D_\parallel \end{bmatrix}$$

(4.60)

where $D_\perp$ and $D_\parallel$ correspond to perpendicular and parrallel directions relative to the magnetic field **B**, and D_H is called the Hall term. In terms of plasma parameters the diffusion tensor is

$$[D_{\alpha\beta}] = \frac{3kT\tau}{m} \begin{bmatrix} \dfrac{1}{1 + v_c\tau} & \dfrac{v_c\tau}{1 + (v_c\tau)^2} & 0 \\ \dfrac{v_c\tau}{1 + (v_c\tau)^2} & \dfrac{1}{1 + (v_c\tau)^2} & 0 \\ 0 & 0 & 1 \end{bmatrix}$$

(4.61)

where we have used for the particle velocity $\bar{v}_F = (3kT/m)^{1/2}$. The mobility tensor is related to the diffusion tensor by the Einstein relation

$$[\mu_{\alpha\beta}] = \frac{eZ}{kT}[D_{\alpha\beta}]$$

(4.62)

It is useful to note that for $D_\parallel$ the characteristic step size in the diffusion process [Equation (4.58)] is the mean free path. In the direction perpendicular to **B**, this step size is reduced by $[1 + (v_c\tau)^2]^{-1}$. When $(v_c\tau)^2 >> 1$, the plasma is very anisotropic and the magnetic field significantly influences transport. In this case, noting that

$$v_c\tau = \frac{\lambda}{r_c}$$

(4.63)

we find

$$D_\perp \approx D_\parallel \left(\frac{r_c}{\lambda}\right)^2$$

(4.64)

that is, diffusion perpendicular to **B** has a characteristic step size equal to the gyro radius rather than the mean free path.

More rigorous analysis [20,21] yields Equation (4.58) with a numerical coefficient of the order of unity. We will take the coefficient as unity (the other transport properties have slightly different numerical coefficients, but they all are approximately one), and we will use appropriate plasma collision times. In this way, using Equations (4.9), (4.18), and (4.61),

$$D_{\parallel} = \frac{25\sqrt{3}\pi\varepsilon_0^2(kT)^{5/2}}{Z^4 m^{1/2} n e^4 \ln\Lambda} \tag{4.65}$$

or

$$D_{\parallel} = 2.81 \times 10^{11} \frac{T^{5/2}}{Z^4 A^{1/2} n \ln\Lambda} \tag{4.65a}$$

which gives, for electrons

$$D_{\parallel}^e = 1.20 \times 10^{13} \frac{T_e^{5/2}}{n_e \ln\Lambda} \tag{4.65b}$$

and for ions

$$D_{\parallel}^i = 2.81 \times 10^{11} \frac{T_i^{5/2}}{Z^4 A^{1/2} n_i \ln\Lambda} \tag{4.65c}$$

If the magnetic field is straight and uniform, using Equation (4.64), we obtain

$$D_{\perp}^e = 3.71 \times 10^{-21} \frac{n_e \ln\Lambda}{B^2 T_e^{1/2}} \tag{4.66}$$

$$D_{\perp}^i = 1.59 \times 10^{-19} \frac{Z^2 A^{3/2} n_i \ln\Lambda}{B^2 T_i^{1/2}} \tag{4.66a}$$

Note that for the same plasma physical conditions $D_{\parallel}^e \gg D_{\parallel}^i$ and $D_{\perp}^e \ll D_{\perp}^i$. The first inequality results from the fact that, at the same temperature, electrons have a higher thermal speed than ions. The second inequality occurs because the ions have a larger gyro radius than electrons, when the temperatures and magnetic field are the same. The units of D are m^2/s.

Over distances larger than the Debye length, electric neutrality $n_e \approx n_i$ is maintained. Therefore, the flux rate of electrons and ions must maintain charge neutrality. Such diffusion is called ambipolar diffusion, and, when analyzed, if $D_{\perp}^i/T_i >> D_{\perp}^e/T_e$, and neutral atoms are present, then one finds [3] that the particle flux perpendicular to **B** is

$$nv = - D_{\perp}^e\left(1 + \frac{T_i}{T_e}\right)\frac{dn}{dx} \qquad (4.67)$$

or when $T_i = T_e$, $D_{\perp} = 2D_{\perp}^e$. Diffusion is determined by the smaller of the ion and electron diffusion coefficients.

A perpendicular diffusion rate found in many gaseous-discharge experiments and early fusion experiments seemed to follow the Bohm formula, namely,

$$D_{\perp,\text{Bohm}} = \frac{kT}{16eB} = 5.40 \times 10^{-6}\frac{T}{B} \qquad (4.68)$$

which is an empirical result. Its functional form can be obtained if, in Equation (4.58), τ is replaced by τ_c. $D_{\perp,\text{Bohm}} \sim B^{-1}$, which is poor for fusion plasma confinement, but most fusion confinement experiments have fortunately shown that Equation (4.68) does not apply.

4.7.2. Thermal Conductivity

In a way completely analogous to the discussion of the diffusion coefficient, elementary kinetic theory predicts that the coefficient of thermal conduction $\varkappa$ is

$$\varkappa_0 = nc_v D_0 = \tfrac{3}{2}k\, nD_0 = \tfrac{3}{2}\, nkv^2\tau \qquad (4.69)$$

where c_v is the constant volume specific heat which, for particles with three degrees of freedom, is $c_v = 3k/2$.

The energy flux by thermal conduction $\overline{\mathbf{q}}$ is

$$\mathbf{q} = - \varkappa_{\parallel}\nabla_{\parallel}T - \varkappa_{\perp}\nabla_{\perp}T \pm \frac{\varkappa_{\parallel}}{\omega_c\tau}(\hat{\mathbf{b}} \times \nabla T) \qquad (4.70)$$

where $\hat{\mathbf{b}} = \mathbf{B}/|\mathbf{B}|$.

The third term results from a thermal force that is perpendicular to both **B** and ∇T, resulting from the gyration of particles about **B**. This effect is discussed by Braginskii [21]. Using Equations (4.65b), (4.65c), and (4.69), we obtain

$$\varkappa_{\parallel}^e = 2.48 \times 10^{-10}\frac{T_e^{5/2}}{\ln\Lambda} \qquad (4.71)$$

$$\varkappa_{\parallel}^{i} = 5.82 \times 10^{-12} \frac{T_i^{5/2}}{Z^4 A^{1/2} \ln\Lambda} \tag{4.71a}$$

and

$$\varkappa_{\perp}^{e} = 7.68 \times 10^{-44} \frac{n_e^2 \ln\Lambda}{B^2 T_e^{1/2}} \tag{4.72}$$

$$\varkappa_{\perp}^{i} = 3.29 \times 10^{-42} \frac{Z^2 A^{3/2} n_i^2 \ln\Lambda}{B^2 T_i^{1/2}} \tag{4.72a}$$

Again note that for the same plasma conditions $\varkappa_{\parallel}^{e} \gg \varkappa_{\parallel}^{i}$ and $\varkappa_{\perp}^{i} \gg \varkappa_{\perp}^{e}$ assuming Z and A are not too different from one. The reasons for this are the same as those given in the corresponding diffusion coefficient discussion. The units of $\varkappa$ are W/m K.

4.7.3. Viscosity

The viscous stress tensor $\Pi_{\alpha\beta}$ in a fluid-like plasma in the absence of a magnetic field is

$$\Pi_{\alpha\beta} = \eta_0 W_{\alpha\beta} \tag{4.73}$$

where elementary kinetic theory gives for the coefficient of viscosity

$$\eta_0 = nmD_0 = 3nkT\tau \tag{4.74}$$

and the rate of strain tensor $W_{\alpha\beta}$ is

$$W_{\alpha\beta} = \frac{\partial V_\alpha}{\partial x_\beta} + \frac{\partial V_\beta}{\partial x_\alpha} - \frac{2}{3} \delta_{\alpha\beta} \frac{\partial V_l}{\partial x_l} \tag{4.75}$$

where here, V represents the macroscopic fluid velocity and $\delta_{\alpha\beta}$ is equal to 1 when $\alpha = \beta$ and equals 0 when $\alpha \neq \beta$. Using Equations (4.65b), (4.65c), and (4.74), we obtain

$$\eta_0^e = 1.09 \times 10^{-17} \frac{T_e^{5/2}}{\ln\Lambda} \tag{4.74a}$$

$$\eta_0^i = 4.66 \times 10^{-16} \frac{A^{1/2} T_i^{5/2}}{Z^4 \ln\Lambda} \tag{4.74b}$$

In a strong magnetic field ($\omega_c \tau \gg 1$) the components of the viscous stress tensor $\Pi_{\alpha\beta}$ have the following form in a coordinate system with the Z axis parallel to the magnetic field:

$$\Pi_{zz} = -\eta_0 W_{zz} \tag{4.76}$$

$$\Pi_{xx} = -\frac{\eta_0}{2}\left(W_{xx} + W_{yy}\right) - \frac{\eta_\perp}{6}\left(W_{xx} - W_{yy}\right) \mp \frac{\eta_0}{2\nu_c\tau}W_{xy} \tag{4.76a}$$

$$\Pi_{yy} = -\frac{\eta_0}{2}\left(W_{xx} + W_{yy}\right) - \frac{\eta_\perp}{6}\left(W_{yy} - W_{xx}\right) \pm \frac{\eta_0}{2\nu_c\tau}W_{xy} \tag{4.76b}$$

$$\Pi_{xy} = \Pi_{yx} = -\frac{\eta_\perp}{3}W_{xy} \pm \frac{\eta_0}{6\nu_c\tau}\left(W_{xx} - W_{yy}\right) \tag{4.76c}$$

$$\Pi_{xz} = \Pi_{zx} = -\frac{\eta_\perp}{3}W_{xz} \mp \frac{\eta_0}{\nu_c\tau}W_{yz} \tag{4.76d}$$

$$\Pi_{yz} = \Pi_{zy} = -\frac{\eta_\perp}{3}W_{yz} \pm \frac{\eta_0}{\nu_c\tau}W_{xz} \tag{4.76e}$$

Where the terms that contain $(\nu_c\tau)^{-1}$ have a double sign, use the upper sign for ions and the lower sign for electrons. The reader is referred to Braginskii [21] for further discussion. The ordinary coefficient of viscosity η_0 is given by Equation (4.74). The perpendicular viscosity coefficient $\eta_\perp$ is

$$\eta_\perp = \frac{\eta_0}{\left(\nu_c\tau\right)^2} \tag{4.77}$$

or for electrons and ions, respectively:

$$\eta_\perp^e = 3.38 \times 10^{-51}\frac{n_e^2\ln\Lambda}{B^2 T_e^{\frac{1}{2}}} \tag{4.77a}$$

$$\eta_\perp^i = 2.64 \times 10^{-46}\frac{Z^2 A^{\frac{5}{2}} n_i\ln\Lambda}{B^2 T_i^{\frac{1}{2}}} \tag{4.77b}$$

4.7.4. Electrical Conductivity

The electrical conductivity coefficient σ is the transport coefficient related to the flux of electrical charge. Conservation of momentum for electrons and ions leads to the generalized Ohm's Law [2], which for a plasma containing a magnetic field is

$$\mathbf{E} = \frac{\mathbf{j}_\parallel}{\sigma_\parallel} + \frac{\mathbf{j}_\perp}{\sigma_\perp} + \mathbf{V} \times \mathbf{B} + \frac{\nu_{ce}\tau_{ei}}{B}\left[\mathbf{j} \times \mathbf{B} + \nabla p_e\right] \qquad (4.78)$$

where electron inertia and thermoelectric terms have been neglected. The first part of the last term of Equation (4.78) that concerns $\mathbf{j} \times \mathbf{B}$, the vector cross product between current density $\mathbf{j}$ and the magnetic field, is called the Hall term; it is significant when $\nu_{ce}\tau_{ei} \gg 1$. Simple kinetic theory gives [21]:

$$\sigma_\parallel = \frac{2n_e e^2 \tau_{ei}}{m_e} \qquad (4.79)$$

and

$$\sigma_\perp = \frac{\sigma_\parallel}{2} \qquad (4.80)$$

The magnetic field does not have much effect on the friction produced by electrons colliding with ions. Using Equations (4.19a) and (4.79), gives

$$\sigma_\parallel = \frac{50\pi^{1/2}\varepsilon_0^2(kT_e)^{3/2}}{m_e^{1/2}e^2 Z \ln\Lambda} \qquad (4.81)$$

or

$$\sigma_\parallel = 1.45 \times 10^{-2}\frac{T_e^{3/2}}{Z \ln\Lambda} \qquad (4.81a)$$

and

$$\sigma_\perp = 7.25 \times 10^{-3}\frac{T_e^{3/2}}{Z \ln\Lambda} \qquad (4.82)$$

The units of σ are $\Omega^{-1}\,\mathrm{m}^{-1}$. The resistivity coefficient $\rho = \sigma^{-1}$, or

$$\rho_\parallel = 6.87 \times 10^1 \frac{Z \ln\Lambda}{T^{3/2}} \qquad (4.83)$$

and

$$\rho_\perp = 1.37 \times 10^2 \, \frac{Z \ln\Lambda}{T^{3/2}}$$

(4.84)

When impurity ions are present, they influence the electrical resistivity. In such circumstances, it is convenient to use an effective Z, namely,

$$Z_{\text{eff}} \equiv \frac{\Sigma n_i Z_i^2}{n_e} = \frac{\Sigma n_i Z_i^2}{\Sigma n_i Z_i}$$

which was introduced in Section 4.4.1 for use in calculations involving momentum exchange.

4.8. DIMENSIONLESS PARAMETERS

As the theory of a scientific discipline matures, dimensional analysis and the analytical solution of physical problems lead to the establishment of dimensionless parameters whose values indicate the relative importance of some physical phenomena. These dimensionless groups, when known, help in modeling the behavior of the media and greatly reduce the quantity of experimental data required for prediction. Dimensionless parameters sometimes have associated with them the name of the person who first introduced them into the literature. Among the more important dimensionless groups in plasma physics and fluid phenomena are the following:

$$\text{Magnetic Reynolds number} = \text{Re}_M = \mu\sigma VL$$

(4.85)

where μ is the permeability, σ is the electrical conductivity, and V and L are the characteristic velocity and length scales associated with the phenomenon. The magnetic Reynolds number is a measure of the ratio of the motion-induced current density to the total current density. When $\text{Re}_M \gg 1$, motion-induced currents are significant and important. The magnetic Reynolds number was introduced into the literature by W. H. Elasser in 1956 in a paper on hydromagnetic dynamo theory.

$$\text{Hartmann Number} = \text{Ha} = BL\left(\frac{\sigma}{\eta}\right)^{1/2}$$

(4.86)

where B is the magnetic induction, L is the characteristic length scale, σ is the electrical conductivity, and η is the viscosity. The Hartmann number is a measure of the ratio of the Lorentz force to the viscous force, or the ratio of joule heating to viscous heating. When Ha $\gg$ 1, electromagnetic force or heating effects are large compared to the corresponding effects caused by viscosity.

$$\text{Lundquist Number} = \text{Lu} = \sigma BL\left(\frac{\mu}{\rho}\right)^{\frac{1}{2}}$$

(4.87)

where σ is the electrical conductivity, B, is the magnetic induction, L is the length scale, μ is the permeability, and ρ is the density. When Lu $\gg$ 1, magnetohydrodynamic phenomena are important and the magnetic field does not significantly diffuse through the media, at least over lengths of the order of L.

$$\text{Interaction parameter} = \text{In} = \frac{\sigma B^2 L}{\rho V}$$

(4.88)

where σ is the electrical conductivity, B is the magnetic induction, L is the length scale, ρ is the density, and V is the velocity of the media. The interaction parameter is a measure of the ratio of the Lorentz force to the inertia force. When In $\gg$ 1, Lorentz forces are more important than inertia forces. The interaction parameter In $= \text{Ha}^2/\text{Re}$, where Ha is the Hartmann number and Re is the Reynolds number.

$$\text{Reynolds Number Re} = \frac{VL}{\nu}$$

(4.89)

where V is the characteristic velocity, L is the characteristic length, and ν is the kinematic viscosity ($= \eta/\rho$). This is a famous fluid dynamics parameter which measures the relative ratio of inertia to viscous forces. In fluids, under many circumstances, the transition from laminar to turbulent flow is a function of only the Reynolds number. When Re $\ll$ 1, viscous forces are very large compared with inertia forces.

$$\text{Beta} = \beta = \frac{2\mu_0 nkT}{B^2}$$

(4.90)

where nkT is the plasma pressure and $B^2/2\mu_0$ is the magnetic pressure. When $\beta \ll$ 1, the magnetic field has nearly its vacuum value since the plasma pressure is relatively very small. The parameter β was introduced in this text in Equation (3.40) of Chapter 3.

$$\text{Prandtl Number} = \text{Pr} = \frac{c_v \eta}{\varkappa}$$

(4.91)

where c_v is the specific heat at constant volume, η is the viscosity, and $\varkappa$ is the thermal conductivity. The Prandtl number is a measure of the ratio of the viscous dissipation to the thermal conduction. When Pr $<<$ 1, conduction is more significant than viscous dissipation.

$$\text{Peclet Number} = \text{Pe} = \frac{\rho c_v VL}{\varkappa} \tag{4.92}$$

where ρ is the density, c_v is the specific heat at constant volume, V and L are the characteristic velocity and length scales, and $\varkappa$ is the thermal conductivity. The Peclet number is a measure of the ratio of convection to conduction energy transfer.

These are some of the dimensionless parameters which arise in plasma physics and fluid dynamics. The fact that in plasma physics there are very many length and time scales, and hence many possible dimensionless parameters, is what makes this subject rich. Particular physical situations often are a function of several dimensionless parameters. Radiation, partially ionized gases, impurities, and so on make the number of possible dimensionless parameters in plasma physics very large.

REFERENCES

1. F. F. Chen, *Introduction to Plasma Physics*, Plenum Press, New York, 1974.
2. L. Spitzer, Jr., *Physics of Fully Ionized Gases*, 2nd rev. ed., Wiley, New York, 1962.
3. K. Miyamoto, *Plasma Physics for Nuclear Fusion*, MIT Press, Cambridge, MA, 1980.
4. National Bureau of Standards, *The International System of Units (SI)*, Special Publication 330, U.S. Dept. of Commerce, Washington, D.C., 1977.
5. M. N. Saha, "Ionization in Solar Chromosphere," *Philos. Mag.* **40**, 472 (1920).
6. L. H. Aller, *Astrophysics*, Ronald Press, New York, 1963.
7. C. W. Allen, *Astrophysical Quantities*, 2nd ed., University of London, Athlone Press, 1964.
8. H. Goldstein, *Classical Mechanics*, rev. ed., Addison Wesley, Reading, MA, 1980.
9. L. Spitzer and R. Härm, *Phys. Rev.* **89**, 977 (1953).
10. T. H. Stix, *Plasma Physics* **14**, 367 (1972). V. I. Pistunovitch, *Sov. J. Plasma Phys.* **2**, 1 (1976).
11. T. H. Stix, *The Theory of Plasma Waves*, McGraw-Hill, New York, 1962.
12. V. L. Ginzburg, *Propagation of Electromagnetic Waves in Plasma*, Gordon and Breach, New York, 1961.
13. H. Alfvén, *Cosmical Electrodynamics*, Oxford University Press, England, 1950.
14. A. Jeffrey and T. Taniuti, *Non-Linear Wave Propagation with Applications to Physics and Magnetohydrodynamics*, Academic Press, New York, 1964.
15. C. L. Longmire, *Elementary Plasma Physics*, Interscience, New York, 1963.
16. T. G. Northrup, *The Adiabtic Motion of Charged Particles*, Interscience, New York, 1963.
17. G. Bekefi, *Radiation Processes in Plasmas*, Wiley, New York, 1966.
18. S. Maxon, *Phys. Rev. A* **5**, 1630 (1972).

At the third IAEA International Conference in Novosibirsk in 1968, the Russians reported that they had achieved in their T-3 tokamak a stable hot plasma with $T_e \approx 1$ keV, $B \approx 3$ T, and $n\tau \sim 10^{18}$ m^{-3} s. The T-3 plasma had an energy confinement time of several milliseconds, which was about 30 times larger than would have been achieved if Bohm diffusion had been present. These were very significant achievements. They were, however, viewed with skepticism by some plasma physicists who had seen nothing but Bohm-like diffusion and relatively low plasma temperatures ($T \lesssim 100$ eV) in their plasma devices.* Soon thereafter Artsimovich invited the British to send a team of their scientists to Moscow to measure by laser−Thomson scattering, the electron temperature in the T-3 plasma [3]. These measurements confirmed the Russian claims and the tokamak was then widely acknowledged as the most successful, plasma confinement concept of that time.

Tokamaks were built in laboratories around the world, and after some learning effort, all of them were able to duplicate the Soviet tokamak results. Larger tokamaks were constructed in the mid 1970s; for example, T-10 in the U.S.S.R. and PLT in the United States. Additional heating power, in particular powerful neutral beams, were injected into tokamaks, further raising their plasma temperature. For example, PLT achieved $T_i \approx 7$ keV and increased its energy-confinement time to nearly 0.1 s. The high-field ($B \approx 10$ T) Alcator** tokamak in the United States achieved an $n\tau_E$ value of above 10^{19} m^{-3} s, albeit with a plasma temperature $T \approx 1$ keV. In October 1983, MIT scientists using pellet injection in Alcator C, finally achieved an $n\tau_E$ value of $(6-8) \times 10^{19}$ m^{-3}/s, a value slightly in excess of breakeven criteria (see Section 3.5.). Larger tokamaks are being built: TFTR in the United States, T-15 in the U.S.S.R., JET in Europe, and JT-60 in Japan. TFTR began operation in late 1982 and JET started about 6 months later. By late 1983 TFTR had achieved an energy confinement time of nearly 0.2 s and appeared to be following the "volume" energy confinement scaling law, $\tau_E \sim n_e R^2 a$ [see Section 5.8 and Equation (5.93a)]. The TFTR and JET are designed to use tritium along with deuterium fuel. These new tokamaks operating in the mid 1980s are generally expected to at least achieve energy breakeven. Even larger tokamaks are being designed with the expectation that they will be built to obtain both increased plasma performance (e.g., ignition) and engineering experience with large, nearly power-reactor-size machines. More time and effort have been devoted to tokamaks than for any other single controlled fusion concept. There are comprehensive reviews of tokamak theory and experimental data by Artsimovich in 1972 [4], by Furth in 1975 [5], by a committee in 1979 [6], Sheffield in 1981 [7], and by E. Teller in 1981 [8].

---

* An exception to this statement is the classical diffusion observed experimentally in a straight theta-pinch plasma with $T \sim 1$ keV.

** Alcator is an acronym derived from the Latin, altus campus torus, which means high−field−torus.

5.2. TOROIDAL GLOSSARY

Consider the sector of a torus as shown in Figure 5.1. The toroidal coordinates are r, θ, ϕ; the toroidal direction is ϕ and the poloidal direction is θ. The major radius of the torus is R_0, and the minor radius is r. The torus has a width of $2w$ and a height $2h$, and the corresponding plasma dimensions are $2a$ and $2b$. If the plasma minor cross section is circular, its outer boundary is designated by $r = a$. The plasma minor cross section is smaller than the torus vacuum-vessel dimensions. This is usually achieved by means of a mechanical aperture known as a limiter, which limits the radial dimension of the plasma. The toroidal magnetic field B_ϕ is produced primarily from currents flowing in the toroidal field coils. The poloidal field B_θ in a tokamak is produced primarily by a toroidal current flowing in the plasma. An equilibrium or vertical field B_v is needed to maintain the plasma column in radial equilibrium.

In an air-core-induction tokamak, equal currents in the same direction in the inner and outer coils induce the plasma ohmic current. Currents in the opposite direction, in these coils, induce the vertical or equilibrium field. This is strictly true for an infinite-aspect-ratio machine; the low-aspect-ratio case is more complicated.

The torus aspect ratio A is defined as

$$A = \frac{R_o}{w} = \varepsilon^{-1}$$

(5.1)

and the inverse aspect ratio $\varepsilon = A^{-1}$, which is a number less than 1. The plasma aspect ratio is R_0/a.

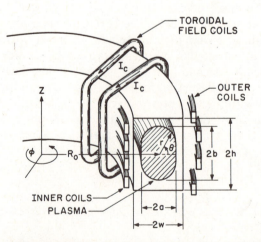

FIGURE 5.1. Toroidal geometry: R_o = major radius; r = minor radius; θ = poloidal direction; ϕ = toroidal direction; Z = toroidal major axis of symmetry.

A magnetic field can have any three-dimensional form that is divergence free. That is

$$\nabla \cdot \mathbf{B} = 0 \tag{5.2}$$

which is one of Maxwell's laws.

A magnetic field line, or line of magnetic force, is a line in space that is everywhere tangent to the magnetic induction vector $\mathbf{B}$; that is,

$$\frac{dx}{B_x} = \frac{dy}{B_y} = \frac{dz}{B_z} \tag{5.3}$$

A field line that is not closed (i.e., does not exactly "bite its tail") will generally cover a surface or fill a volume ergodically. This means that by following a field line long enough, it will eventually get arbitrarily close to any point on a surface or in a volume where it is ergodic.

The magnetic flux ψ is the amount of magnetic field lines passing through a given surface, that is,

$$\psi = \iint \mathbf{B} \cdot \hat{\mathbf{n}} ds \tag{5.4}$$

The toroidal flux ψ_{tor} is made up of toroidal magnetic field B_ϕ passing through a minor cross-sectional area. Similarly, the poloidal flux ψ_{pol} is made up of poloidal field B_θ passing through the equitorial plane.

A magnetic surface $\psi = $ const is one on which all the lines of magnetic induction lie, that is,

$$\nabla\psi \cdot \mathbf{B} = 0 \tag{5.5}$$

Or, one ergodic line of magnetic induction can define a flux surface. Magnetic flux surfaces are useful because density, temperature, pressure, and electrostatic potential are, for a steady-state situation, all constant on flux surfaces.*

An important property of tokamaks is the q value, or safety factor. The q value is the number of times a magnetic field line winds the long way around the torus divided by the number of times it winds the short way around, in the limit of an infinite number of turns. The q value is a property of the field lines, being everywhere the same on a magnetic surface. A rational magnetic surface is one which has the property

$$q_{rat} = \frac{n}{m} \tag{5.6}$$

*See Section 5.4 for a proof of this statement.

where n and m are integers. On a rational magnetic surface the field line makes n transit the long (ϕ) way around, for exactly every m transits the short (θ) way. Field lines that lie on rational magnetic surfaces exactly "bite their tails."

For a circular, minor cross-section tokamak

$$q(r) = \frac{r}{R} \frac{B_\phi}{B_\theta}$$

(5.7)

where R is the radial coordinate measured from the axis of symmetry and r is the radial coordinate measured from the center of the minor cross section, and for a noncircular, minor cross-section tokamak

$$q(\psi) = \frac{RB_\phi}{2\pi} \oint \frac{dl}{R^2 B_\theta} = \frac{l_\theta}{2\pi R_0} \frac{B_\phi}{\langle B_\theta \rangle}$$

(5.8)

where the line integral is evaluated over a closed flux surface whose field line length is l_θ, $\langle B_\theta \rangle$ is the value of B_θ averaged over that flux surface, R_0 is the major radius of the torus.

Consider Figure 5.2, which shows flux surfaces for the NUWMAK fusion reactor design [9]. The nearly eliptically shaped surfaces, nested within one another, surround a single field line called the magnetic axis. The radial distance between consecutive flux surfaces contains the same quantity of

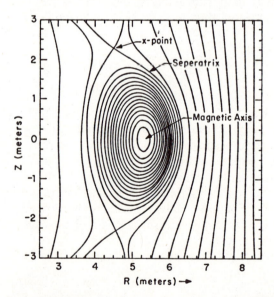

FIGURE 5.2. Magnetic surfaces for the NUWMAK fusion reactor design [9]; note the two x-points and the separatrix. Each closed surface about the magnetic axis encloses a fixed quantity of poloidal flux ψ_{pol}. The NUWMAK design has $B_o = 6.05$ T, $R_o = 5.125$ m, $b/a = 1.64$, $a = 1.125$ m, $R/a = 4.55$, $q(0) = 1.09$, $q(a) = 2.63$, and plasma current $I_\phi = 7.2$ MA.

poloidal flux. This flux plot has two x-points that identify a separatrix flux surface. The flux surfaces on one side of the separatrix are topologically different from those on the other side. In Figure 5.2, flux surfaces inside the separatrix are closed, nested contours; those outside are open contours. These latter surfaces intercept the first wall of the reactor chamber.

The rotational transform angle i is a measure of the poloidal angle a field line has rotated after tranversing 2π radians in the toroidal direction. Measured in radians, it is defined as

$$i = \frac{2\pi}{q} \tag{5.9}$$

Because the q value may vary from one ψ surface to the next, the magnetic line of force will point in a different direction as one proceeds radially to different flux surfaces. This results in a shear in the magnetic field where the shear parameter Θ is defined as [5]

$$\Theta = \frac{r^2}{2\pi R}\left(\frac{di}{dr}\right) = \frac{r^2}{R}\frac{d(q^{-1})}{dr} \tag{5.10}$$

The vacuum toroidal magnetic field in the torus results from the current that threads the hole of the torus. From Maxwell's laws, if I is the total current which threads the torus hole,

$$I = \oint \frac{\mathbf{B} \cdot d\mathbf{l}}{\mu_0} = \frac{2\pi R B_\phi}{\mu_0} \tag{5.11}$$

or

$$B_\phi(R) = \frac{\mu_0 I}{2\pi R} = \left(\frac{R_0}{R}\right) B_\phi(R_0) \tag{5.12}$$

The vacuum value of the toroidal magnetic field varies inversely with R. When a plasma is in the torus, and there is poloidal current, Equation (5.11) still applies, with B_ϕ the local toroidal field and I the sum of both the toroidal-field coil currents and the poloidal plasma current.

The toroidal current I_ϕ is generated in the torus plasma by creating a toroidal electric field E_ϕ. This is achieved by altering in time the magnetic flux that penetrates the hole in the torus. Considering a closed loop around the torus, from Maxwell's equations

$$\oint \mathbf{E} \cdot d\hat{\mathbf{l}} = -\frac{d\psi_{pol}}{dt} = -\frac{d}{dt}\int\int \mathbf{B} \cdot \hat{n}\,ds \tag{5.13}$$

The flux change can be achieved by either an air-core or an iron-core transformer whose flux links the torus core.

A tokamak may be defined as a toroidal device that has a large toroidal plasma current and a strong toroidal magnetic field such that

$$B_\phi > B_\theta > B_v \tag{5.14}$$

where B_v is the vertical or equilibrium field. It also has the property

$$q(\psi) \geqslant 1 \tag{5.15}$$

This latter property, that the safety factor be equal to or larger than 1, results from stability criteria discussed later (see Section 5.5). The reader interested in more detailed discussion of toroidal magnetic field properties may consult the texts by Miyamoto [10] and by Bateman [11].

5.3.   PARTICLE MOTIONS IN A TOKAMAK

It is instructive and revealing to consider the motion of a single charged particle in a tokamak magnetic field. This simplification of a single particle obviously neglects first, plasma currents, and second, many-body Coulomb-collision scattering effects. The first approximation is valid in the low-β limit ($\beta = 2\mu_0 p/B^2 \to 0$), and the second approximation is removed in a rigorous treatment of plasma transport in toroidal geometry [12]. It is instructive because it shows why a simple toroidal field cannot confine a plasma, and the orbits reveal important magnetic geometry effects on plasma kinetic transport.

Consider first a simple vacuum toroidal magnetic field generated only from current in the toroidal field coils. This field has components

$$\mathbf{B} = [B_r, B_\phi, B_\theta] = \left[0, \frac{R_0 B_\phi(0)}{R}, 0 \right] \tag{5.16}$$

as given by Equation (5.12), where $B_\phi(0)$ is the toroidal field strength at $R = R_0$. A charged particle will move along a line of induction with speed $v_\parallel$, and drift radially in the Z direction, across the lines of force, with a guiding center drift velocity (see Section 4.55),

$$\mathbf{v}_D = \frac{m}{ZeB^2} \left[v_\parallel^2 \frac{(\mathbf{R} \times \mathbf{B})}{R^2} + \frac{v_\perp^2}{2} (\hat{\mathbf{b}} \times \nabla B) \right] \tag{5.17}$$

But for a simple vacuum toroidal magnetic field,

$$\frac{\mathbf{R} \times \mathbf{B}}{R^2} = \hat{\mathbf{z}} B_\phi(0) \frac{R_0}{R^2}$$

$$\hat{\mathbf{b}} \times \nabla B = \hat{\mathbf{z}} \, \frac{B_\phi(0) R_0}{R^2}$$

and

$$B_\phi = \frac{R_0}{R} \, B_\phi(0)$$

so that

$$\mathbf{v}_D = \hat{\mathbf{z}} \left(\frac{m}{Ze} \right) \frac{1}{R_0 B_\phi(0)} \left[v_\parallel^2 + \frac{v_\perp^2}{2} \right] \tag{5.18}$$

where $\hat{\mathbf{z}}$ is the unit vector in the Z direction, Ze is the particle charge ($-e$ for electrons and $+Ze$ for ions), and $B_\phi(0)$ is the toroidal field at $R = R_0$. Electrons and ions drift in opposite Z directions. Equation (5.18) is called the toroidal drift. The result of these drifts is that the plasma rapidly becomes polarized, with an electric field in the Z direction (see Figure 5.3). The resulting electric field causes both electrons and ions to $\mathbf{E} \times \mathbf{B}$ drift [Equation (4.49)] in the outward radial direction with a speed of E/B. The bulk plasma in a simple toroidal field simply moves radially outward until it strikes the wall. There is no radial equilibrium and hence no plasma confinement in a simple toroidal field. The macroscopic equivalent of this result can be recognized by the fact that the toroidal magnetic field has a radial gradient $\nabla B_\phi = -R_0/R^2/B_\phi(0)$. Noting that $B^2/2\mu_0$ is a magnetic pressure implies that the plasma feels an outward radial pressure force for which there is, in a simple torus, no counterbalancing force.

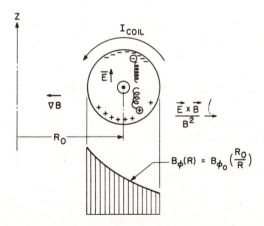

FIGURE 5.3. In a simple toroidal field, toroidal drift causes an electric field in the Z direction. The entire plasma $(\mathbf{E} \times \mathbf{B})/B^2$ = drifts radially outward. A simple toroidal field does not provide plasma containment.

The polarization that results from a simple toroidal magnetic field can be short-circuited by connecting the upper and lower (Z) regions of the torus with magnetic lines of force. Particles, spiraling along these lines of force, will no longer remain at $\pm Z$, but can transit back and forth. A simple way to achieve this effect is to introduce a poloidal magnetic field B_θ. The vector addition of the toroidal and poloidal fields results in helical magnetic lines of force with a rotation transform i. This can provide nested flux surfaces, or ergodic field structure.

But, even with toroidal helical lines of magnetic induction, the magnetic field strength must be larger on the inside region of the torus than on the outer regions. This provides an important feature to the drift orbit of a charged particle. We discuss this next.

An important fundamental property of a charged particle moving in a magnetic field is that its magnetic moment μ is constant (see Reference 10):

$$\mu \equiv \frac{mv_\perp^2}{2B} = \text{const} \tag{5.19}$$

This relationship is valid so long as the particle has constant total kinetic energy (its motion is adiabatic) and the magnetic field strength does not change too much over the distance of a gyro radius. A particle moving along a line of magnetic induction, entering a region of increasing field strength, because of the constancy of μ, can reach a point where $v_\parallel = 0$; that is, it is "stopped" and all its kinetic energy at that point resides in $w_\perp$. Its only recourse is to move back into the region of smaller B. The particle has been reflected by the magnetic field. The condition for such a magnetic mirror reflection [10] can be shown to be

$$\frac{v_\parallel}{v_\perp} \leqslant \left(\frac{B_{\max}}{B_{\min}} - 1\right)^2 \tag{5.20}$$

or

$$\left(\frac{v_\perp}{v}\right)^2 \geqslant \frac{B_{\min}}{B^{\max}} = \sin^2\alpha_{\text{cr}} \tag{5.20a}$$

where $B_{\max}$ is the maximum strength of the magnetic mirror. This reflection is illustrated in Figure 5.4a. When a particle has $(v_\perp/v)^2 > B/B_{\max}$, then it is reflected; if it is less (i.e., has a large $v_\parallel$), then it passes through the mirror. The angle α_{cr} measures that conical angle in phase $(v_\perp, v_\parallel)$ space that contains those particles which pass through the mirror. α_{cr} is called the loss-cone angle.

Consequently, as particles spiral along helical lines of force in a tokamak magnetic geometry, some of them will be reflected by the stronger mirrorlike field that exists on the inner regions of the torus. The magnitude of the toroidal field is approximately

$$B \approx |B_0| \, (1 - \varepsilon \cos \theta) \tag{5.21}$$

where

$$R = R_0 \, (1 + \varepsilon \cos \theta)$$

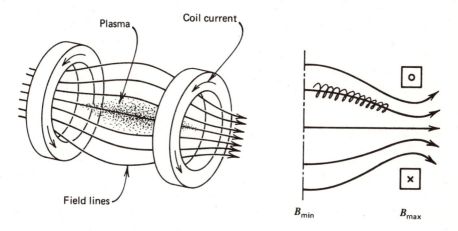

Simple magentic mirror

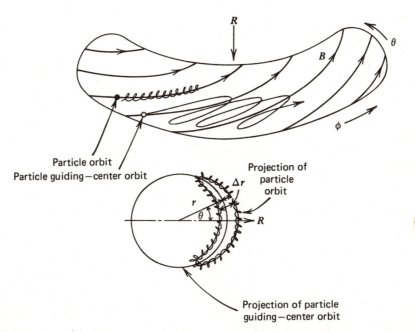

FIGURE 5.4. (*a*) Illustration of simple mirror trapping and particle reflection. (*b*) Mirror reflections produce complex orbits in a tokamak.

ε is the inverse aspect ratio r/R_0 and θ is the poloidal angle. For $(v_\perp/v_\parallel)^2 > \varepsilon^{-1}$, these particles may be trapped in the outer, weaker region of the magnetic field. Such particles are called trapped particles. Owing to the toroidal drift, these trapped particles do not reflect back along the same line of force on which they entered the mirror. They oscillate as illustrated in Figure 5.4b. The orbit has a banana shape, and the particle executes a series of banana-shaped orbits as it migrates in the toroidal direction.

The width Δr_b of the drift surface made up of banana orbits can be shown to be (see Section 5.7)

$$\Delta r_b \approx \frac{2\varepsilon^{1/2}m\bar{v}}{qB_\theta} = 2\varepsilon^{1/2}r_{c\theta} \tag{5.22}$$

which is proportional to a gyro radius based on the poloidal magnetic field strength. Since $B_\theta < B_\phi$, $r_{c\theta} >> r_c$, and the banana width Δr_b affords a larger step size in radial diffusion than would classical diffusion. Banana orbits play an important role in plasma-diffusion and particle-confinement times in toroidal magnetic systems. These are discussed in Section 5.7. The fact that the radial width of a drift surface in a tokamak is proportional to the gyro radius based on the poloidal magnetic field indicates the importance of the plasma toroidal current on confinement of alpha particles. By setting $\Delta r_b = a$, the plasma radius, and using Equation (5.22) together with Ampere's law, $\mu_0 I_\phi = 2\pi a B_\theta$ results in an equation for the magnitude of I_ϕ necessary to contain fusion alpha particles in a tokamak. It is

$$I_\phi \geqslant \frac{4\pi\varepsilon^{1/2}m_\alpha v_\alpha}{e\mu_0} = 3.1 \times 10^6 \text{ A} \tag{5.22a}$$

In a tokamak there is a toroidal electric field E_ϕ needed to generate the toroidal current. There is, as a result of this field, a radial drift velocity of the banana orbit center, which can be shown to be [10,13].

$$v_{\text{drift}} = \frac{E_\phi}{B_\theta} \tag{5.23}$$

The sign of B_θ, produced by the current induced by E_ϕ, is opposite to the sign of E_ϕ, so that $v_{\text{drift}} < 0$, and the banana center moves radially inward. This inward velocity is larger than the E_ϕ/B_ϕ particle guiding center drift. Therefore, the banana orbits should pinch the toroidal column inward. This theoretically predicted phenomenon is called the Ware pinch [13]; it has, however, not yet been observed in experiments.

5.4. TOROIDAL EQUILIBRIUM

5.4.1. Basic Equations

To maintain a hot plasma away from the container wall, an equilibrium state of plasma and magnetic fields must exist. The equilibrium equations of plasma physics are described and discussed in the book by Bateman [11], which also contains many thought-provoking questions.

The ideal magnetohydrodynamic theory and its application to magnetic fusion systems are reviewed by Freidberg [14]. The basic equilibrium equation is:

$$\nabla p = \mathbf{j} \times \mathbf{B} \qquad (5.24)$$

where it is assumed that within the plasma $n_e = n_i$, and, therefore, the electric field force is negligible. From Maxwell's equations, for steady-state conditions:

$$\mu_0 \mathbf{j} = \nabla \times \mathbf{B} \qquad (5.25)$$

$$\nabla \cdot \mathbf{B} = 0 \qquad (5.26)$$

$$\nabla \cdot \mathbf{j} = 0 \qquad (5.27)$$

It follows from Equation (5.24) that

$$\mathbf{B} \cdot \nabla p = 0 \qquad (5.28)$$

$$\mathbf{j} \cdot \nabla p = 0 \qquad (5.29)$$

so $\mathbf{B}$ and ∇p are orthogonal. Therefore, flux surfaces are constant-pressure surfaces, and the current density vector $\mathbf{j}$ is parallel to flux surfaces. Combining Equations (5.24) and (5.25), we obtain another form of the magnetohydrodynamic equilibrium equation:

$$\nabla\left(p + \frac{B^2}{2\mu_0}\right) = (\mathbf{B} \cdot \nabla)\frac{\mathbf{B}}{\mu_0} \qquad (5.30)$$

This equation illustrates the role of the plasma pressure p and the magnetic pressure $B^2/2\mu_0$. When the radius of curvature of $\mathbf{B}$ is much larger than the length over which $\mathbf{B}$ changes, then the right-hand side of Equation (5.30) is negligible and the equation can be integrated, giving

$$p + \frac{B^2}{2\mu_0} \approx \frac{B_0^2}{2\mu_0} \qquad (5.30a)$$

where $B_0^2/2\mu_0$ is the magnetic pressure where $p = 0$. For an axisymmetric system with $\partial/\partial z = 0$, Equation (5.30) is

$$\frac{\partial}{\partial r}\left(p + \frac{B_\phi^2 + B_\theta^2}{2\mu_0}\right) = -\frac{B_\theta^2}{r\mu_0} \tag{5.31}$$

Integrating by parts,

$$\left(p + \frac{B_\phi^2 + B_\theta^2}{2\mu_0}\right)_{r=a} = \frac{1}{\pi a^2}\int_0^a\left(p + \frac{B_\phi^2}{2\mu_0}\right)2\pi r\,dr \tag{5.31a}$$

or

$$\frac{B_\phi^2(a)}{2\mu_0} + \frac{B_\theta^2(a)}{2\mu_0} = \langle p\rangle + \frac{\langle B_\phi\rangle^2}{2\mu_0} \tag{5.31b}$$

where $\langle\,\rangle$ indicates volume averaged quantities, and it is assumed that $p(a) = 0$. This simple algebraic equation is useful to estimate tokamak relations.

5.4.2.   Axisymmetric Equilibrium

A straight cylinder, an axisymmetric toroid, and a configuration with helical symmetry are three examples of geometry that have at least one ignorable coordinate. Consider an axisymmetric toroid with cylindrical coordinates R, ϕ, Z with ϕ the ignorable angle coordinate. Equation (5.2) in cylindrical coordinates is

$$\frac{1}{R}\frac{\partial}{\partial R}(RB_R) + \frac{\partial B_z}{\partial Z} = 0 \tag{5.32}$$

This equation is satisfied by

$$RB_R = -\frac{\partial\psi}{\partial Z} \tag{5.33}$$

$$RB_Z = \frac{\partial\psi}{\partial R} \tag{5.34}$$

where ψ is the poloidal flux function defined by Equation (5.4). Indeed, $\nabla\psi\cdot\mathbf{B} = 0$ is satisfied. From Equation (5.28), namely, $\nabla p\cdot\mathbf{B} = 0$, we have:

$$-\frac{\partial p}{\partial R}\frac{\partial\psi}{\partial Z} + \frac{\partial p}{\partial Z}\frac{\partial\psi}{\partial R} = 0 \tag{5.35}$$

and

$$p = p(\psi) \qquad (5.36)$$

Similarly, from Equation (5.29), $\nabla p \cdot \mathbf{j} = 0$, and Equation (5.25), $\mu_0 \mathbf{j} = \nabla \times \mathbf{B}$:

$$-\frac{\partial p}{\partial R}\frac{\partial}{\partial Z}(RB_\phi) + \frac{\partial p}{\partial Z}\frac{\partial}{\partial R}(RB_\phi) = 0 \qquad (5.37)$$

Defining $I(\psi)$ as the total current in the hole of a flux surface $\psi = $ const, Equation (5.11) gives

$$RB_\phi = \frac{\mu_0}{2\pi} I(\psi) \qquad (5.38)$$

The radial component of the basic equilibrium equation $\mathbf{j} \times \mathbf{B} = \nabla p$ can now be written in terms only of the function ψ as

$$\frac{\partial^2\psi}{\partial R^2} - \frac{1}{R}\frac{\partial\psi}{\partial R} + \frac{\partial^2\psi}{\partial Z^2} + \mu_0 R^2\frac{\partial p(\psi)}{\partial\psi} + \frac{\mu_0^2}{8\pi^2}\frac{\partial I^2(\psi)}{\partial\psi} = 0 \qquad (5.39)$$

or

$$-L(\psi) = \mu_0 R^2 p' + \frac{\mu_0^2}{4\pi^2}II' \qquad (5.40)$$

where

$$L(\psi) \equiv \left(R\frac{\partial}{\partial R}\frac{1}{R}\frac{\partial}{\partial R} + \frac{\partial^2}{\partial Z^2} \right)\psi$$

Equation (5.39) is the equilibrium equation for axisymmetric systems and is named the Grad−Shafranov equation after the two plasma scientists who independently derived the equation.

For calculating equilibria, using Equation (5.39), the functions $p(\psi)$ and $I(\psi)$ may be chosen arbitrarily. For other than quadratic functions of ψ, Equation (5.40) is nonlinear, and therefore difficult to solve. The left-hand side of Equation (5.40) comes from the toroidal current crossed with the poloidal magnetic field, while the II' term on the right-hand side corresponds to the poloidal current crossed with the toroidal magnetic field.

Axisymmetric plasma equilibria must satisfy Equation (5.40), and hence the Grad−Shafranov equation has been and is studied in great detail. By assuming usually quadratic forms of the pressure function $p(\psi)$ and the current function

$I(\psi)$, and solving numerically Equation (5.39), for $\psi(r, z)$, one can then construct flux plots, such as shown in Figure 5.2. The current density is given in terms of $I(\psi)$ as follows:

$$j_R = -\frac{1}{2\pi R}\frac{\partial I(\psi)}{\partial Z} \tag{5.41}$$

$$j_\phi = -\frac{1}{\mu_0}\left(\frac{\partial}{\partial R}\frac{1}{R}\frac{\partial \psi}{\partial R} + \frac{1}{R}\frac{\partial^2 \psi}{\partial Z^2}\right) = \frac{1}{\mu_0 R}\left(\mu_0 R^2 p' + \frac{\mu_0^2}{8\pi^2}(I^2)'\right) \tag{5.41a}$$

$$j_Z = \frac{1}{2\pi R}\frac{\partial I(\psi)}{\partial R} \tag{5.41b}$$

There are three interesting limits of the Grad−Shafranov equation:

1.   There is no poloidal current in the plasma; therefore, $I(\psi) = 0$, and the toroidal field is exactly the vacuum field $B_\phi = B_\phi(0)R_0/R$. The plasma pressure is confined entirely by the toroidal current crossed with its own poloidal field. This case corresponds to $\beta_\theta = 1$, that is,

$$\beta_\theta = \frac{2\mu_0 p}{B_\theta^2} = 1 \tag{5.42}$$

As will be shown later, this corresponds to a "low-β tokamak" equilibrium.

2.   The current density is everywhere parallel to **B**; that is, $\mathbf{j} = \alpha\mathbf{B}$. Therefore, $\nabla p = p'(\psi) = 0$ and there is no magnetic confinement of plasma. Where this condition applies, the region is called "force free." Scientists at the Jutphaas Laboratory in Holland have designed plasma toroids in which force-free regions exist between the plasma column and the vacuum wall.

3.   The plasma is confined almost entirely by the poloidal current crossed with the toroidal field; that is, $|II'(\psi)| \gg L(\psi)$. This is the case of the "high-β tokamak," and $\beta_{pol} > 1$, as is discussed in Section 5.6.

5.4.3.   Tokamak Equilibrium

Situations that have complete toroidal equilibria can be obtained by computational solution of the Grad−Shafranov equation. How to arrange for this solution to occur in a real situation is not trivial. In tokamak experiments, equilibrium along the major radius is achieved by adjusting the magnitude of a vertical magnetic field, so that the cross product of the toroidal plasma current and this applied vertical field produces a radially inward force to just balance the radial outward forces.

In early tokamak experiments, a highly conducting shell surrounded the plasma. As the hot plasma moved radially outward, it produced image currents

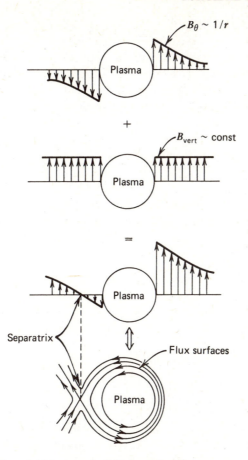

FIGURE 5.5. The poloidal magnetic field from the plasma current contributes an outward radial force on the plasma column. A vertical magnetic field interacting with the toroidal plasma current provides an inward radial force. The combined magnetic field of a circular plasma has a separatrix with one x-point.

in this shell which produced the required vertical field for equilibrium. The plasma magnetic axis is shifted outward. Eventual Ohmic decay of these image currents make this technique satisfactory only for relatively short times. Another method to achieve radial equilibrium makes use of an independent, applied external vertical field.

Consider the poloidal magnetic field from the plasma current as shown in Figure 5.5. Because of the topology of a torus, the poloidal flux inside the hole is more dense than on the outside. Therefore, the poloidal magnetic pressure is larger on the inside of the torus than on the outside, resulting in a radial outward force for a diamagnetic plasma. Or, what is equivalent, the toroidal current crossed with the poloidal magnetic field results in a radial outward force. In addition, the poloidal current crossed with the toroidal magnetic field also gives a radial outward force. And finally, the plasma pressure in a torus results in an outward force that, from the particle point of view, is the centrifugal force produced by the toroidal velocity component of the plasma particles.

The calculation of the magnitude of the applied vertical field required to provide radial equilibrium was first developed by Shafranov [15]. Its derivation can also be found in the books by Miyamoto [10] and by Bateman [11] and will not be repeated here. For a plasma torus, the vertical equilibrium field B_v is found to be

$$B_v = \frac{-\mu_0 I_\phi}{4\pi R}\left[\ln\frac{8R}{a} + \frac{l_i}{2} + \beta_\theta - \frac{3}{2}\right] \tag{5.43}$$

where, l_i is the internal inductance of the plasma loop per unit length, and its value depends on the current distribution within the plasma. For a circular cross section:

$$l_i = \frac{\displaystyle\int_0^a B_\theta^2(r)r\,dr}{a^2 B_\theta^2(a)} \tag{5.44}$$

where

$$B_\theta(a) = \frac{\mu_0 I_\phi}{2\pi a} \tag{5.44a}$$

For a uniform current $j_\phi = $ constant, $l_i = \frac{1}{2}$, and for a peaked current $l_i > \frac{1}{2}$. The poloidal beta β_θ is defined for noncircular tokamaks, as

$$\beta_\theta = \frac{2\mu\langle p\rangle}{(\oint d\hat{l}\cdot\mathbf{B}_\theta(a)/\oint d\hat{l})^2} = \frac{2\mu\langle p\rangle}{\overline{B}_\theta^2(a)} \tag{5.44b}$$

where $\overline{B}_\theta^2$ is the square of the mean value of the poloidal field averaged around the outer plasma poloidal flux surface, and $\langle p\rangle$ is the volume averaged pressure:

$$\langle p\rangle = \frac{\displaystyle\int_0^a p(2\pi r)dr}{\pi a^2} \tag{5.44c}$$

A high-β tokamak has $\beta_\theta \sim A > 1$.

If the vertical field B_v were uniform and straight, the plasma column would be neutrally stable to vertical displacement. To avoid this potential stability problem, the vertical field is slightly curved, bowing radially inward (positive radius of curvature) at the top and bottom of the torus. This curvature of the externally applied vacuum magnetic field is described by the decay index n, defined as

$$n \equiv -\frac{R}{B_v}\left(\frac{dB_v}{dR}\right)$$

(5.45)

The condition for vertical stability is $0 < n$. For horizontal stability, the restoring force provided by B_v must decrease with radius slower than the expansion forces. In the limit of large aspect ratio ($\ln 8R/a \gg 1$), the condition for horizontal stability of a circular tokamak turns out to be $n < \frac{3}{2}$ [11].

Although it is convenient and usual to assume that there is no variation in B_ϕ in the toroidal direction, the fact is that there is space between the toroidal field magnets that cause toroidal field ripple. This field ripple should be kept small ($\Delta B_\phi/B_\phi(0) < 1\%$) to achieve good plasma confinement.

5.5. TOROIDAL STABILITY

5.5.1. Introduction

Achieving a plasma equilibrium that is stable has been recognized as a fundamental plasma problem since the earliest days of magnetic-fusion research. Plasma-stability study is a field of primary research importance, and there now exists a very large and sometimes complex literature dealing with this topic. For example, the entire book by Bateman [11] is devoted to, and deals exclusively with, one important aspect of this research, namely, the magnetohydrodynamic (mhd) class of instabilities of toroidal tokamaks. The macroscopic, or mhd-fluid approach to analyzing the stability of a plasma configuration, has been, by-in-large, very successful in predicting the most serious large scale instabilities observed in actual plasma experiments. What is even more important, this body of theory has helped provide a group of rules, which when adhered to in experiments, does indeed provide a macroscopically stable plasma.

There are also important microscopic plasma-stability problems which require plasma kinetic theory to interpret. For example, instabilities arise as the result of some non-Maxwellian particle distributions, and such microscopic instabilities influence transport properties, confinement, heating, wave properties, etc. And there are stability aspects of magnetic confinement that do not neatly fit either macro- or microscopic categorization, but at times are a mixture of both, for example, current-driven instability. Stability studies continue to be a lively area of research in fusion.

There have been a great number of instabilities predicted for various plasma-magnetic-field configurations. So many, in fact, that just listing them and their properties fills a handbook–dictionary [16]. A good introduction to tokamak equilibrium, a large subject in itself, can be obtained by referring to the books by Bateman [11] and Miyamoto [10], tokamak review papers [4–6], and the review of mhd equilibrium and stability by Wesson [17], and by Freidberg [14].

It does appear, however, that the subject of tokamak stability, while not yet complete, is mature. Large-scale macroscopic stability can be predicted fairly well, even for rather complex minor cross-section-shaped toroids. On the other hand, many mhd and kinetic instabilities that are theoretically predicted have yet to be seen in actual experiments. Perhaps they do not grow, because of some form of damping, or they degenerate into plasma turbulence that manifests itself in increased transport. It should also be remembered that stability theory starts with the assumption of a precise equilibrium, a situation sometimes not achieved in actual experiments.

5.5.2. Mhd-Stability Fundamentals

The classical theoretical approach to study stability of a plasma is to perturb slightly the mhd-equilibrium configuration and examine whether the amplitudes of the perturbations grow or decay with increasing time. To do this, the mhd equations are linearized, and the equilibrium solution is perturbed to first order in terms of its Fourier components:

$$\xi(\mathbf{r},t) = \xi(r)\exp(-i\omega t) \tag{5.46}$$

where $\omega = \omega_r + i\omega_i$; the subscripts refer to real and imaginary parts of ω. In the linearized case, each frequency component can be treated independently. This results in a system of linear algebraic equations, whose solutions are contained in a dispersion equation. The dispersion equation yields eigenvalues of ω. Stability depends upon the sign of the imaginary part, ω_i. If $\omega_i > 0$, the perturbation is unstable, and when $\omega_i < 0$, it is stable. When $\omega_r \neq 0$, the perturbation is oscillatory, and when $\omega_r = 0$, it grows or damps monitonically. Solution of the linearized equations has a form proportional to $\exp(\gamma t)$, where γ is the growth rate. When there is more than one unstable mode, the fastest growing mode is considered the most dangerous.

It is usually very difficult to solve the plasma-stability eigenvalue problem except in some relatively simple, but nonetheless informative, physical cases. There is another, and very useful approach, to test for stability: multiply the mhd equation of motion by the time derivative of the plasma displacement (from equilibrium) and integrate, subject to boundary conditions, over the volume of the plasma. The total perturbed energy is constant in time, and consists of the sum of the potential and kinetic energies. Any perturbation of the potential energy, δW, that decreases its value must produce a corresponding increase in the kinetic energy; that is, the system is linearly unstable. The perturbation need not be an eigenfunction of the equations nor the fastest growing mode. Any test function that satisfies the boundary conditions and can be integrated is satisfactory. This is an important advantage of the energy principle. Alternately, if all perturbations increase the potential energy (i.e., δW is positive), then the system is linearly stable. This energy principle was first developed by Bernstein, Frieman, Kruskal, and Kulsrud [18] and is widely used to test for mhd plasma stability.

Present-day stability techniques usually involve very large computer codes and are relatively complex. Within the discipline of plasma physics, stability analysis has become a specialty, and because of the size and richness of this subject, will probably remain so. Scientists and engineers who deal with many other important aspects of fusion will most probably have to rely upon a stability specialist for advice, but remember that experimental observations are the ultimate facts upon which real reactor designs must depend.

It is very helpful to know the language of plasma-stability analysis, and some of the most important results. To accomplish this, we first consider mhd stability of a straight circular cylinder of plasma. Then we examine tokamak toroidal stability, and finally we review some important kinetic instabilities of toroids. The mathematical details can be found, for example, in References 10 and 11.

5.5.3.   Mhd Instabilities of a Straight Circular Cylinder

The stability of a straight plasma circular cylinder was studied intensively early in the history of fusion. Laboratory creation of straight plasma cylinders is usually done rapidly, accompanied by shock waves, and when most of the plasma current is in the direction of the cylindrical axis (Z-direction), it is called a Z-pinch. When most of the plasma current is poloidal (θ direction), it is called a θ-pinch, and when the Z and θ currents are similar in magnitude, it is called a screw-pinch.

First consider a sharp-boundary plasma cylinder (coordinates r, θ, Z) magnetic field B_{iZ} inside the boundary and B_{eZ} outside the plasma. Both are assumed constant. An azimuthal magnetic field $B_\theta = \mu_0 I / 2\pi r$ is outside the plasma column. The plasma is assumed perturbed by the displacement, $\xi(r,t) \exp\left[\gamma t + i(m\theta - kZ)\right]$, where k is the propagation constant. There are two driving mechanisms (e.g., sources of energy) for producing mhd instabilities in the circular cylinder case: (1) pressure gradients and (2) currents parallel to the magnetic field. Instabilities caused by the former are called interchange modes and the latter are called kink modes.

When there is no external longitudinal magnetic field, that is, $B_{eZ} = 0$, the plasma cylinder is found to be unstable. In particular, the $m = 0$ mode is called the sausage instability because of the shape of the plasma, which is illustrated in Figure 5.6a. This instability can be overcome, that is, stabilized, by the addition of a small longitudinal field. That, however, gives rise to a new instability, the $m = 1$ kink mode. The kink mode is illustrated in Figure 5.6b. The plasma

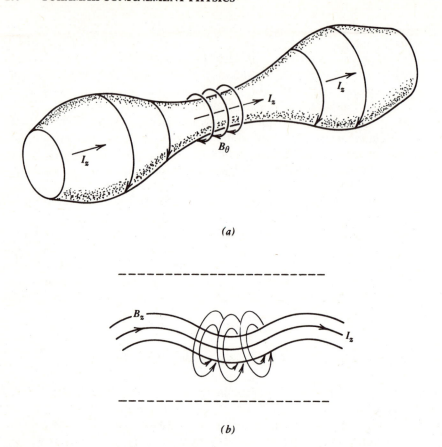

(a)

(b)

FIGURE 5.6.   Examples of mhd instabilities of a straight plasma column: (a) sausage instability: $B = 0$, $m = 0$; (b) kink instability: $B_z \neq 0$, $m = 1$.

column will, however, be stable, even for $m = 1$, when the ratio of the poloidal
to toroidal fields obey the Kruskal–Shafranov condition:

$$\left|\frac{B_\theta}{B_z}\right| < \frac{2\pi a}{L} \qquad (5.47)$$

where L is the column length and a is the cylinder radius. It is perhaps the most
important stability criterion for pinches. Because of the presence of the strong
B_z field, the instability is restricted to long wavelengths. The criterion $q(a) > 1$
can be viewed as a condition for prohibiting the formation of potentially
unstable long-wavelength modes because of periodicity requirements. It sets a
limit on the magnitude of plasma current, as a function of the strength of the
longitudinal magnetic field. The instability appears as a nearly rigid helical
deformation of that volume of plasma that violates the Kruskal–Shafranov
condition.

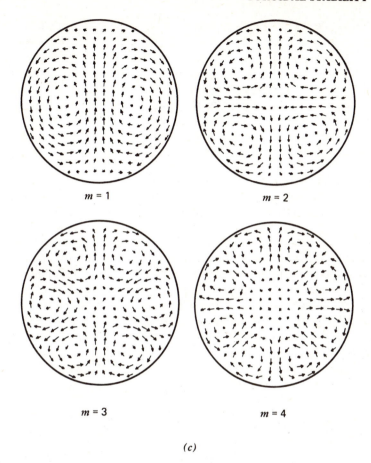

$m = 1$ $m = 2$

$m = 3$ $m = 4$

(c)

FIGURE 5.6. (*continued*) Examples of mhd instabilities of a straight plasma column: *(c)* fixed-boundary instabilities with uniform current equilibrium with $ka = 1$ [from G. Bateman, W. Schneider, and W. Grossman. *Nuclear Fusion* **14**, 669 (1974)].

For a diffuse plasma in which the current permeates the plasma, and which is usually zero at the boundary, similar results are found. For $m \geq 2$, instabilities are found that are localized (called local modes) near the mode rational surfaces, $m/n = (2\pi a/L)(B_\theta/B_z)$ where m and n are integers. These local modes can be stabilized by providing sufficient magnetic shear. This stability criterion, discovered by Suydam, is

$$\frac{r}{4}\left(\frac{\bar{\mu}'}{\bar{\mu}}\right)^2 > \frac{2\mu_0}{B_z^2}\left(\frac{dp}{dr}\right) \tag{5.48}$$

where $\bar{\mu} \equiv B_\theta/rB_z$ and the prime represents the radial derivative. A fixed-boundary plasma may exhibit unstable internal modes, as illustrated in Figure 5.6c. If, however, a conducting wall surrounds the plasma column, not only does it alter the boundary conditions, but it generally aids in stabilizing the column.

It is important to keep in mind that stability analysis nearly always concerns small-amplitude linear perturbations. It says nothing about nonlinear growth and the eventual plasma behavior. Linearized instability predictions need not always be the cause for rejection of a plasma confinement system. If the nonlinear consequences can be tolerated, that is, they are not too damaging, then the reactor designer may still accept them if the rewards are sufficient, for example, more compact design, higher beta, etc.

5.5.4.   Mhd Tokamak Stability

The hydrodynamic stability of large-aspect-ratio, low-β, circular cross-section tokamaks is fairly well understood, and it appears that completely mhd-stable configurations are possible. The cases of small-aspect ratio, shaped cross-sections, and higher-beta tokamaks are less well understood and are the topics of current research. The status of mhd tokamak stability is described in review papers by Wesson [17] and Freidberg [14].

The methods used, and results found, in analyzing the stability of tokamak toroids are similar to those found for cylinders. There are of course effects caused by toroidal curvature, and the cylinder length is replaced by $2\pi R_0$. To be realistic, diffuse current profiles must be assumed and their shape affects stability and growth rates. Finite resistivity of the plasma and noncircular cross-section shapes add to the degree of complexity as more reality is introduced into the analysis.

The Kruskal−Shafranov condition necessary for stability of kink modes of a circular tokamak is

$$q(r) = \frac{r}{R} \frac{B_\phi}{B_\theta} > 1$$

(5.49)

where q is the safety factor. When $r < a$, an instability is referred to as internal, and when $r = a$, as external. Much of the basic insight concerning external kink instabilities is due to Shafranov. The most dangerous mhd instabilities in a straight tokamak are the external kink modes that occur when $q(a) < 1$. The internal kink plays an important role in tokamak operation because it limits the value of the toroidal current density on axis. Equivalent expressions for noncircular tokamaks involve Equation (5.8). The destabilizing effect for kink modes of diffuse current tokamaks is found to be the current-density gradient dj_ϕ/dr. It is the torque arising from the dj_ϕ/dr term in the δW expression which drives this instability. For a free-boundary plasma, it is further found that when $q(a) > m/n$, and $q(a) > 1$, it is stable. The minimum value of n, the toroidal mode number, is unity. Stability is obtained for $q(a) > m$. For a given $q(r)$, stability can be described in terms of $nq(a)$ only. Growth rates are a function of the local Alfvén speed and a characteristic length, for example, the plasma radius a. Stability and growth rates depend on the plasma current profile, or $q(r)$ profile. Fortunately, the current gradient also provides strong stabilizing effects through shear of the magnetic field. It is found that complete stability against kink modes may be obtained for $q(a) > 1$ by peaking the current distribution, and the maximum current allowable is obtained for the lowest allowable value of $q(0)$. Even when internal modes become unstable, as it presumably does during saw-tooth oscillations, it is not catastrophic since its activity mainly consists of a redistribution of plasma energy within a relatively small region internal to the $q = 1$ surface. In effect, the oscillations anchor the average value of the safety factor on axis to its threshold value, $q(0) = 1$. The current density at the origin is related to the safety factor at the origin, for a circular cross-section tokamak, by

$$q(0) = \frac{2B_\phi}{\mu_0 R_0 j_\phi(0)} \tag{5.50}$$

Internal modes, often called interchange modes, are localized about resonant surfaces $q = m/n$. The stability criteria for these internal modes is given by Mercier's criterion, which is the toroidal generalization of the Suydam criteria for a straight cylinder [see Equation (5.48)]. The Mercier-tokamak stability criterion is

$$\frac{1}{4}\left(\frac{\tilde{\mu}'}{\tilde{\mu}}\right)^2 + \frac{p'/a}{B_\phi^2/2\mu_0}(1 - q^2) > 0 \tag{5.51}$$

Since $p' < 0$, internal modes are stabilized for $q > 1$. For current profiles $j_\phi(r)$ that decrease with r, the q value increases with r, the minimum of q occurring at $r = 0$. Thus a tokamak with $p' < 0$ is stable to localized modes provided $q(0) > 1$. A stability diagram illustrating the regions in which ideal kink and internal modes are stable is shown in Figure 5.7.

This discussion of tokamak stability has, to this point, been based on an ideal (perfect conductivity) plasma. The introduction of finite conductivity in the analysis results in new phenomena, since magnetic field lines can break and rejoin. This effect plays a special role in a narrow layer around the mode rational surfaces, $q = m/n$. It is found that a resistive plasma may, around resonant surfaces, break up into helical filaments called magnetic islands. These resistive internal kink modes around resonant surfaces are commonly called tearing modes. Tearing modes have lower growth rates than ideal modes, they seem to saturate, and are believed to be more tolerable than ideal kink modes. If this is so, tearing modes may not be too damaging, but they can cause increased transport. One of the operating limits of present-day tokamaks, a major disruption (see Section 5.9), is believed to be related to m/n tearing-mode phenomena. Disruptions impose even stronger restrictions on $q(a)$.

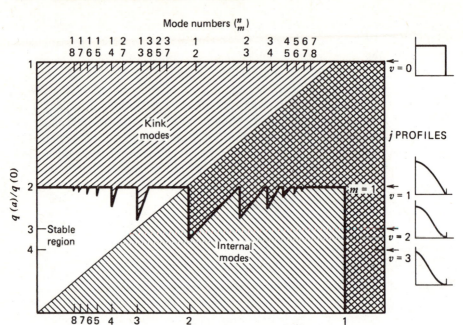

FIGURE 5.7.   Stability diagram for ideal kink and internal modes of a large-aspect-ratio circular cross-section tokamak. The current distribution is $j_\phi(r) = j_\phi(0)[1 - r/a)^2]^\nu$. Complete stability against kink modes may be obtained for any $q(a) > 1$ by sufficient peaking of the current profile (larger value of ν). The maximum stable current for this model is obtained with a parabolic current distribution with $q(0) = 1$ and $q(a) = 2$. (Data from J.A. Wesson [17].)

Two basic requirements for a tokamak are a large plasma current, relative to the current in the toroidal field coils, needed for good confinement, and a high value of beta for high fusion-power density. To optimize the plasma current, it is desirable to have a relatively low aspect ratio ($R/a \sim 4$), because of q limitations. There appears to be advantages to achieving both requirements by going to noncircular cross-section tokamaks. Typical shapes are the ellipse, the dee, the doublet, and the bean illustrated in Figure 5.8. The circular tokamak, when heated to high pressure and high-β, is predicted to naturally evolve into a dee shape. And dee-shaped toroidal field coils have some mechanical stress advantages. The doublet has large magnetic shear. The bean cross section is calculated to be stable at high beta. Such elongated plasmas introduce the possibility of axisymmetric mode instabilities ($n = 0$, any m) but these are usually avoided by appropriate choice of the decay index (see Section 5.4.3). Most present-day reactor designs choose a dee-shaped cross section with an elongation of about 1.5.

The maximum obtainable stable tokamak beta is not known and is a topic of research. It is presently believed that there is an upper limit to beta, for tokamaks, set by ballooning instability. In considering the terms that constitute

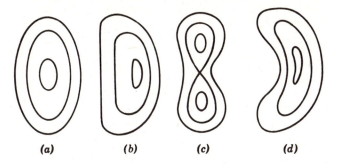

FIGURE 5.8.   Noncircular tokamak minor cross-sections: *(a)* ellipse, *(b)* dee, *(c)* doublet, *(d)* bean. The axis of toroidal symmetry is on the left of each figure.

W, one of the negative terms involves a pressure gradient ∇p. For higher-β tokamaks, it is this ∇p term which dominates δW and which becomes more significant than the current-driven term. The instability that is driven by the ∇p occurs only around the outer edge of the torus, a region of bad field curvature, and along a length $\sim 1/Rq$ of the field line. The plasma bulges out most where both the pressure gradient and curvature are strongest and both point roughly in the same direction. The fastest growing ballooning instabilities are high-m modes. A flow pattern for a ballooning mode is illustrated in Figure 5.9. Ballooning modes are expected even for a collisionless plasma and also one with resistivity. Some tokamak experiments, for example, those with intense neutral-beam heating and some pinch tokamaks, have operated with $\langle \beta \rangle \approx 3\%$ and seem to be macroscopically stable. Theory also predicts a second, stable, high-beta regime [19].

5.5.5.   Microinstabilities

Microinstabilities are instabilities associated with kinetic models of a plasma, and they are often associated with non-Maxwellian velocity distributions. Fluid models are limited to phenomena of relatively low frequency, usually less than the ion gyro frequency. Kinetic or microinstabilities often affect transport properties and can evolve into plasma turbulence. Microinstabilities are a large and complex topic of plasma physics, and here we shall discuss just a few of those instabilities believed to be relevant for fusion. The interested reader may consult the later chapters of the book by Miyamoto [10].

When a beam of particles propagates through a background plasma, waves can be excited. The energy to excite a wave comes from the kinetic energy of the beam. When some fluctuation occurs in the beam motion, charged particles may be bunched and an electric field induced. If the electric field acts to amplify the bunching, the disturbance can grow. Such an instability is called a "two-stream" instability. Electron or ion beams injected into a plasma can cause this instability.

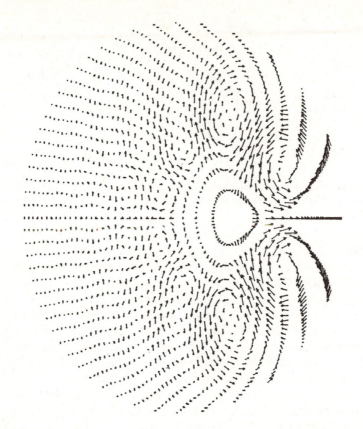

FIGURE 5.9.   Flow pattern for a tokamak ballooning mode with $n = 3$ at $\langle\beta^*\rangle = 3\%$. Beta star is defined as

$$\langle\beta^*\rangle^2 \equiv 4\mu_0^2 \langle p^2\rangle / B_\phi^4(0)$$

Toroidal axis is on the left. [Figure from the paper by A.M.M. Todd et al., *Phys. Rev. Lett.* **38**, 826 (1977).]

Where there is a density gradient in a plasma, an instability may arise because of electron drift caused by the gradient. This instability is called the drift or universal instability, and is expected at the edge of a plasma where the density gradient may be steep. It can be stabilized by magnetic shear, which enhances Landau damping. Landau damping is a collisionless mechanism by which waves and some particles with the appropriate speeds exchange energy.

A velocity space anisotropy occurs in a plasma confined by mirror fields. This occurs because of the particle loss-cone, namely, those particles which have a large $v_\parallel$ [see Equation (5.20)] escape from the plasma. The loss of these particles is the source of a class of instabilities that are thought to be important in tokamaks, because of the trapped particles which exist in the toroidal helical magnetic field. Trapped particles are confined in the local mirrors, and move back and forth in the local mirror fields with a bounce frequency ω_b. When a disturbance frequency is less than the bounce frequency of either electrons or ions, a trapped particle instability can occur. Dispersion equations have been studied for the trapped ion mode ($\omega \ll \omega_{bi} \ll \omega_{be}$) and the trapped electron mode ($\omega_{be} \gg \omega \gg \omega_{bi}$). The trapped ion mode has bulk fluid or fluid-like motions. The effects of collisions on trapped-particle instabilities have been extensively studied, and the dissipative trapped-particle instability is a result of trapped electrons becoming untrapped by virtue of collisions. Various models of anomolous diffusion have been developed to attempt to account for the effects of trapped-particle instability in tokamaks at fusion-plasma conditions; see, for example, the review papers by Kadomtsev and Pogutse [20] and Coppi [21], and the monograph by Manheimer [22].*

These are but a few of many kinetic instabilities that have been analyzed and that may occur under appropriate plasma conditions. The trapped-particle instability is predicted to substantially increase cross-field diffusion in a fusion tokamak, but it has not yet been observed in tokamak confinement experiments.

5.6. HIGH-BETA TOKAMAK

The importance of operating a fusion power tokamak at a relatively high value of beta was discussed in Section 3.3. Most fusion tokamak reactor-design studies indicate that to be economically competitive with other forms of energy for electric power plants, a tokamak should have a volume averaged beta $\langle \beta \rangle \gtrsim 5\%$, and the higher, the better, within the metallurgical limits of the first wall to sustain the neutron flux.

The high-beta tokamak is discussed by Freidberg [14]. To understand physically what constitutes a high-beta tokamak, consider a sharp-boundary plasma tokamak, where equilibrium at the plasma surface requires

$$B_{\phi e}^2 + B_\theta^2 = 2\mu_0 p + B_{\phi i}^2 \qquad (5.52)$$

where $B_{\phi e}$ and $B_{\phi i}$ refer to the toroidal magnetic field strength just exterior to, and just inside, the plasma surface, respectively.

The conventional low-β tokamak has its toroidal field equal to the vacuum field, $B_\phi = R_0 B_{\phi 0}/R$. The difference between $B_{\phi e}$ and $B_{\phi i}$ is of the order ε^2, which is considered negligible. Consequently, from Equation (5.52):

$$p = \frac{B_\theta^2}{2\mu_0} \qquad (5.52a)$$

---

*The first experimental observation of trapped-particle instability was reported by S. Prager, A. Sen, and T. Marshall [*Phys. Rev Lett.* **33**, 692 (1974)].

or

$$\beta_\theta \equiv \frac{2\mu_0 p}{B_\theta^2} = 1 \tag{5.52b}$$

where β_θ is the poloidal beta [see Equations (5.42) and (5.44b)]. The toroidal beta β_ϕ, which is the beta of economic interest in a tokamak because $B_\phi > B_\theta$, is

$$\beta_\phi \equiv \frac{2\mu_0 p}{B_\phi^2} = \beta_\theta \frac{B_\theta^2}{B_\phi^2} = \beta_\theta \left(\frac{\varepsilon}{q}\right)^2 \tag{5.52c}$$

Since $\beta_\theta \approx 1$ and the lowest allowable q value is 1, we find

$$\beta_\phi \approx \varepsilon^2 \tag{5.52d}$$

The plasma inverse aspect ratio ε may be of the order $\varepsilon \sim \frac{1}{5}$, giving $\beta_\phi \sim 4\%$, and usually $\beta_\phi < 4\%$. A conventional low-β tokamak plasma is confined by its poloidal field [Equation (5.52b)], and has a $\beta_\phi \sim \varepsilon^2$. These results are illustrated in Figure 5.10.

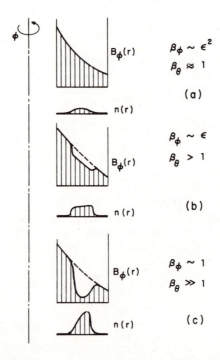

FIGURE 5.10. Beta ordering for circular tokamaks: *(a)* low-beta tokamak, *(b)* high-beta tokamak, and *(c)* pinch tokamak.

Note that in a low-beta tokamak the energy density of the plasma ($\frac{3}{2}nkT$) is proportional to β_θ, ε^2, and B_ϕ^2, that is,

$$nkT = \beta_\theta \left(\frac{\varepsilon}{q}\right)^2 \frac{B_\phi^2}{2\mu_0}$$

An upper limit of β_ϕ is believed to be a function of

$$\left(\frac{\varepsilon}{q}\right)^2 = \left(\frac{a}{Rq}\right)^2$$

see, for example, the arguments in Reference 10, p. 191.

A high-β tokamak has a sufficient poloidal current that the toroidal magnetic field profile has a modest magnetic well, as illustrated in Figure 5.10b. The difference between the exterior and interior toroidal field is of the order ε, that is,

$$B_{\phi i} = B_{\phi e} \left[1 - O(\varepsilon)\right] \tag{5.52e}$$

Combining Equation (5.52), (5.52a), and the definition of q, and neglecting terms of order ε^2 results in

$$\beta_\phi \approx 2\varepsilon \tag{5.52f}$$

Thus, a high-beta tokamak can be defined as one where the toroidal-beta is of the order ε, which means that beta will be between 10% and 50%. The poloidal beta is

$$\beta_\theta > 1$$

the plasma is diamagnetic, and the toroidal-field well provides the plasma confinement. A high-beta tokamak has, according to Freidberg [14], the following properties:

$$\frac{B_\theta}{B_\phi} \sim \varepsilon$$

$$\beta_\phi \sim \varepsilon$$

$$q \sim 1$$

and

$$\beta_\theta \sim \frac{1}{\varepsilon}$$

This last criteria, namely, $\epsilon\beta_\theta \geq 1$, is a good test for whether high-beta tokamak physics conditions have been achieved in experiments.

Finally, there is the toroidal screw pinch; if the cross section is long and thin, it is called a belt pinch. The poloidal current is very large and the depth of the toroidal-field well is large, namely,

$$B_{\phi i} = B_{\phi e} [1 - O(1)] \tag{5.52g}$$

This case, the limit of the high-beta tokamak, has a very deep magnetic well, with $\beta_\phi \sim 1$ and $\beta_\theta \gg 1$. This is illustrated in Figure 5.10c.

There are several ingredients that are helpful in obtaining a high-beta tokamak. They are:

1. Elongated minor cross section, that is, dee, doublet, bean, or ellipse.
2. Radially nonsymmetric current and pressure profiles, $q(r)$ and $p(r)$.
3. Force-free currents external to the plasma column.
4. Significant poloidal plasma current.

There are two processes, so far distinct, which are used to obtain a high-beta tokamak. The flux-conserving tokamak (FCT) begins with a low-beta tokamak equilibrium, and then is subjected to intense heating by neutral beams or RF. The time scale of this heating is such that the magnetic flux surfaces are frozen in the plasma. This heating is on a time scale faster than the resistive magnetic diffusion time, but slower than the Alfvén transit time. The separatrix does not move into and touch the plasma, which would permit plasma to escape. A circular tokamak plasma evolves into a dee-shaped plasma during FCT heating. This is shown in the computer similations of Dory and Peng, illustrated in Figure 5.11. The second process uses high-power pulsed implosion to obtain either the high-beta, or the pinch tokamak. This process involves high-voltage technology and is not favored by reactor designers. It is useful in research devices for studying the plasma-physics properties of higher-beta tokamaks.

There is no limit (other than $\beta_\phi \leq 1$) to the value of beta, set by plasma equilibrium. Solutions of the Grad–Shafranov equation have been found for very high beta tokamaks [23]. The question of their stability is as yet unanswered. The ballooning mode is predicted to occur when beta is increased and is typically expected when $\langle \beta \rangle$ is from 2 to 5%. Coppi and co-workers theoretically predict the existence of a second, stable tokamak regime [19].

Coppi chooses to illustrate tokamak stability in $\hat{S} - G$ space where $\hat{S}$ is the shear, $\hat{S} = d\ln q/d\ln r$ and G = pressure gradient,

$$G = -\frac{2\mu_0 R_0 q^2}{B_0^2}\left(\frac{dp}{dr}\right)$$

FCT FLUX SURFACES ($q_0 = 1$, $A = 4$)

• INTERNAL FLUX SURFACES EVOLVE INTO
D−SHAPE AT HIGH β

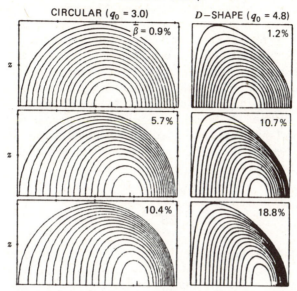

FIGURE 5.11. Flux-surface profiles for two sequences of flux-conserving-tokamak (FCT) equilibria. The torus axis is on the left. Note that the profile tends to dee shape, the magnetic axis shifts radially outward, and the flux contours become more tightly packed on the outside; that is, the toroidal current is peaked on the outside. [From R.A. Dory and Y-K. M. Peng, *Nuclear Fusion* **17**, 21 (1977).]

The pressure-gradient function G can be shown to be equivalent to

$$G = \varepsilon \langle \beta_\theta \rangle$$

where we have used the approximation $(dp/dr) \approx \bar{p}/a$ and Equation (5.52c). The second, high-beta stability boundary is a function of shear and $\varepsilon\beta_\theta$. δW stability analyses for tokamaks also find stability a function of $\varepsilon\beta_\theta$; thus, $\varepsilon\beta_\theta \gtrsim 1$ is a simple and useful criterion for higher-beta tokamaks, as was pointed out earlier in this section. Flux-conserving tokamaks have achieved stable operation for $\langle \beta \rangle \approx 3\%$, with no signs of ballooning modes. Perhaps this instability, if it occurs, saturates and the beta limit of tokamaks will be set by energy transport processes. The interested reader can find more details about high-beta tokamak prospects in References 6, 11, 23, 24, and 25.

5.7. NEOCLASSICAL TOKAMAK PLASMA TRANSPORT

Classical plasma transport theory was summarized in Section 4.7. The characteristic step size for plasma diffusion normal to a magnetic field is the gyro radius [Equation (4.64)]. The kinetic theory of plasma transport in tokamaks was initiated by Galeev and Sagdeev [26] who showed that trapped particles with banana orbits are responsible for a significant enhancement of the transport properties at small values of the collision frequency (i.e., in the plasma regime where fusion rectors must operate). The enhancement was explained qualitatively on the basis of a random walk of the banana orbits, with a step size characterized by the width of a banana orbit rather than a gyro radius. Following this pioneering paper there was a decade of kinetic analysis of tokamaks, and the development of what is called neoclassical transport theory. A large literature now exists on this subject, and it is comprehensively reviewed by Hinton and Hazeltine [12]. We summarize here the essential features of neoclassical transport theory; the interested readers should consult Reference 12.

Recall from Section 5.3, particle motions in a tokamak, that the magnetic moment μ is

$$\mu \equiv \frac{mv_\perp^2}{2B} = \text{const} \tag{5.19}$$

The particle kinetic energy E is

$$E = \frac{mv_{th}^2}{2} = \frac{mv_\parallel^2}{2} + \mu B \tag{5.53}$$

and $v_\parallel$ is therefore

$$v_\parallel = \left[\frac{2}{m} (E - \mu B) \right]^{\frac{1}{2}} \tag{5.54}$$

Particles have $v_\parallel = 0$ when $\mu B = E$. In a conventional low-β circular tokamak, the toroidal magnetic field is

$$B_\phi = \frac{B_\phi(0)}{1 + \frac{r}{R_0} \cos \theta} \approx B_\phi(0)\left(1 - \frac{r}{R_0} \cos \theta\right) \tag{5.55}$$

See Figure 5.12 for the geometry. Although $B_\theta \ll B_\phi$ and is neglected in estimating the magnitude of B, the poloidal field is important in studying the particle trajectories. The minimum value of magnetic field B_{min} occurs on the outside of the torus ($\theta = 0$) and the maximum is on the inside ($\theta = \pi$); that is, $B_{max} \approx B_\phi(0)(1 + \varepsilon)$ and $B_{min} \approx B_\phi(0)(1 - \varepsilon)$ where the poloidal field has been neglected in this estimate. Particles are trapped when

$$\frac{v_\parallel}{v_\perp} \leq \left(\frac{B_{max}}{B_{min}} - 1 \right)^{\frac{1}{2}} \tag{5.56}$$

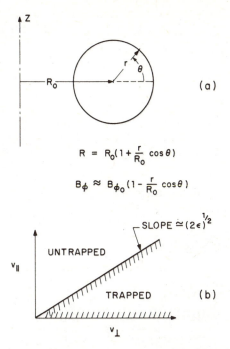

FIGURE 5.12.   (a) Geometry useful for trapped particles and neoclassical transport discussions. (b) Phase space showing regime of trapped particles in a tokamak.

or

$$\frac{v_\parallel}{v_\perp} \leqslant \left(\frac{R+r}{R-r} -1\right)^{1/2} \approx \left(\frac{2r}{R-r}\right)^{1/2}$$

$$\leqslant \left(\frac{1+\varepsilon}{1-\varepsilon} -1\right)^{1/2} = (2\varepsilon)^{1/2} \tag{5.57}$$

where ε is the inverse aspect ratio (r/R_0). The pitch angle that delineates trapped particles from untrapped ones is $(2\varepsilon)^{1/2}$. A scattering that removes a particle from the trap is one which changes the pitch angle in phase (velocity) space into the untrapped region (see Figure 5.12b). The fraction of particles which are trapped f_T is given by the volume percentage of phase space, namely, $(2\varepsilon)^{1/2}$; that is,

$$f_T \approx (2\varepsilon)^{1/2} \tag{5.58}$$

and since a compact, low-aspect ratio is desirable, say $\varepsilon \sim \frac{1}{3}$, then $f_T \sim 80\%$; a large fraction of the particles is trapped. The parallel velocity $v_\parallel$ is related to the thermal velocity v_{th} by

$$v_{\text{th}}^2 = \frac{2kT}{m} = v_{\parallel}^2 + v_{\perp}^2 = v_{\parallel}^2 \left(1 + \frac{v_{\perp}^2}{v_{\parallel}^2} \right)$$

$$v_{\text{th}}^2 \approx v_{\parallel}^2 \left(1 + \varepsilon^{-1} \right) \approx \frac{v_{\parallel}^2}{\varepsilon}$$

or

$$v_{\parallel} = \varepsilon^{1/2} v_{\text{th}} \tag{5.59}$$

The trajectory length l for a particle moving along a helical path depends upon the safety factor q, and is

$$l = 2\pi q R \tag{5.60}$$

l is called the connection length. The collisionless bounce time τ_B and bounce frequency ν_B are defined as

$$\tau_B = \frac{qR}{v_{\parallel}} = \frac{qR}{\varepsilon^{1/2} v_{\text{th}}} = \left(\frac{R}{r} \right)^{1/2} \frac{qR}{v_{\text{th}}} \tag{5.61}$$

and

$$\nu^B \approx \frac{\varepsilon^{1/2} v_{\text{th}}}{qR} \tag{5.62}$$

Trapped particles while bouncing between their mirror points (see Figure 5.4b) experience the ∇B drift in the Z direction; see Equation (5.18). Their drift velocity v_d is approximately

$$v_d \approx \frac{m v_{\text{th}}^2}{eBR} \approx \left(\frac{r_c}{R} \right) v_{\text{th}} \tag{5.63}$$

The width of a banana orbit $\Delta r_b = 2\Delta r$ (see Figure 5.4b) is approximately

$$\Delta r_b \approx \frac{2v_d}{\nu_b} \tag{5.64}$$

where ν_b is the bounce frequency $= \tau_B^{-1}$. Using Equations (5.62) and (5.63) we obtain

$$\Delta r_b \approx 2\varepsilon^{1/2} r_{c\theta} = 2r_c \frac{q}{\varepsilon^{1/2}} \tag{5.65}$$

Since $2q/\varepsilon^{1/2}$ is a number of the order of 10, the banana width is an order of magnitude larger than the particle gyro radius; Δr_b is the important step size in tokamak-transport theory.

Accordingly, perpendicular diffusion in a tokamak should have a neo-classical diffusion coefficient $D_{\perp,NC}$:

$$D_{\perp,NC} \approx \nu_{\text{eff}} \,(\Delta r_b)^2 \, f_T \qquad (5.66)$$

and the effective collision frequency ν_{eff} is, for electrons, the collision frequency ν_{ei} modified to account for the change in pitch angle needed to get out of the trapped region of phase space:

$$\nu_{\text{eff}} = \frac{\nu_{ei}}{(\Delta\theta)^2} \approx \frac{\nu_{ei}}{\left(\dfrac{v_{\parallel}}{v_{\perp}}\right)^2} = \frac{\nu_{ei}}{\varepsilon}$$

$$(5.67)$$

To escape from the trapped region, the pitch angle $\Delta\theta$ must change by approximately $\varepsilon^{1/2}$ rather than $\pi/2$. Using Equations (5.67), (5.65), and (5.58) in (5.66), we obtain for electrons

$$D_{\perp,NC} \approx \frac{q^2}{\varepsilon^{3/2}} \, \nu_{ei} r_c^2 = \frac{q^2}{\varepsilon^{3/2}} \, D_{\perp,0} \qquad (5.68)$$

Thus, the neoclassical perpendicular diffusion coefficient is found to be $q^2/\varepsilon^{3/2}$ larger than the classical plasma diffusion coefficient. Since $q \sim 3$ and $\varepsilon \sim \frac{1}{3}$, the value of $q^2/\varepsilon^{3/2}$ can be of the order of 50.

The nontrapped particles diffuse classically with a coefficient $D_{\perp,0}$. Thus, when the effective collision frequency is less than the bounce frequency, the neoclassical electron-diffusion coefficient is

$$D_{\perp,NC}^b = \nu_{ei} r_c^2 \left(1 + \frac{q^2}{\varepsilon^{3/2}} \right) \qquad (5.69)$$

when

$$\nu_{ei} < \frac{\varepsilon^{3/2} v_{\text{th}}}{Rq} \qquad (5.70)$$

The quantities ν_e^* and ν_i^* are defined as

$$\nu^* \equiv \left(\frac{R}{r} \right)^{3/2} \frac{\nu q R}{v_{\text{th}}} \qquad (5.70a)$$

where v on the right-hand side of Equation (5.70a) is either v_{ei} (for electrons) or v_{ii} (for ions); that is,

$$v_{ei} = 3.88 \times 10^{-6} \frac{Zn_e \ln\Lambda}{T_e^{3/2}} \tag{4.19c}$$

and

$$v_{ii} = 8.85 \times 10^{-8} \frac{Z^4 n_i \ln\Lambda}{A^{1/2} T_i^{3/2}} \tag{4.18d}$$

When $v^* < 1$ trapped-particle effects are important; if $v^* \gg 1$, they cannot play a significant role. For a plasma with equal electron and ion temperature v_e^* and v_i^* will be about the same. When $v^* < 1$, Equation (5.69) applies, and this condition is called the banana regime.

When a trapped particle executes a significant fraction of a banana orbit but is scattered into the loss-cone before a bounce orbit is complete, that is, for electrons

$$\frac{\varepsilon^{3/2} v_{th}}{Rq} < v_{ei} < \frac{v_{th}}{Rq} \tag{5.71}$$

then the probability of an electron completing a banana orbit is

$$\frac{v_b}{v_{eff}} \approx \frac{(2\varepsilon)^{1/2} v_{th}/Rq}{v_{ei}/\varepsilon} \tag{5.72}$$

In this regime of collision frequency, the transport coefficient is

$$D^p_{\perp,NC} \approx D^b_{\perp,NC} \frac{v_b}{v_{eff}} \tag{5.73}$$

or

$$D^p_{\perp,NC} \approx r_c^2 v_{th} \frac{q}{R} \tag{5.74}$$

This regime where the neoclassical coefficient is independent of collision frequency is called the plateau regime.

Finally, for collision-dominated toroidal diffusion, where, for electrons

$$v_{ei} > \frac{v_{th}}{Rq} \tag{5.75}$$

the characteristic time for electron radial diffusion τ_{PS} is:

$$\tau_{PS} = \frac{(Rq)^2}{D_{\parallel}}$$

(5.76)

where

$$D_{\parallel} = \frac{v_{th}^2}{v_{ei}}$$

(5.77)

Particles experience a ∇B drift during the τ_{PS} time, so that the diffusion coefficient is

$$D^{PS} = v_d^2 \tau^{PS}$$

(5.78)

Using Equations (5.63), (5.76), and (5.78), we obtain for electrons

$$D^{PS} = r_c^2 v_{ei} q^2$$

(5.79)

Including the classical rate of diffusion gives, finally

$$D_{\perp NC}^{PS} = r_c^2 v_{ei} (1 + q^2)$$

(5.80)

which is the electron classical coefficient increased by $(1 + q^2)$. This result was first obtained by Pfirsch and Schlüter and is called the Pfirsch−Schlüter regime.

These results are, for electron neoclassical diffusion, summarized in Figure 5.13. The boundaries of each regime have been analyzed and rounded shoul-

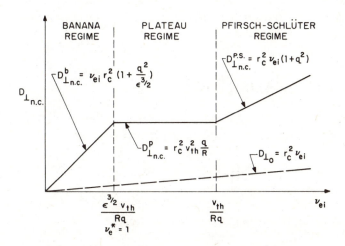

FIGURE 5.13. Neoclassical electron transport for tokamaks.

ders are predicted where we, for simplicity, have indicated sharp changes. These and many more subtle details can be found in the very thorough review by Hinton and Hazeltine [12] and in the text by Miyamoto [10]. It is anticipated that in the banana regime and the lower part of the plateau regime, trapped-particle instabilities will substantially increase the actual transport rate. Tokamak experiments have now covered most of the neoclassical diffusion regime, and the experimental evidence indicates that for electrons, there is enhanced, anomolous diffusion, very much larger than the predicted neoclassical amounts. The ions behave nearly neoclassically, that is, within a factor of 3−5. Neoclassical-transport theory is important because it provides a theoretical foundation with which one can judge actual results, and it represents the best that can be expected.

As indicated in Section 4.7, all the classical transport properties are related. Accordingly, there are neoclassical predictions for the other properties such as thermal conductivity, etc. The reader should refer to References 6 and 12 for further discussion.

An interesting new effect is predicted for plasma in the banana regime. Radial neoclassical diffusion can induce a toroidal current. When a radial density gradient exists, there is a difference in the number of particles on neighboring banana orbits passing through a given point. The current density that results from these trapped, banana electrons is predicted to be

$$j_b \approx - \frac{\varepsilon^{1/2}}{B_\theta} \left(\frac{\partial p}{\partial r} \right) \tag{5.81}$$

This current, called a "bootstrap current," has not yet been observed in experiments. Should it exist, it would offer a possibility for a steady-state tokamak. The poloidal confining magnetic field would be provided by the bootstrap current, and the toroidal electric field is, in this circumstance, zero. More extensive discussion and further references on the bootstrap current can be found in References 10 and 12.

5.8.   TOKAMAK SCALING LAWS

The diffusion equation, based on the conservation of matter, relates the diffusion coefficient D to the number density $n(\mathbf{r},t)$. It is

$$\frac{\partial n(\mathbf{r},t)}{\partial t} = \nabla \cdot [D \nabla n(\mathbf{r},t)] \tag{5.82}$$

if we approximate the left-hand side of Equation (5.82) by

$$\frac{\partial n(\mathbf{r},t)}{\partial t} \approx - \frac{n(\mathbf{r})}{\tau_n} \tag{5.83}$$

and assume D is a constant, then the solution of Equation (5.82) has the form

$$n(\mathbf{r},t) = n_0(\mathbf{r})\exp\left(\frac{-t}{\tau_n}\right)$$
(5.84)

For a straight cylinder of radius a, these assumptions lead to the equation

$$\frac{1}{r}\frac{\partial}{\partial r}\left(r\frac{\partial n}{\partial r}\right) + \frac{n}{D\tau_n} = 0$$
(5.85)

whose solution for the boundary condition $n(a) = 0$ [10] is

$$n(r) = n_0 J_0\left(\frac{2.4r}{a}\right)\exp\left(\frac{-t}{\tau_n}\right)$$
(5.86)

where

$$\tau_n = \frac{a^2}{(2.4)^2 D} = \frac{a^2}{5.8D}$$
(5.87)

and J_0 is the zeroth-order Bessel function. The quantity τ_n is a measure of the particle-confinement time resulting from diffusion. When there are no particle sources, such as can occur from recycling from the wall, neutral-beam injection, or gas puffing, then Equation (5.87) may be used to estimate the diffusion coefficient from measurements of $n(\mathbf{r},t)$.

A widely used measure is the gross energy confinement time τ_E, which was introduced in Section 3.5 in the discussion of power breakeven. If W is the internal energy of the entire plasma volume and Q is the total of all forms of energy power input to the plasma (such as ohmic, neutral beams, RF power), then τ_E is defined by

$$\frac{dW}{dt} = Q - \frac{W}{\tau_E}$$
(5.88)

When the total plasma internal energy W is constant in time and the only source of energy input is ohmic heating by the toroidal plasma current I_ϕ, then for a circular tokamak

$$\tau_E = \frac{4\pi R^2 \frac{3}{2}k \int_0^a (n_e T_e + n_i T_i)r\,dr}{VI_\phi}$$
(5.89)

where V is the total loop voltage change in the toroidal direction. Tokamak experimentalists measure V, I_ϕ, and $n(r)$ and $T(r)$, and by using Equation

(5.89), τ_E is evaluated. This measure of the energy confinement includes all energy exchange mechanisms such as radiation, charge exchange, or conduction. It is the most widely used, experimentally measured parameter, which permits comparison of different tokamaks and comparison with theoretical predictions.

Another interesting measure is the particle confinement time τ_p. It is defined, for the central plasma region ($r \approx 0$), as

$$\tau_p(0) = \frac{n_e(0)}{S_b(0) + S_W(0) - \dfrac{dn_e(0)}{dt}}$$

(5.90)

where $S_b(0)$ is the beam source rate, $S_W(0)$ is the particle source rate from neutrals from the wall and limiter, and $dn_e(0)/dt$ is the measured rate of change of the central electron density. This is more difficult to determine because S_b and S_w are often not well known.

Other experimental measurements that are often compared are the central (maximum) electron temperature $T_e(0)$, the maximum stable central electron number density $n_e(0)$, and the maximum stable plasma pressure that can be sustained, implied from values of $\langle \beta \rangle$ and β_θ. Many differences between theory and experiment and different experiments are observed. These are discussed in the reviews of tokamaks [5, 6, 7] and in much of the current published literature, as exemplified by the recent French effort to obtain tokamak scaling laws [27].

For example, some tokamak scaling laws that have been proposed are: "Alcator" gross energy-confinement-time scaling [28]:

$$\tau_E = 6 \times 10^{-21} \bar{n}_e a^2$$

(5.91)

or the modified Alcator scaling [27,29]:

$$\tau_E = 5 \times 10^{-21} \bar{n}_e q^{1/2}(a) a^2$$

(5.92)

By regression analysis of data from many tokamaks, Hugill and Sheffield propose the scaling [30]:

$$\tau_E = 9.1 \times 10^{-14} \langle n_e \rangle^{0.61} a^{1.57} B_\phi^{0.88}$$

(5.93)

and the MIT group propose [31] a "volume" ($R^2 a$) law,

$$\tau_E = 1.15 \times 10^{-21} \bar{n}_e R^{2.3} a^{0.8}$$

(5.93a)

Another example of a scaling law, for the central ion temperature, first proposed by Artsimovich and expanded upon by the TFR group [27], is

$$T_i(0) = 5 \times 10^{-2} \left[\bar{n}_e B_\phi I_\phi R^2\right]^{\frac{1}{3}} A_i^{-\frac{1}{2}} \left(\frac{\bar{n}_e}{\bar{n}_i}\right)^{\frac{1}{3}} \tag{5.94}$$

This expression is based on neoclassical transport in the plateau regime.

There are many scaling relations and they all make use of experimental data. Some of the relations are purely empirical, some are based on theoretical models of the plasma behavior, and others employ dimensional analysis and regression analysis. The electron energy confinement time τ_{Ee} has been empirically scaled by Daugheney* as

$$\tau_{Ee} = 3 \times 10^{-17} \langle n_e \rangle \langle T_e \rangle^{\frac{1}{2}} a^2 / I$$

Transport coefficients have also been empirically scaled using experimental data and computer simulation codes. For example, the electron thermal conductivity $\kappa_\perp^e$ is predicted from Alcator A data[†] to scale as,

$$\kappa_\perp^e = 5.0 \times 10^{19} / \bar{n}_e$$

More extensive data analysis of many tokamaks has lead to the following scaling laws by Coppi and Mazzucato,[‡]

$$\kappa_\perp^e = 5.8 \times 10^{22} \cdot \frac{B_\phi}{\bar{n}_e T_e q(a)}$$

Scaling laws are valuable, but to extrapolate them beyond their experimental base is risky.

There are also semiempirical relations used to provide maps of tokamak operation so that experimentalists can be guided in parameter space to those regions where tokamaks have been found to operate well. For example, a line in $q(a)$ vs $\bar{n}_e R / B_\phi$ space (see Reference 32) is thought to define the regime where tokamaks can operate without experiencing a major disruption. Murakami et al. [33] proposed an empirical scaling law observed to relate a density limitation for ohmically heated tokamaks, namely,

$$\bar{n}_e \leq 2 \times 10^{19} \frac{B_\phi}{R_0} \tag{5.95}$$

Some theoretical basis for the form of Equation (5.95) can be found in Section 6.5, which discusses ohmic heating in a tokamak. Such relationships often

*C.Daugheney, *Nuclear Fusion* **15**, 967 (1975).
†E. Apgar et al., IAEA-6, 1977, Vol. 1, p. 247.
‡B. Coppi and E. Mazzucatto, *Phys. Lett. A* **71**, 337 (1979).

change as physical understanding is increased and as experimentalists find improved ways to operate tokamaks.

Why should there be differences in tokamak performance from one device to another? There are geometric differences between different devices, differences in plasma impurity levels, different values of magnetic field ripple, different methods of supplementary heating, and different start-up and operating procedures. It is a complex and difficult task to obtain all the data needed, for example, to account for the energy flow in a real tokamak plasma. There are many physical phenomena occurring in tokamaks. Because of this complexity, it will be necessary to make full-size, full-power, and full-time duration tests of tokamaks to determine the actual plasma properties and confinement characteristics that exist in future fusion-power reactors. Eventually, as understanding increases, the appropriate dimensionless scaling parameters should become known, and the need for full scale testing will be greatly reduced.

5.9. EXPERIMENTAL OBSERVATIONS

The status of experimental tokamak experience can be assessed by referring to reviews, for example references 6, 7, and the most recent proceedings of the international IAEA fusion conferences; see Reference 26 of Chapter 1.

The equilibrium shape and position of a low-beta tokamak plasma conform very well with mhd theory as exemplified by two-dimensional solutions of the Grad—Shafranov equation. Tokamak plasmas nearly always exhibit oscillations such as magnetic field fluctuations, soft-x-ray-sawtooth signals, and density fluctuations. Those signals, with frequencies typically $5-15$ kHz, are believed to be caused by internal helical (kink) structures with mode numbers like $m = 2$ or 3. When $q(0) \leq 1$, poloidal flux loops, x-ray signals, and electron synchrotron radiation detectors all indicate fluctuations whose periods are of the order of 10 ms in large ohmically heated tokamaks. The origin of these signals seems to be on the $q = 1$ and $q = 2$ surfaces with modes such as $m/n = 1/1, 2/1, 3/2$, etc. The properties of these oscillations are described rather well by tearing-mode theory. As the modes grow, there is mode coupling, and new modes sometimes appear.

The stable tokamak operating regime is determined in part by the disruptive instability. A disruptive instability is a sudden, large disturbance that develops rapidly ($10-100$ μs in ohmically heated tokamaks) and is characterized by hard-x-ray bursts, flattening of the toroidal current profile, large negative loop voltage spikes, some loss of energy, and the magnetic axis moving radially inward. Small disruptions may repeat; large ones terminate the discharge. Although $m = 2$ oscillations are a common feature, and tearing modes seem involved, there is as yet no commonly accepted detailed explanation of the disruptions. A large disruption deposits high-power density on the first wall, limiter, etc., and is an important consideration in the design of new, large tokamaks.

Large computer codes have been developed to predict the properties and structure of tokamak plasmas. To reproduce experimental observations it appears that the ions are behaving nearly (within a factor of $2-5$) like neo-classical theory predicts. The electrons seem to behave very anomalously, with electron thermal conduction very different (one or two orders of magnitude larger) from neoclassical predictions. Most of these observations come from PLT data [34], where reactor-grade plasmas have been obtained ($v_i^* \leq 10^{-1}$) over most of the radial profile. Neutral-beam heating does not seem to significantly alter transport properties [35]. Energy-confinement times up to $\tau_E = 100$ ms and particle-confinement times $\tau_p(0) \approx 150$ ms have been obtained in PLT. There is evidence of plasma rotation, particularly when there is unidirectional injection of neutral beams. At low densities ($n_e \leq 10^{19}\ m^{-3}$) intense x-rays result from runaway electrons. Most tokamaks exhibit some form of plasma turbulence with relatively large density fluctuations usually occurring near or on the plasma boundaries. Operation with $Z_{\text{eff}} \approx 1$ can be achieved in tokamaks (pulse length ≤ 1 s) by careful low-power discharge cleaning and gettering techniques. Migration of impurities within the tokamak plasma is not very well understood.

The Alcator tokamak has stably confined plasmas with $n \leq 10^{21}\ m^{-3}$, but at relatively low temperatures and with $v^* \sim 8$. The ISX-B and TOSCA tokamaks have operated with $\langle \beta \rangle \leq 3\%$ with no significant indications of ballooning instabilities. These and other tokamak parameters are being actively explored in fusion programs around the world. Details about preionization of the gas, current distribution, and build-up, and so on, during a tokamak start-up transient are important. Each machine seems to have its own start-up characteristics, but all well designed low-beta tokamaks settle down to equilibrium independent of the variations that take place during start-up. The current literature should be examined to determine recent experimental observations.

REFERENCES

1. A. D. Sakharov and I. E. Tamm, *Plasma Physics and the Problem of Controlled Thermonuclear Reactors*, M. A. Leontovich (ed.), Pergamon, New York, 1959–1960, Vol. I, first three articles.

2. A. H. Morton, "Tokamaks in Australia," *The Australian Physicist* **18**, 195 (1981).

3. N. J. Peacock et al., Nature **224**, 488 (1969); M. J. Forrest, N. J. Peacock, D. C. Robinson, V. V. Sannikov, and P. D. Wilcock, Culham Report CLM-R 107, July 1970.

4. L. A. Artsimovich, "Tokamak Devices," *Nuclear Fusion* **12**, 215 (1972).

5. H. P. Furth, "Tokamak Research," *Nuclear Fusion* **15** 487 (1975).

6. J. M. Rawls (ed.), *Status of Tokamak Research*, U.S. Department of Energy Report. DOE/ER-0034, Oct. 1979.

7. J. Sheffield, *Status of the Tokamak Program*, ORNL/TM-7778 (1981), *Proc. IEEE* **69**, 885 (1981).

8. E. Teller (ed.), *Fusion*, Academic Press, New York, (1981) Vol. I, Pts. A and B, Chapters 2 and 3.

9. University of Wisconsin Design Team *NUWMAK, A Tokamak Reactor Design Study*, UWFDM-330, March 1979.

10. K. Miyamoto, *Plasma Physics for Nuclear Fusion*, MIT Press, Cambridge, MA, 1980.

11. G. Bateman, *MHD Instabilities*, MIT Press, Cambridge, MA, 1978.

12. F. L. Hinton and R. D. Hazeltine, "Theory of Plasma Transport in Toroidal Confinement Systems," *Rev. Mod. Phys.* **48**, 239 (1976).

13. A. A. Ware, *Phys. Rev. Lett.* **25**, 15 (1970).

14. J. P. Freidberg, "Ideal Magnetohydrodynamic Theory of Magnetic Fusion Systems," *Rev. Mod. Phys.* **54**, 801 (1982).

15. V. D. Shafranov, *Rev. Plasma Phys.* **2**, 103 (1966).

16. F. F. Cap, *Handbook of Plasma Instabilities*, Academic Press, New York, 1976.

17. J. A. Wesson, "Hydromagnetic Stability of Tokamaks" (review paper), *Nuclear Fusion* **18**, 87 (1978).

18. I. B. Bernstein, E. A. Frieman, M. D. Kruskal, and R. M. Kulsrud, *Proc. Roy. Soc. London Ser. A* **244**, 17 (1958).

19. B. Coppi, A. Ferreira, J. W-K. Mark, and J. J. Ramos. *Nuclear Fusion* **19**, 715 (1979).

20. B. B. Kadomtsev and O. P. Pogutse, *Nuclear Fusion* **11**, 67 (1971).

21. B. Coppi, *Nuovo Cimento* **1**, 357 (1969).

22. W. M. Manheimer, *An Introduction to Trapped-Particle Instabilities in Tokamaks*, ERDA Critical Review Series TID-27157, National Technical Information Service, U.S. Dept. Commerce, Washington, D.C., 1977.

23. P. C. Van der Laan and L. W. Mann, "Tokamak Equilibria with Beta Close to One," *Proceedings of the Finite Beta Theory Workshop*, Varenna, Italy Summer School of Plasma Physics, U.S. Dept. of Energy, CONF-7709167, Sept. 1977, p. 5.

24. R. A. Gross and T. C.Marshall, "Implications of Recent High-Beta Tokamak Research," *Comments on Plasma Physics and Controlled Fusion* **5**, 233 (1980).

25. J. T. Hogan, "Beta Limits in Beam-Heated Tokamaks," *Nuclear Fusion* **20** 1119 (1980).

26. A. A. Galeev and R. Z. Sagdeev, *Sov. Phys. JETP* **26**, 233 (1968).

27. TFR Group, "Tokamak Scaling Laws, with Special Emphasis on TFR Experimental Results," *Nuclear Fusion* **20**, 1227 (1980).

28. E. Apgar et al., *6th IAEA Conference on Plasma Physics and Controlled Nuclear Fusion Research*, Vol. 1. IAEA, 1977, p. 247

29. D. R. Cohn, R. R. Parker, and D. L. Jassby. *Nuclear Fusion* **16**, 31 (1976); *Nuclear Fusion* **16**, 1045 (1976).

30. J. Hugill and J. Sheffield, *Nuclear Fusion* **18**, 15 (1978).

31. B. Blackwell et al., *9th International Conference on Plasma Physics and Controlled Nuclear Fusion Research*, IAEA, Baltimore, MD, Sept. 1982, paper CN-41/I-3.

32. J. W. M. Paul, "Review of Results from DITE Tokamak." Proceedings of the 8th European Conference on Controlled Fusion and Plasma Physics, Vol. II. IAEA, 1977, p. 49.

33. M. Murakami, J. D. Callen, and L. A. Berry; *Nuclear Fusion* **16**, 347 (1976).

34. W. Stodiek et al. "Transport Studies in the Princeton Large Torus," paper A-1, Proceedings of the 8th IAEA Conference on Plasma Physics and Controlled Nuclear Fusion Research, IAEA, Brussels, 1980.

35. H. Eubank et al., "PLT Neutral Beam Heating Results, "Paper C-3, *Proceedings of the 8th IAEA Conference on Plasma Physics and Controlled Nuclear Fusion Research*, IAEA, Brussels, 1980.

6 | FUSION TECHNOLOGY

6.1. INTRODUCTION

The success of fusion energy depends as much on progress in technology as it does on advances in plasma physics. One of the first studies to recognize some of the special technology problems associated with fusion was the preliminary design, published in 1954, of a fusion power plant by Spitzer and coworkers [1]. This pioneering study focused attention on a number of areas of technology that would require substantial development before fusion could become a reliable, safe, and useful energy resource. Among the technology areas highlighted in that study are the development of large magnets, the nucleonics (shielding, breeding, tritium handling) of fusion, lithium-technology problems, divertors needed to scrape off impurities, and methods to achieve new fuel injection. Other early magnetic fusion technology studies and reviews are those of Post [2], Rose [3], Mills et al. [4], Ribe [5], Carruthers et al. [6], Emmert et al. [7], and Steiner [8]. Fusion technology topics are described in the proceedings of meetings and symposia [9–11] and journals devoted to fusion technology [12–14]. In this chapter we discuss the technology of fusion power reactors, particularly that needed for tokamaks.

6.2. POWER BALANCE AND DESIGN OPERATING POINT

It is useful and instructive to examine the power flow in a fusion-energy system
to reveal the relationship among different components of the power plant. This
is done by characterizing each major subsystem with an efficiency and doing
either an energy or power accounting, which is sometimes called a power
balance. These studies help the fusion-reactor designer choose the operating
design point for the power plant. The operating design point is usually determined by a consideration of the physics and the economics resulting from the
interaction of the various components. Power balances have been performed
for many different fusion concepts and can be found, for example, in the paper
by Nozawa and Steiner [15], the book by Kammash [16], the review by Steiner
[8], the parametric system studies by the Argonne National Laboratory [17],
and the review by Conn [18].

    As an example of a power balance, consider a tokamak power plant,
separated into several subsystems as illustrated in Figure 6.1. An example of
the sequence of events of the fusion plant are as follows: (1) the gaseous fuel in
the tokamak is ionized and ohmically heated; (2) neutral-beam supplementary
heating is turned on for a time t_0, during which the tokamak plasma is brought
to ignition; (3) after ignition, there is a burn phase; (4) the pulse is completed
and some of the magnetic energy from the toroidal current is recovered and
stored for the next cycle.

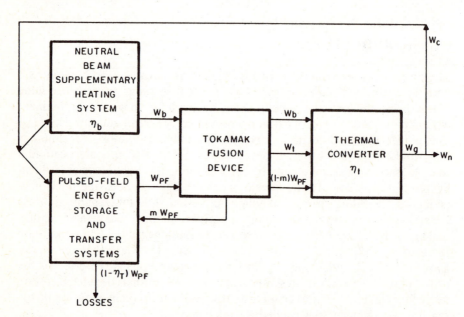

FIGURE 6.1. Energy-flow diagram for a tokamak fusion power plant with neutral-beam supplementary heating.

The net electrical energy output W_n, based on the energy flow in Figure 6.1, is

$$W_n = [W_b + W_t + (1 - m)W_{\text{PF}}]\eta_t$$
$$- \left[\frac{W_b}{\eta_b} + W_{\text{PF}} + (1 - \eta_T)W_{\text{PF}}\right] + mW_{\text{PF}} \qquad (6.1)$$

where W_b = energy delivered by the neutral-beam injector system with an energy efficiency η_b.

W_t = total nuclear-energy released.

W_{PF} = energy delivered by the pulsed-field energy storage and transfer system.

m = fraction of the pulsed-field energy W_{PF} that is returned to the energy storage and transfer system during toroidal current shutdown.

η_T = energy transfer efficiency of the pulsed-field energy storage and transfer system.

η_t = energy efficiency of thermal convertor.

The energy delivered by the neutral-beam system W_b depends on the time t_0 that this system is used; for example, $W_b = P_b t_0$, where P_b is the beam power. The orders of magnitude used by Steiner [8] are $P_b \sim 20$ MW and $t_0 \sim 10$ s, but more precise values will depend on the actual energy-confinement time of the tokamak plasma. The technology of neutral-beam heating is discussed later, in Section 6.6.

The overall fusion plant efficiency η is defined as

$$\eta \equiv \frac{W_n}{W_t} \qquad (6.2)$$

which can be expressed as

$$\eta = [Q_t^{-1} + 1]\eta_t - \frac{R_b}{\eta_b} - R_{PF}[1 + (1 - \eta_T) - m(1 - \eta_t)] \qquad (6.3)$$

The total amplification factor Q_t, a term used in Equation (6.3), is defined as

$$Q_t \equiv \frac{W_t}{W_b + W_{\text{PF}}} \qquad (6.4)$$

The dimensionless energy ratios for the neutral beam and the pulsed energy storage and transfer system are designated by

$$R_b \equiv \frac{W_b}{W_t} \tag{6.5}$$

$$R_{PF} \equiv \frac{W_{PF}}{W_t} \tag{6.6}$$

The blanket that surrounds the tokamak plasma not only converts the neutron flux from the plasma into thermal energy, but it may also produce energy amplification that results from further nuclear reactions such as $^1n + {}^6Li \rightarrow {}^4He + {}^3T + 4.8$ MeV (see Section 2.1), a reaction employed to breed tritium. Steiner assumed a blanket energy amplification factor $M = 1.2$, where M is the ratio of the energy released within the blanket to the energy released by fusion events in the plasma. Using tokamak reactor numbers from the early design study by the University of Wisconsin group [7], Steiner estimated $Q_t \approx 490$, $R_b = 5.8 \times 10^{-6}$; and $R_{PF} = 2.0 \times 10^{-3}$.

Because the neutral-beam and pulsed-field energies are, in this example, such a small fraction of the total nuclear-energy released, the overall plant efficiency is essentially the efficiency of the thermal convertor; that is, from Equation (6.3), $\eta \approx \eta_t$. The pulsed-field energy in the UWMAK-I tokamak design is ≈ 60 GJ, which can be transferred with an estimated efficiency $\eta_T \approx 0.90$.

The relatively high overall plant efficiency $\eta \approx 0.40$ is, in this model, a consequence of the large plasma-energy amplification factor of $Q_t/M \approx 410$. Two assumptions are critical to achieving this high energy amplification in tokamaks: (1) the power requirements for fueling are relatively small and (2) long burn times (about 90 min) can be achieved. If we assume that refueling takes place by using neutral beams and that $R_b \gg R_{PF}$, then

$$R_b \approx \frac{1}{Q_t} \tag{6.7}$$

and the overall plant efficiency is approximately:

$$\eta = (Q_t^{-1} + 1)\eta_t - \frac{1}{Q_t \eta_b} \tag{6.8}$$

If the neutral beams are operated for the duration of the plasma burn, for example, to also provide for refueling, then

$$Q_t = \frac{P_t}{P_b} \tag{6.9}$$

where P_t is the total nuclear power, P_b is the beam power, and

$$P_t = \frac{W_t}{t_b} \tag{6.10}$$

where t_b is the burn time. Figure 6.2a shows the variation of η with the ratio P_t/P_b for a beam efficiency $\eta_b = 0.7$ and a thermal convertor efficiency $\eta_t = 0.4$. On the basis of the figure, power for fueling by neutral beams should not exceed about 10% of the nuclear power. For the reactor design used in Steiner's study, this means $P_b \lesssim 500$ MW.

If, on the other hand, fueling power can be neglected by, for example, successful pellet refueling, then

$$P_{PF} \gg R_b \tag{6.11}$$

and

$$R_{PF} \approx \frac{1}{Q_t} \tag{6.12}$$

The overall plant efficiency for this case is approximately

$$\eta = (Q_t^{-1} + 1)\eta_t - Q_t^{-1}[1 + (1 - \eta_T) - m(1 - \eta_t)] \tag{6.13}$$

The total energy amplification can be expressed in terms of the burn time t_b (in seconds), which for the Steiner study, using an early tokamak reactor design, gives

$$Q_t = \frac{490t_b}{5400} \tag{6.14}$$

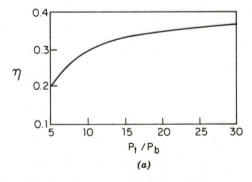

FIGURE 6.2. (a) Variation of the overall tokamak plant efficiency as a function of the ratio of the nuclear power P_t to the neutral-beam power P_b. From Steiner [8].

The duty factor f is defined as

$$f \equiv \frac{t_b}{t_b + t_d} \tag{6.15}$$

where t_d is the downtime per cycle. The downtime for the early UWMAK-I tokamak design [7] was 400 s. Figure 6.2b shows the overall plant efficiency η and duty factor f as a function of burn time t_b for $\eta_T = 0.9$, $m = 0.8$, and $\eta_t = 0.4$. It appears, from these data, that burn times as short as 40 s might be acceptable. However, the associated downtime for such short burns would also have to be extremely short ($t_d \lesssim 10\,\text{s}$) in order to achieve reasonably economic duty factors. To start up and shut down the large toroidal current ($I \approx 21\,\text{MA}$) in this tokamak model, in a few seconds, would involve very high voltages.

This example of a tokamak power balance shows quantitatively the importance of high Q_t, low fueling power, and long burn times. The latter will require very good impurity control for a long time. Methods of fueling and impurity control are discussed in later sections of this chapter.

The importance of a high Q_t can be seen in examining the magnitude of the fraction of circulating energy W_c/W_n. For $Q_t \lesssim 40$, the fraction of recirculating energy rises rapidly [16] to values of $W_c/W_n \gtrsim 0.20$, which are considered impractical since the losses associated with large power recirculation become excessive. The incorporation of fissile material within the blanket relaxes the economic constraints on the pulsed-field system, but because a tokamak has the potential for very large power amplification, it is of less significance than for some other fusion concepts.

It is common practice to use computer codes to simulate the various components of a fusion reactor system and to use them to study the system performance. For example, the design team at the Argonne National Laboratory

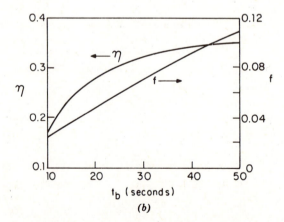

FIGURE 6.2. (b) Variation of the overall tokamak plant efficiency and duty factor f as a function of burn time t_b. From Steiner [8].

made many parametric system studies of tokamak reactors [17] before choosing the design point and operating characteristics of a fusion reactor design.

6.3. PLASMA – WALL INTERACTIONS: IMPURITY CONTROL

Magnetic confinement has as its principal purpose isolation of the hot plasma from the relatively cold first wall which surrounds the plasma. However, some plasma always does come into contact with the first wall and other surfaces such as limiters. The resulting interactions are complex and can have very significant ramifications. The two major areas of concern are that (1) the first wall of the reactor will be damaged by the plasma and by the reaction products and (2) impurities freed from the wall will enter the plasma and result in additional radiative energy losses. For the latter, as shown in Chapter 3 (see Figure 3.8), small concentrations of high-Z impurities can make it impossible to achieve fusion ignition. An extensive review of plasma – surface interactions has been written by McCracken and Stott [19], and there are periodic international conferences devoted to this subject [20, 21].

The plasma chamber or vacuum vessel must, under normal conditions, sustain a vacuum base pressure of less than 10^{-7} torr ($n_0 \approx 4 \times 10^{15}$ m^{-3} or a few parts per million of the fusion plasma density). All wall material surfaces should initially be chemically cleaned and preferably baked or discharge cleaned. Heating the wall will reduce the surface concentration of absorbed molecules by thermal desorption. But, heating only releases the weakly bound molecules, so that at any practical baking temperature the more tightly bound molecules and atoms will remain. They can subsequently be desorbed by particle impact in discharge cleaning. Metal first-wall surfaces are preferred, although some fusion systems have employed glass and ceramic liners. The total quantity of gas desorbed from a stainless steel surface prepared by conventional machining and vapor degreasing was found to be of the order of 35 torr $\cdot$ liter/m^2, which was equivalent to 114 monolayers of absorbed gas. Baking vacuum components to 450°C prior to final assembly is recommended. The main outgassing species from stainless steel are usually water vapor, with lesser concentrations of oxygen, carbon monoxide, carbon dioxide, and hydrogen.

Most of the absorbed impurity species not removed by baking can be removed by various forms of discharge cleaning. The physical processes involved in discharge cleaning are sputtering, desorption, chemical interactions, and electronic excitation. There are many variables involved in tokamak discharge cleaning, which include the plasma species (hydrogen or inert gases such as neon), the plasma current, the toroidal field strength, the frequency of the discharge, etc. The interested reader should consult a summary review of discharge cleaning [21] for details. In many cases a low-temperature discharge is found more effective than glow discharges. A popular method to reduce impurities further is to trap and bury them on the wall by gettering. Using an evaporable titanium getter source, a layer of metal is deposited on the wall just prior to a plasma discharge. The source is usually mounted on a retractable probe. The main process is assumed to be that impurities are trapped chemically in the titanium and then buried under subsequent evaporated layers. These techniques when carefully applied have permitted plasma discharges of 1 s duration to be formed with $Z_{eff} \approx 1$. A good description of the conditioning of the large ADSEX tokamak vessel can be found in the paper by Neidermeyer et al. [21].

The reflection coefficient for light ions and neutrals (backscatter plus re-emission) is close to unity for nongettered wall surfaces during steady-state operation. It may be smaller than unity during the start-up transient of a machine with initially very clean and desorbed walls. Heavy ions have a small reflection coefficient. The influence of recycled wall gases on plasma properties is most pronounced for low-density systems ($n < 10^{19}$ m^{-3}) with a large surface-to-volume ratio. The impurities usually observed in hydrogen plasmas are oxygen, carbon, nitrogen, and the metal from which the limiter is made. The radiation power loss from a typical tokamak plasma, with less than 1% (by number) of impurities, is often of the order of 30% or more of the ohmic power input.

The plasma, after it is formed, is separated from the surface by a nonneutral ($n_e \neq n_i$) region called a sheath. Usually the plasma has a positive potential relative to the wall, because initially the electron flux to the wall is larger than the ion flux. To maintain plasma charge neutrality and a zero net current flow to the wall, a sheath potential V_s is formed, which is approximately

$$V_s \approx \frac{kT_e}{2e} \ln \left(\frac{m_i}{2\pi m_e} \right) \tag{6.16}$$

a formula first derived by Langmuir and Tonks [22] in 1929. For a deuterium plasma at a temperature of 10 keV, the sheath potential would be, according to this simple model, about 30 keV. Ions accelerated through such a sheath potential would, upon striking the surface, produce secondary electrons which will reduce the potential. Hobbs and Wesson [23], for example, have developed a sheath equation that takes into account secondary electrons. It is

$$V_s = \frac{kT_e}{2e} \ln \left[\frac{m_i(1 - \Gamma)}{2\pi m_e} \right] \tag{6.16a}$$

where Γ is a measure of the secondary electron emission.

If, despite secondary emission, the sheath potential gets large, an arc may strike between the plasma and the surface, resulting in copious electron emission. When this occurs from a localized spot, it is referred to as a unipolar arc [24]. Localized surface melting often occurs as a result of unipolar arcs, and the arc tracks have been observed on metal surfaces in many fusion devices. Estimates of the amount of metal removed by arcs are sufficient to account for metal concentrations observed in some plasmas. In the presence of a magnetic field, the arc is driven by the Lorentz force $\mathbf{j} \times \mathbf{B}$ in a direction at right angles to the current vector, although at low pressures there is some observed discrepancy to this simple concept. The simplest way to avoid unipolar arcs would appear to be to grade the potential and to have the plasma edge at relatively low temperature.

Ion and neutral-atom sputtering is a source of impurity release from the first walls of a fusion reactor. Energetic ions diffuse out of the confining magnetic field and charge-exchange neutrals bombard the first wall, causing sputtering. The basic physical mechanism involved in sputtering is well understood and is predominantly a momentum-transfer process. When an energetic ion or neutral is incident on a solid surface, it produces a collision cascade by collisions with the lattice atoms. Sputtering takes place when a surface atom receives sufficient energy to exceed the binding energy. The emitted wall material is ionized in the plasma. The study of sputtering has developed into a subject of considerable magnitude. The sputtering yield depends primarily on the primary mass, energy, and incident angle of the incoming particle, as well as the mass and the surface binding energy of the wall material. The total yields at normal incidence have been measured for nearly all ion–target combinations of interest in the energy range from 100 eV to 10 keV [25]. The dimensionless yield ratio for kilovolt light ions impinging upon stainless steel is between 10^{-1} and 10^{-3}. The yield from light-ion sputtering is only weakly dependent on the wall temperature. Semiempirical relations have been developed [25] that, when used with plasma transport codes, can estimate the amount of wall sputtering to be expected in reactor situations. Recycling of wall-surface particles is an important factor in controlling the plasma density and purity.

Another phenomena resulting from a heavy dose of high-energy incident particles, such as hydrogen or helium impinging upon metals, is the formation of surface blisters. There is a critical dose needed to cause blisters and for helium this is found to be between 10^{21} and 10^{22} ions/m^2. Empirical relations have been developed for the blister diameter and blister thickness. The blisters weaken the material and flake, producing macroscopic damage to the wall.

Hydrogen (or deuterium) embrittlement can be a source of difficulty in first-wall materials that are subject to hydrogen bombardment. Thermal shock and evaporation of the first-wall material can be serious when there is a plasma disruption or loss of control of plasma confinement. And finally, runaway electrons impinging upon limiters or the first wall can produce local cracking or melting. The runaway electron phenomenon results from the fact that the electrons' collision cross section decreases with increasing energy [see Equation (4.12)] so that there is a critical energy, $m_e v_{cr}^2/2e$, for a given electric field strength E, above which the velocity of the electron increases without limit. This critical energy is

$$\frac{m_e v_{cr}^2}{2e} = \frac{e^2 n \ln \Lambda}{4\pi \varepsilon_0^2 E} \approx 5 \times 10^{-16} \frac{n}{E} \tag{6.17}$$

For $n = 10^{20}$ m^{-3}, $E \approx 100$ V/m. A detailed review of runaway electrons in toroidal discharges has been given by Knoepfel and Spong [26].

The majority of first-wall designs for fusion reactors chose austenitic steels or Inconels as the material because of their availability, fabricability, weldability, together with good magnetic and vacuum properties. Conn [27] has reviewed the choice of first-wall materials, including such factors as radiation damage, induced radioactivity, thermomechanical properties, fabricability, cost, and availability. He concluded that for the first generation of fusion reactors stainless steels will most likely be the first choice. To help reduce the effects of impurities from sputtering, the limiter is sometimes constructed of graphite, and it has been proposed that the first wall be coated with a thin layer of carbon, beryllium, or boron.

6.4. DIVERTORS AND LIMITERS

A magnetic divertor can be used to greatly reduce the interaction between a plasma and the surrounding material surfaces, reducing both impurity-atom and recycled-gas flows to the plasma. A divertor is a device that provides for magnetic field lines near the plasma edge to guide the escaping plasma into a separate (burial) chamber where the particles are neutralized, pumped away, and prevented from returning to contaminate the main plasma volume. All divertors are based on the idea of generating a null at some point, or along a line, in a component of the magnetic field. This generates a separatrix, and field lines on the outside of this separatrix are carried away from the plasma surface as they pass the null point. The separatrix, sometimes called a magnetic limiter, becomes the boundary of the toroidally confined plasma.

There are three basic divertor geometries that are considered for a toroidal machine such as a tokamak. Consider Figure 6.3 where they are illustrated. The toroidal field divertor produces a null in the toroidal field by means of a current-carrying coil encircling the minor radius of the torus, with its current flowing in the opposite direction to that in the toroidal field coils. This type of divertor was first used with success in the C-stellerator in 1965. It is considered incompatible for a tokamak because of the tight space requirements in the hole of the tokamak. A toroidal divertor also spoils the axisymmetry of a tokamak.

The bundle divertor uses a pair of relatively small current-carrying loop coils, located outside the torus, which divert a toroidal flux bundle outside the main torus. It spoils the axisymmetry of the tokamak concept, but it is electrically and mechanically convenient, since it is located on the outer radial side of a tokamak. A bundle divertor has been successfully operated on the DITE tokamak at Culham [28].

The poloidal-field divertor is based on one or more current conductors, concentric with the tokamak plasma current, which produce nulls in the poloidal field. It has the advantage of maintaining axisymmetry, and can be used with a system whose minor cross section is noncircular. The major disadvan-

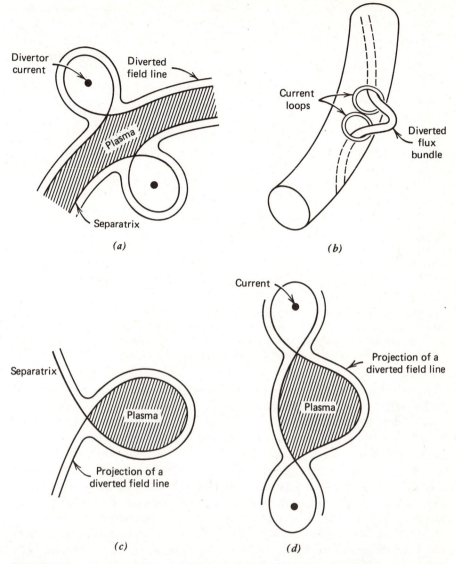

FIGURE 6.3. Schematic divertor configurations: *(a)* toroidal divertor (top view of torus); *(b)* bundle divertor; *(c)* poloidal divertor—single null point (currents not shown); *(d)* poloidal divertor—double null points (not all currents shown).

tage is the need to have the conductors inside the vacuum vessel and, consequently, linking the toroidal-field coils. This presents a rather complex and difficult assembly and maintenance problem. Poloidal-field divertors have been successfully operated on several tokamaks including the large machines PDX (Princeton) and ASDEX (Garching). A cross section of the ASDEX tokamak, illustrating the poloidal divertor, is shown in Figure 6.4.

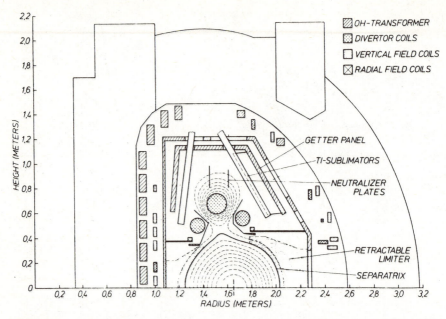

FIGURE 6.4. The ASDEX tokamak in the Max-Planck Institut für Plasmaphysik, EURATOM Association, Garching, Germany. Note the limiter, the poloidal divertor, including the burial chamber with its plates and the vacuum-getter panels. It is interesting to note the location of the coils for ohmic heating (OH), vertical field, divertor, etc., and the magnetic flux surfaces.

Experimental results for operation using a divertor with the PDX and ASDEX tokamaks are reported in the proceedings of the 8th International Conference (Brussels, 1980) [29]. Satisfactory plasma equilibrium and stability were achieved despite the added complexity of the cross section configurations. In PDX the divertor captures about 70% of the input power while 30% is radiated to the walls. Plasmas with $Z \approx 1$ were produced. The ASDEX tokamak provides very useful data on the individual effects caused by the stainless steel limiter, gettering in the divertor chamber, divertor operation without pumping, and divertor operation with pumping. In general, it was found that divertor operation considerably extends the machine operating parameter range, enabling routine discharges up to 3 s. Both these tokamaks have been operated with substantial neutral-beam heating during which the divertor operated satisfactorily. Bundle-divertor experimental results in the DITE tokamak, together with gettering, are reported to be quite successful [28], producing $Z_{\text{eff}} \approx 1$ plasma. Both poloidal and bundle divertors thus appear to be satisfactory from a physics point of view. Which geometry may be preferable for a fusion power reactor is not yet clear. There are substantial heat-load and particle-pumping technological requirements common to all divertors. It is anticipated that most of the helium ash from a D−T burning reactor will be removed by the divertor.

A mechanical limiter is one means by which plasma can be kept from directly interacting with the primary vacuum wall. A limiter may consist primarily of a metal ring in the poloidal direction (poloidal limiter) providing a plasma aperture; or it may be a toroidal ring or belt on the outer side of the torus; or it may consist of simple straight rail segments. The limiter must be capable of withstanding large thermal fluxes without melting or cracking. A major plasma disruption that can terminate the discharge current may deposit a large fraction of the plasma thermal and magnetic energy onto the limiter in a short ($\sim$ 1 ms) time. The area of the limiter that may be involved in receiving the energy is uncertain. Intense and localized thermal fluxes onto the limiter can also result from runaway electrons, particularly during low-density start-up activities. Because of the need to withstand thermal shock, the materials chosen for limiters are, for example, tungsten, tantalum, molybdenum, refractory alloys, or various forms of carbon or graphite. Because some limiter materials will enter the plasma because of sputtering, there is the desire to use low-Z materials like carbon, or to coat a metal limiter with a low-Z material such as beryllium or boron.

It is possible to have a mechanical divertor, somewhat like a scoop, into which the edge-region plasma flows. The scoop requires intensive vacuum pumping and cooling. Sketches of a simple mechanical limiter and a pumped-limiter are illustrated in Figure 6.4A. In a recent fusion-reactor design [30], the

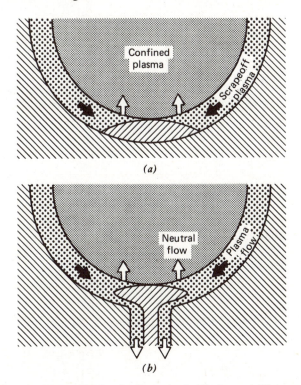

FIGURE 6.4A.   Illustration of *(a)* a simple mechanical limiter and *(b)* a pumped limiter.

limiter and divertor functions are combined. The plasma impurity control and exhaust system are, in this design, combined in a toroidal belt limiter, behind which is the vacuum system. This concept, employed in the STARFIRE tokamak design, is described further in Chapter 7.

6.5. PLASMA HEATING

Plasma temperature will increase when the rate at which energy is put into it exceeds the rate at which it can leak out. Methods of energy input include ohmic heating, neutral beams, electromagnetic waves, and compression. Energy leaves a plasma via radiation, conduction, and particle losses. In a fusion system, heating and confinement are often physically dependent on each other in rather complicated ways. We consider here some of the principal methods of heating a tokamak.

6.5.1. Ohmic Heating

The power density deposited in a plasma by ohmic heating is:

$$P_\Omega = \frac{j^2}{\sigma}$$

(6.18)

where j is the current density and σ is the electrical conductivity. In a tokamak, the plasma current is primarily toroidal and the appropriate conductivity is $\sigma_\parallel$, where, here, trapped-electron effects are neglected. Using Equation (4.81a), we obtain

$$P_\Omega = 6.87 \times 10^1 \frac{j_\phi^2 Z \ln \Lambda}{T^{3/2}}$$

(6.19)

In a tokamak there is a limit, set by mhd stability, on the value of j_ϕ. For a circular cross section tokamak, at $r = 0$, the safety factor $q(0)$ and $j_\phi(0)$ are related [see Equation (5.50)] by

$$j_\phi(0) = \frac{2B_\phi}{\mu_0 R_0 q(0)}$$

(6.20)

Setting $q(0) = 1$, upon combining Equations (6.19) and (6.20), we obtain

$$P_\Omega^{max}(0) = 1.75 \times 10^{14} \frac{Z B_\phi^2 \ln\Lambda}{R_0^2 T_e^{3/2}}$$

(6.21)

Equation (6.21) predicts the maximum ohmic heating power density at the tokamak center consistent with mhd kink stability [i.e., $q(0) = 1$].

By equating the ohmic power to the bremsstrahlung radiation loss at $r = 0$ in a tokamak, that is, $P_\Omega(0) = P_{ff}$, we obtain an equation for the electron temperature at $r = 0$:

$$T_e(0) = 1.05 \times 10^{27} \frac{B_\phi}{R_0 n_e(0)} (\ln\Lambda)^{1/2}$$

(6.22)

For a D−T power reactor with $B_\phi = 5$ T, $R_0 = 5$ m, $n_e = 0.5 \times 10^{21}$ m^{-3}, we find $T_e(0) = 8.6 \times 10^6$ K or about 740 eV. This suggests that pure ohmic heating may not be suitable, since ignition requires at least 5 keV. By examining Equation (6.22), higher center temperatures can be achieved by employing higher values of toroidal field B_ϕ and smaller values of R_0 and $n_e(0)$.

Other forms of energy loss such as particle diffusion and impurity radiation make it difficult to achieve the desired high temperature from pure ohmic heating. For example, another estimate of $T(0)$ in a tokamak can be obtained using the empirically observed Alcator gross energy-confinement-time scaling [see Equation (5.91)]. Thus:

$$P_\Omega = \frac{j^2}{\sigma_\parallel} = \frac{3/2(n_e + n_i)kT}{\tau_E} \approx \frac{3n_e kT}{6 \times 10^{-21} \bar{n}_e a^2}$$

(6.23)

Combining Equations (6.21) and (6.23), and solving for $T(0)$, we obtain

$$T(0) = 2.76 \times 10^6 \left(\frac{B_\phi a}{qR}\right)^{4/5} (\ln\Lambda)^{2/5}$$

(6.24)

Equation (6.24) predicts the maximum temperature at the center of a tokamak, subjected to ohmic heating and Alcator energy-confinement scaling. For $B_\phi = 5$ T, $a = 1$ m, $\ln\Lambda \approx 15$, $q = 1$, $R = 5$ m, we find that $T(0) \approx 9 \times 10^6$ K or about 800 eV.

Equations (6.22) and (6.24) suggest that higher values of $T(0)$ might be achieved by increasing the toroidal field strength, going to lower electron density, and having a low-aspect-ratio device. Whether a tokamak might achieve ignition by pure ohmic heating depends very much on the real scaling laws for energy-confinement time. Prediction of ignition requires a careful consideration of radiation and conduction losses, diffusion, alpha-particle confinement, and energy exchange with the background plasma. Large computer codes are used to perform ignition studies, and they usually use a mixture of theory and empirical observations to describe energy-transfer processes. Most plasma physicists believe heating methods, in addition to ohmic, must be used to obtain tokamak ignition.

There are, however, some scientists who believe that a very high magnetic field (16−20 T) can be developed which will ignite by purely ohmic heating. The high plasma density is employed with the expectation of Alcator energy-confinement scaling [see Equation (5.91)] continuing far beyond present data. The high magnetic field permits a high plasma current density $j_\phi \sim 8$ MA/m^2. The resulting design is a compact, tokamak geometry with a very high neutron flux to the wall, and very difficult magnet technology. One design for a 355 MW$_e$ ohmically ignited tokamak is called a Riggatron (see Chapter 9).

6.5.2. Neutral-Beam Heating

A major method for supplementary plasma heating in tokamak reactors is the use of neutral beams. The most developed method to date to heat to fusion ignition conditions consists of the injection and capture of energetic neutral hydrogen isotopes. The injected particles traverse the vacuum and external magnetic field unhindered and become trapped in the confinement region as a result of charge exchange or ionizing collisions. These trapped high-energy ions then share their energy by Coulomb collisions with the background plasma. The technology of neutral-beam-heating sources has undergone a remarkable development, which has been reviewed by Kunkel [31] and by Memon [32].

The neutral-particle-beam energy must be sufficiently large to ensure adequate penetration into the plasma before ionization and trapping takes place, but it should not be so large that a substantial fraction of the beam completely traverses the plasma and impinges upon the far side wall. The power deposited by the beam must be sufficient to achieve the desired heating effect, and the beam should ideally be free of impurities.

The intensity I of a beam of neutral atoms passing through a plasma attenuates proportional to the beam intensity, the distance traversed, and the probability α of electron loss per unit distance; that is,

$$dI = -\alpha \, I \, dx \tag{6.25}$$

The effective cross section for electron loss σ_{eff} (i.e., trapping) is given [33] by

$$n_e \sigma_{\text{eff}} = n_i \sigma_i + n_i \sigma_{cx} + \frac{n_e \langle \sigma_e v_e \rangle}{v_b} + \sum_j n_j (\sigma_{ji} + \sigma_{jcx}) \tag{6.26}$$

where σ_i and σ_{cx} are the ionization and charge exchange cross sections, σ_e is the electron ionization cross section, n_e and n_i are the electron and ion densities in the plasma, v_e and v_b are electron and beam velocities, and $\langle \sigma_e v_e \rangle$ represents a Maxwellian average. The last term in Equation (6.26) accounts for impurity effects. It is assumed that $v_i < v_b < v_e$. Values of individual cross sections, and overall effective cross sections, can be found in References 31−33. An approximate expression for α is

$$\alpha = n_e \sigma_{\text{eff}} \approx 1.8 \times 10^{-18} \frac{n_e}{E} \tag{6.27}$$

where E is the energy in keV per nucleon of the neutral particle. The characteristic length for trapping, or e-folding attenuation of the neutral beam, is

$$\lambda_i = \frac{1}{\alpha} = \frac{1}{n_e \sigma_{\text{eff}}} \tag{6.28}$$

and the beam attenuation is given by

$$\frac{I}{I_0} = e^{-\alpha x} = e^{-x/\lambda_i} \tag{6.29}$$

A schematic neutral-beam system is illustrated in Figure 6.5. A plasma source or generator is required to provide a deuterium plasma whose density is $n_i > 10^{18}$ m^{-3} with uniform (in space and time) properties over several hundred square centimeters of extraction surface. Ions are extracted at a current density of about 3×10^3 A/m^2 and accelerated electrostatically to the desired energy. The large ion current (~ 100 A) requires the use of multiple apertures to form the beam. Multigrid accel−decel arrangements are employed together with careful ion focusing to produce a low-angular-divergence (less than 1°) beam.

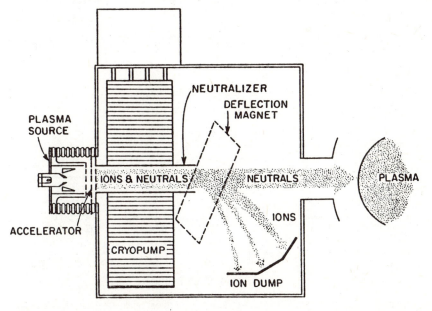

FIGURE 6.5. Schematic illustration of a neutral-beam injection system (from Reference 32).

This ion beam, typically several hundred square centimeters in cross section, is closely coupled to a conductance-limited duct or neutralizing cell in which a fraction of the ions is converted into energetic neutrals by charge exchange with the neutral background gas in the cell. Unfortunately, the efficiency of neutralization of positive ions decreases with beam energy, as indicated in Figure 6.6, so that above about 150 keV, it is impractical. Negative-ion sources are being developed for higher-energy neutral beams. The unneutralized component of the beam, consisting for example of D^+, D_2^+, and D_3^+, is magnetically deflected into a dump.

The neutrals stream straight ahead through a drift duct that connects the injector to the plasma target. The power per unit area delivered to the torus port is between 10^4 and 4×10^4 kW/m². The pressure beyond the gas neutralizer cell must be kept below a pressure of the order of 10^{-5} torr. The gas efficiency (ions out/gas particles in) of the source is typically about 50% and the gas throughput is several tens of torr·liter/s. Cryocondensation panels with pumping speeds of the order of 10^5 liter/s·m² are used to achieve the required vacuum.

The power flow along a PDX beamline is shown in Figure 6.7, illustrating where power is lost from the beam. A view of the PDX beam-injector system is shown in Figure 6.8.

The fast neutrals from the beam injector enter the plasma target volume and become ionized. Their subsequent trajectories depend on the direction of injection with respect to the direction of the magnetic field and the discharge current [34]. There are two classes of orbits: trapped and untrapped or passed. The trapped ions mirror in a tokamak field and have a characteristic cross

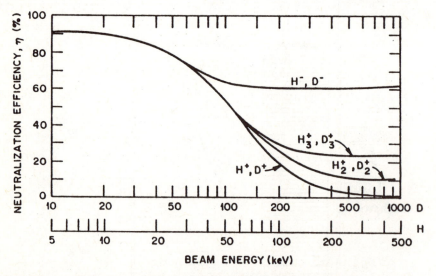

FIGURE 6.6.   Neutralization efficiency for different beam energies; efficiency equals power in neutral atoms out divided by power in ion beam entering neutralizer cell.

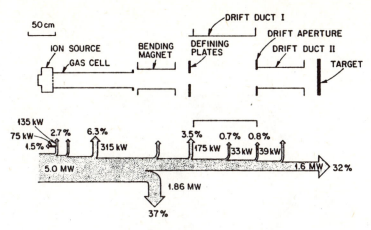

FIGURE 6.7.   Power flow along a PDX neutral-beamline injector (from Reference 32).

section banana-shape trajectory. The untrapped ions circumnavigate the major and the minor axes of a tokamak. For low injection energy ($\leqslant 30$ keV), coinjection (injection parallel to the plasma current direction) is more advantageous than counterinjection. The difference between co- and counterinjection is less serious in reactor-size plasmas because the poloidal gyro-radius, which is the same order as the size of the banana width, is small compared to the minor radius of the plasma. Perpendicular injection avoids the problem of plasma toroidal rotation that occurs if nonsymmetrical toroidal injection is used.

The injected particles, if confined in stable orbits, transfer their energy by Coulomb collisions with the background plasma. The rate at which the injected particles slow down and transfer their energy can be predicted by the slowing-down time and equilibration time formulas given in Section 4.4.2.

Neutral-beam injection and plasma heating have been performed in many laboratories and in many different plasma confinement devices. Heating at several times the ohmic power in tokamaks is a successful technique, having achieved, for example, ion temperatures of above 7 keV in PLT, with no observed deleterious effects.

The ion temperature increase ΔT_i, resulting from neutral-beam injection, scales with beam power and electron density $P_b/\bar{n}_e$. For PLT at 2.1 MW of beam power, when $\bar{n}_e \approx 2 \times 10^{19}$ m^{-3}, $\Delta T_i/P_b \approx 2$ eV/kW. A summary of neutral-beam-injection performance on both the PLT and PDX tokamaks was given by Schelling et al. [35]. The problems associated with coupling of arrays of high-power injectors, beam divergence, and damage of low-level integrated circuits by high-voltage faults are discussed in this summary. It is also observed that the stable high-density limit of tokamak operation [see Section 5.8 and Equation (5.95)] can be substantially extended by supplementary neutral-beam heating. The possibility of using neutral-beam-driven currents has been proposed by Ohkawa [36] to allow for continuous tokamak operation.

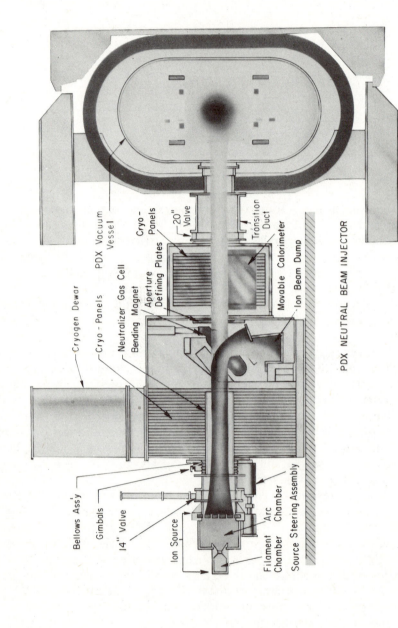

Cryogen Dewar

Cryo - Panels

Neutralizer Gas Cell

Bending Magnet

Aperture Defining Plates

Cryo - Panels

PDX Vacuum Vessel

20" Valve

Transition Duct

Movable Calorimeter

Ion Beam Dump

Bellows Ass'y

Gimbals

14" Valve

Ion Source

Filament Chamber

Arc Chamber

Source Steering Assembly

PDX NEUTRAL BEAM INJECTOR

FIGURE 6.8. PDX neutral-beam injector firing (perpendicular) into PDX plasma.

144

Neutral-beam injectors capable of delivering several megawatts of energy at up to 120 keV and with pulse lengths of about 1 s have been developed. Unfortunately, the efficiency of producing positive-ion-based neutral beams becomes increasingly poorer at higher energies (see Figure 6.6) and the technology becomes physically very large and expensive. For a tokamak reactor ($a \geqslant 1$ m, $n_e \approx 10^{20}$ m^{-3}), beam trapping with good penetration implies beam energies of several hundred kilovolts. Negative-ion-based injectors with the inclusion of internal-energy-recovery subsystems promise improvements in overall efficiency. Straight neutral-beamline penetrations also make neutron shielding of a fusion reactor more complex. If neutral beams are used to supplementarily heat a tokamak reactor, it is estimated [32] that to achieve ignition they would need to be about 200 keV, at an input beam power of about 75 MW, with a pulse length of between 5 to 10 s. The present neutral-beam-injection systems deliver about 5×10^5 J, while a power tokamak reactor would need of the order of 5×10^7 J to achieve ignition.

It is also possible to achieve interesting fusion output without the requirement of ignition, by sustaining the plasma at fusion conditions using intense neutral-beam injection. Jassby has reviewed the status of neutral-beam-driven tokamak fusion reactors [37]. A beam-driven reactor consists of a Maxwellian bulk plasma whose temperature is maintained by a very small population of injected ions, together with charged fusion reaction products. If the population of injected ions is a large percentage, the system is referred to as a two-component torus, and if most of the ions are circulating via injection, it is referred to as a counterstreaming torus. A proposed attractive feature of beam-driven tokamaks is their capability for useful applications (i.e., neutron sources, fuel breeding, etc.) in devices of relatively small size.

6.5.3. Resonance-Frequency Heating

Many types of waves can propagate in fusion plasmas and they have many natural modes. These waves offer a great variety of supplementary-heating possibilities. Although not yet as thoroughly studied or explored as neutral beams, resonance-frequency (RF) heating is perceived as an attractive supplementary-heating method having potential advantages, such as lower cost than neutral beams and easier neutron shielding. Reviews of the status of RF heating for magnetic-fusion devices have been written by Hwang and Wilson [38] and by Porkolab [39], and there are proceedings of international conferences devoted to this subject [40].

RF heating involves several distinct steps: (1) launching the waves, (2) wave propagation and coupling into the plasma, and (3) absorption and heating. Although there are a large number of waves that can propagate in a plasma, only a few types have the physical characteristics and technological base suitable for fusion reactors. Ion-cyclotron (ICRF), lower-hybrid (LHRF), and electron-cyclotron (ECRF) heating methods are being actively pursued for possible fusion applications. For cyclotron frequencies, when the static confinement field varies in space, the frequency-matching condition is localized to regions of space called cyclotron layers. Low-frequency, that is, $\omega < \omega_{ci}$, methods usually require relatively complex coil antennas inside the plasma chamber; these antennas need high-voltage insulators that are not well suited to the fusion-reactor environment; hence, they are not favored.

Ion-Cyclotron Heating.   A wave that is well suited to achieve the objectives of supplementary fusion heating is the compressional hydromagnetic wave, sometimes called the magnetosonic mode, or simply the fast wave. In the cold-plasma approximation, the dispersion relation, for waves with frequencies below ω_{ci} in a magnetized plasma, has two distinct branches—the fast magnetosonic wave and the slow electrostatic wave. Both of these waves have electric fields that are elliptically polarized in the plane perpendicular to **B**, and they rotate in opposite directions. The slow-wave electric field rotates in the ion cyclotron direction, while the fast-wave electric field rotates in the electron cyclotron direction. The slow wave can only propagate below the ion cyclotron frequency, and in cold-plasma theory it has a resonance at $\omega = \omega_{ci}$. In hot magnetized plasma the slow wave is damped as $\omega \rightarrow \omega_{ci}$, and the wave energy is converted to ion motion in the perpendicular degree of freedom. Unfortunately, the slow wave must be launched from the high-magnetic-field side, and this presents difficulty in tokamaks, for there is not much room on the inside of the torus. The characteristics of both the fast and the slow waves are discussed in the book by Stix [41].

The fast magnetosonic wave can propagate above and below ω_{ci} and is used to heat tokamaks with waves launched from the more accessible low-field side of the machine. An illustration of an ICRF half-turn antenna for fast-wave heating is shown in Figure 6.9. Although cold-plasma theory predicts no ion heating for the fast wave, because it is polarized opposite to the ion gyromotion, a finite temperature or the introduction of a second ion species alters the polarization sufficiently to allow wave damping for ions. Wave damping can occur at second or higher harmonics. If there is another type of ion present in the plasma, the ion−ion hybrid resonance layer occurs near the cyclotron layer of the minority species, and this tends to increase wave absorption.

The wavelengths in the desired frequency ($\omega \sim \omega_{ci}$) and density ($n \sim 10^{20}$ m^{-3}) ranges are long enough to allow reasonable mode separation in large tokamak plasmas and good coupling to antenna structures. A good review of fast-wave heating of a two-component plasma has been given by Stix [42]. By proper selection of the fast-wave frequency mode numbers, magnetic field strength, and plasma composition, it is possible to heat ions and/or electrons to vary the deposition pattern of the heating and to modify the ion velocity distribution.

The fast-wave two-ion regime in which minority species are cyclotron damped has been successfully employed in PLT [43−45]. Hydrogen and ^{3}He ions have been used as a minority species in deuterium, and found to absorb the RF power and transfer it to the deuterons and electrons according to theoreti-

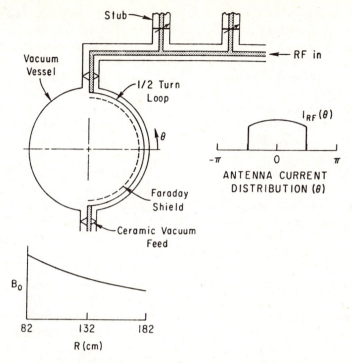

FIGURE 6.9. An illustration of an ICRF half-turn antenna, Faraday shield, ceramic vacuum feed, and tuning stubs used on PLT fast-wave heating experiments (from Reference 43).

cal predictions. The RF frequencies used were 24.6 and 42 MHz, and up to 800 kW for pulse lengths of about 130 ms were delivered to a set of two parallel half-turn loop antennas. About 80% of the power coupled to the plasma. The deuterium heating rates were found to be $\sim 3 \text{ eV} \times 10^{19} \text{ m}^{-3}/\text{kW}$ and $\sim 6 \text{eV} \times 10^{19} \text{ m}^{-3}/\text{kW}$, respectively. For a two-ion mixture, the ion–ion hybrid region is located between the fundamental resonance layers of the individual species, and it strongly affects wave damping and RF power distribution among the ions and electrons. For sufficiently small minority ion concentrations, most of the wave power couples directly to the minority ions. When the minority concentration is sufficiently large, fast waves incident from the low-field side of a tokamak are partially reflected, mode converted to Bernstein waves, and lead to direct electron heating through Landau damping.

Heating experiments using either the first harmonic [46] or the second harmonic [43] have also been quite successful in raising the plasma temperature. The details of wave propagation and the mechanisms of absorption are, at this time, still open to a variety of interpretations. All ICRF experiments carried out in large-size devices have achieved heating efficiencies in excess of 70%. An ICRF-heated design based on the fast magnetosonic mode at $2\omega_{CD}$ (92 MHz) was carried out for the NUWMAK reactor [47]. The

NUWMAK design calculates a need to couple 80 MW for 3 s for ignition, and a primary power requirement of 136 MW for the RF pulse system.

An example of an all-metal antenna successfully employed in the TFR-600 experiment [46] is shown in Figure 6.10. Efficient wave coupling, >90%, is achieved with loop-antenna systems located, for example, in the shadow of the plasma limiter. External component technology is well developed. Megawatt tetrodes and coaxitrons, suitable for pulsed ICRF heating sources exist with efficiencies of about 90% [38, 48]. Transmission of the ICRF power at up to 20 MW per line, for over hundreds of meters, can be done with negligible losses using commercial 23-cm-diameter transmission lines. Double stub trimers are used to match the antenna feed impedance to the source. The RF power is transmitted into the vacuum system through coaxial vacuum breaks, made, for example, from Al_2O_3 insulation. The wave couplers, however, face a severe radiation and thermal environment and must be carefully designed. The thermal flux at the first wall in a tokamak reactor is of the order of 1 MW/m^2 and the neutron flux is greater than 10^{20} n/m^{2}s, with a lifetime dosage of over 10^{27} n/m^2. Reliability of ICRF components is high, and the cost is expected to be significantly less than comparable power neutral-beam systems [48].

Lower-Hybrid Wave Heating.   Wave injection near the lower-hybrid frequency has, for a long time, been proposed as a method for heating tokamaks. This method of heating would have the advantage of using waveguides external to the vacuum vessel for coupling the wave energy into the plasma. Either electron or ion heating can be obtained by a suitable choice of frequency and wave vector. In the cold-plasma approximation the dispersion relation for waves above the lower-hybrid frequency:

$$\frac{1}{\omega_{LH}^2} = \frac{1}{\omega_{ci}^2 + \omega_{pi}^2} + \frac{1}{\omega_{ci}\omega_{ce}}$$

has two branches, the fast wave and the slow electrostatic wave [41]. The slow wave is evanescent near the edge of the plasma where $\omega < \omega_{pe}$, and the wave tunnels through this layer and then propagates along resonance cones. For the wave to reach the lower-hybrid layer, it must satisfy conditions on its parallel index of refraction, and this is called the accessibility condition [38].

The lower-hybrid wave usually utilized for both ion and electron heating is an electrostatic wave, which requires a slow-wave structure for launching. The launched wave is absorbed in a single pass across the plasma (eigenmode structure is not required).

The choice of parallel wavelength is restricted by accessibility and penetration requirements. The imposed parallel refractive index $n_\parallel = k_\parallel c/\omega$ is bounded from below, in order to couple to a lower-hybrid mode, which has access to the plasma center. An upper limit on $n_\parallel$ arises from the need to control linear mode conversion to thermal modes (and/or electron Landau damping) to obtain energy deposition near the plasma center. Analysis of lower-hybrid

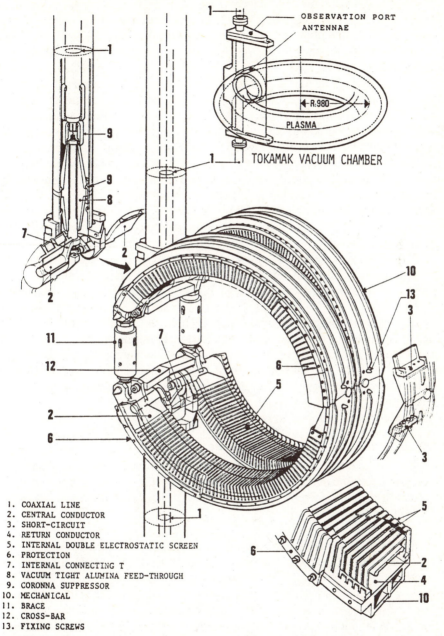

OBSERVATION PORT
ANTENNAE

R.980

PLASMA

TOKAMAK VACUUM CHAMBER

1. COAXIAL LINE
2. CENTRAL CONDUCTOR
3. SHORT-CIRCUIT
4. RETURN CONDUCTOR
5. INTERNAL DOUBLE ELECTROSTATIC SCREEN
6. PROTECTION
7. INTERNAL CONNECTING T
8. VACUUM TIGHT ALUMINA FEED-THROUGH
9. CORONNA SUPPRESSOR
10. MECHANICAL
11. BRACE
12. CROSS-BAR
13. FIXING SCREWS

FIGURE 6.10. Four half-turn antennas used in the TFR-600 ICFR experiments. RF powers to 500 kW at 60 MHz were employed for about 50 ms. All inconnel construction was used with careful Faraday shielding; Reference 46.

wave theory, waveguide–plasma coupling, absorption physics, and wave scattering at the plasma edge is reviewed, for example, by Schuss et al. [49]. A waveguide grill, or phased waveguide array, was proposed by Lallia in 1974, and a theory for its design for lower-hybrid-wave experiments was developed by Brambilla [50].

The technology needed for lower-hybrid heating at fusion-reactor levels is reviewed in References 38 and 48. The ability to remotely site the power supplies and transport the power over long distances with efficient waveguides, together with the ability to have bends and to use radiation-resistant metal, are engineering advantages of the lower-hybrid heating method. Klystron tubes at 2.4 and 2.1 GHz are available at 500-kW power levels. Lower-hybrid heating for $B \approx 5$ T requires frequencies in the range of $2-6$ GHz. The waveguide from the klystron to the launcher should be pressurized with SF_6 gas to avoid electrical breakdown, and the windows separating the pressurized and evacuated regions should avoid direct neutron flux. Waveguides constructed from beryllium-coated copper, where there is exposure to plasma, appear to be a good choice of material. Window materials of BeO, Al_2O_3, or Y_2O_3 are thought to be satisfactory. Power densities in the waveguide-array grills have been tested to power densities up to 100 MW/m^2 without apparent breakdown.

Experiments using lower-hybrid-heating techniques in tokamaks are being actively pursued (see Reference 49 for additional references), and good results have been achieved at power levels up to about 500 kW, with pulse lengths of about 8 ms [48, 51]. The system costs are estimated to be comparable to neutral-beam injectors, and overall system power efficiencies are predicted to exceed 60%.

Another application of lower-hybrid waves is the possibility of driving the plasma current for steady-state tokamak operation. It is proposed for the STARFIRE tokamak reactor design [30]. This use of the lower-hybrid wave is under intense study and future work and experiments will indicate whether it will prove feasible and practical. A review of the theory and the experimental status of electron current drive by lower-hybrid waves has been given by McWilliams and Motley [52].

Electron–Cyclotron Heating. Plasma heating by microwaves at the electron–cyclotron resonance frequency is a well-understood and effective technique. The advantages for electron–cyclotron heating (ECH) are that it can be deposited in a thin, resonant layer at or near the plasma center, the absorption coefficient increases with electron temperature, the damping mechanism is linear and heats the bulk of the electrons (rather than producing a high-energy tail), and the antenna structures are relatively small because the ECH wavelengths are, for most tokamaks, less than 1 cm.

The limits on plasma parameters for ECH are determined by wave-penetration requirements. The extraordinary wave is more heavily damped than the ordinary wave, and for ECH it offers the lowest-frequency requirement; but to avoid cutoff, it must be launched from the inner radius side where access is very difficult. Ordinary wave heating at $\omega = \omega_{ce}$ can be done from the outer radius side when $\omega_{pe}/\omega < 1$. This, however, sets an upper limit on $<\beta>$ [48, 53] and for $\dot{B} \approx 5$ T, requires frequencies of about 140 GHz.

The main obstacle to ECH is the present primitive level of gyrotron tube development for frequencies greater than 35 GHz.

There are gyrotrons available at 28 GHz delivering 200 kW with efficiencies of about 35%. Prototype tubes at 220 kW at 60 GHz for 10 μs with efficiencies of about 50% are being developed. The USSR is reported to have developed a 120 GHz tube. For tokamak reactors, ECH requires high-power gyrotrons in the 120–200 GHz range.

Gyrotron radiation should reflect from mirrors, which will greatly facilitate access to the plasma and prevent neutrons from streaming into the tubes. An analysis of ECH and comparison with neutral-beam heating, to achieve toka-mak ignition, has been given by Cano et al. [53]. ECH experiments have been reported on several tokamaks [48, 54] using up to 80 kW power for times less than about 10 ms. The electron temperature was shown to increase linearly with ECH power.

6.5.4. Other Heating Methods

Other methods of supplementary heating in tokamaks are possible, but none appear to have the attractiveness for reactors that are available using neutral-beam or RF techniques. Some methods that have been considered are: adia-batic compression (major- and/or minor-radius changes), relativistic electron-beam injection, cluster injection (large neutral condensates consisting of $\sim10^5$ molecules), and plasma-gun injection. These alternate methods of heating are discussed briefly in a review of the status of tokamak research [55], and in the proceedings of INTOR workshops [48].

6.6. FUELING AND ASH REMOVAL

As indicated in power-balance and cycle studies (see Section 6.2), long burn times are very important for a fusion power plant to achieve satisfactory overall system efficiency. Although neutral-beam injection can supply fresh fusion fuel, the high energy per injected particle often makes this method of fueling impractical. The injection of frozen pellets of the isotopes of hydrogen is considered the leading candidate for refueling a fusion power reactor based on the tokamak concept. Reviews of the theory and the proposed technology, together with the status of development of pellet fueling, have been written by Chang et al. [56] and by Milora [57].

A 5 GW fusion reactor fueled by equal amounts of deuterium and tritium will consume about 3×10^{21} ions/s. The actual required refueling rate will depend on such factors as the particle-confinement time, the recycling of particles that reach the wall and later reenter the plasma, and the loss rate of particles into divertors. Various refueling methods have been proposed, such as gas puffing, cold-fuel gas blankets, plasma guns, neutral beams, cluster injection, and pellet injection. Pellet injection has the perceived advantage of delivering fresh fuel into the central region of the plasma at energies less than about 1 eV per deuteron. Pellet sizes between 500 μm and 5 mm have been considered, with injection frequencies between about 4000 and 4 pellets per second, respectively.

There is a large literature addressing the theories by which pellets injected into a fusion plasma are perceived to evaporate. Ablation and self-shielding theories have been developed. A conclusion from the theoretical efforts is that a protective blanket consisting of gas and cold plasma will envelope the pellet and partially shield its surface from the intense plasma heat flux. The blanket prolongs the pellet lifetime, but penetration to the plasma center may require injection velocities up to about 10 km/s.

Experiments performed to date have verified the existence of a pellet-shielding mechanism, and pellet-ablation models that use neutral-gas shielding seem adequate to predict what is observed. The tokamak plasma has demonstrated a surprising resilience even to massive density perturbations caused by using large pellets. The overall effect is similar to gas puffing. Large pellets are preferable from the standpoint of attaining deep penetration, and this has important implications for the technology of pellet injection. Injection velocities up to 1 km/s have been achieved with pneumatic gun-type devices. An illustration of an ORNL pneumatic hydrogen pellet injector is shown in Figure 6.11a and a mechanical pellet injector is shown in Figure 6.11b.

Gas puffing at the outer plasma edge of a tokamak has been a successful fueling technique. However, the physical mechanism by which the fresh gas rapidly penetrates the existing plasma is somewhat of a mystery. Both gas-puffing and pellet-injection experiments indicate that mass increases of 10–20% of the total plasma charge are acceptable. No plasma instabilities are observed, nor is particle containment adversely affected. The technology of pellet injection is at an early stage of development, and acceleration of hydrogen ice to about 1 km/s has been achieved. Whether higher velocities and/or deep penetration will actually be required or desired remains to be answered by future experiments.

In a deuterium–tritium fueled reactor, α particles are the ashes that may remain within the burning plasma. The role of alpha particles in tokamak reactors has been extensively reviewed by Kolesnichenko [58]. There are two classical mechanisms by which α particles escape from the plasma. The first is where the orbit of the particle, as a result of toroidal drift, intersects the wall. The second mechanism results from collisional diffusion. In the former case the α particle has nearly its initial energy of 3.5 MeV, while in the latter case it exits the plasma at relatively low particle energy. There have been extensive calculations of α particle orbits and losses [58], including such effects as ripple of the longitudinal magnetic field in tokamaks. Collective processes and possible

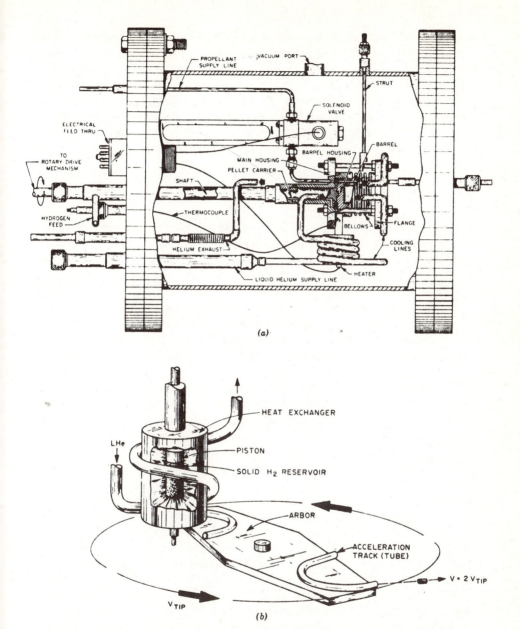

FIGURE 6.11. *(a)* Schematic of ORNL pneumatic hydrogen-pellet injector. This device is capable of injecting single 1.1 mm cylindrical pellets at 1 km/s [57]. *(b)* Schematic of ORNL 30-cm mechanical pellet injector. The high-speed solid-hydrogen extruder, together with the rotating arbor, produces pellets at a rate of 150 per second and injects them at 290 m/s [57].

microinstabilities caused by the nonMaxwellian α particles have also been considered by theorists. The actual rate at which α particles will transfer their energy to the background plasma, and the rate at which they leave the plasma will have to await future experiments with burning and/or ignited plasmas. Whether α particle ash removal will prove to be a difficult technological problem is presently uncertain and it will depend on plasma transport processes, the details of refueling, and the method of edge particle removal, that is, magnetic divertors or mechanical divertors with intense vacuum pumping.

6.7.   TRITIUM PRODUCTION AND HANDLING

Tritium is a radioactive material with a short half-life (12.36 years); it is rare and must be manufactured, usually from lithium. When manufactured in production fission piles, it is quite expensive, about \$9000 per gram. A 1-GW thermal D−T fusion-power plant will consume about 140 g of tritium per day. For D−T fusion it is essential that fusion neutrons be used to breed tritium from the ^{6}Li(n,α)T and ^{7}Li$(n,n'\alpha)$T reactions. The resource aspects of deuterium, tritium, and lithium are discussed in Sections 2.3.1, 2.3.2, and 2.3.3. Because it is radioactive, and because it must be manufactured at the fusion-power plant, using fusion neutrons, tritium handling has a large impact on the design of a D−T fusion-power plant.

The properties of tritium are well known and well documented [59]. There are 2.06×10^{19} tritium atoms per curie (Ci) or, said another way, 1.03×10^{-4} g of tritium will have 3.7×10^{10} disintegrations/s. One gram of tritium is approximately equal to 9600 Ci. The decay of tritium into ^{3}He involves the ejection of an electron (β^- decay) whose maximum energy is 18.6 keV and whose average energy is 5.6 keV. This is soft radiation; there is no gamma radiation from tritium.

Tritium is not considered a significant external radiation hazard because the electron given off during its decay is of low energy and it has very low penetrating power; it does not penetrate the skin. However, tritiated water (HTO or T_2O) and its vapor can be taken into the body by skin penetration. Tritiated water vapor is taken into the body almost as fast by skin penetration as by inhalation, so it presents a special hazard for personnel who must handle high concentrations of tritium. Tritium compounds are used as tracers in biomedical research. The maximum permissible body burden is 1 mCi, which is the highest level for any radionuclide that is recommended by the International Commission on Radiological Protection.

The maximum permissible concentrations for tritium are*:

	Occupational	Population
Air	5 μCi/m^3	0.2 μCi/m^3
Water	0.1 μCi/ml	3×10^{-3} μCi/ml

* Reference: NRC Title 10 CFR part 20, appendix B, Tables 1 and 2, (1982).

There are no significant body reservoirs for tritium, and it is excreted rapidly in urine and sweat. The biological half-life for tritium is about 12 days.

It is essential that fusion neutrons be used to breed tritium for a D–T-fueled fusion reactor. The breeding ratio B is defined as

$$B \equiv \frac{\text{number of T atoms produced}}{\text{number of T atoms burned}}$$

Successful breeding means $B > 1$. Large-scale production of tritium is achieved via the following two neutron reactions with lithium:

$$^1n + {}^6\text{Li} \rightarrow {}^4\text{He} + {}^3\text{T} + 4.785 \text{ MeV} \tag{6.30}$$

$$^1n + {}^7\text{Li} \rightarrow {}^4\text{He} + {}^3\text{T} + {}^1n' - 2.5 \text{ MeV} \tag{6.31}$$

The cross sections for these reactions are given in Figure 2.1. The $^6\text{Li}(n,\alpha)\text{T}$ reaction has an enormous cross section of 941 barns at room temperature, while the $^7\text{Li}(n,n'\alpha)\text{T}$ reaction has a threshold at about 2 MeV and cross-section values of between 0.2 and 0.5 barn between 5 and 14 MeV.

If every 14.1 MeV neutron produced from the D–T reaction produced one triton, then $B = 1$. There are two tritium-producing reactions with lithium, and the breeding process proceeds as follows. A lithium blanket, close to and surrounding the reacting plasma, will intercept the 14 MeV neutrons. The ^7Li $(n,n'\alpha)\text{T}$ reaction occurs with many of these neutrons, producing secondary n' neutrons. These, together with some of the original fusion neutrons, slow down and then undergo the $^6\text{Li}(n,\alpha)\text{T}$ reaction. If every 14 MeV neutron underwent reaction (6.31) and every secondary neutron underwent reaction (6.30), then $B = 2$. Actually, parasitic absorption of neutrons in the structure, first wall, etc., some leakage losses via holes, and other reactions in lithium, make $B < 2$. Neutron multipliers like beryllium, $^9\text{Be}(n,2n)2{}^4\text{He}$, or niobium, $^{93}\text{Nb}(n,2n)^{92}\text{Nb}$, can be used to increase the tritium yield from lithium. However, most $(n,2n)$ reactions are endothermic (about -8 MeV) and deplete the energy yield. Beryllium has the smallest energy deficit and is frequently used for neutron multiplication.

The time that it takes to double the original fuel inventory, t_2, the doubling time, is a measure of the success of breeding. If I is the tritium inventory, P is the tritium production rate, and C is the tritium consumption rate, then

$$\frac{dI}{dt} = P - C = (B - 1)C \tag{6.32}$$

where $B \equiv P/C$. If the excess tritium is removed as it is produced, the doubling time t_2 is that time when the amount produced is equal to the equilibrium inventory I_b. Thus

$$I_b = (B - 1)Ct_2$$

or

$$t_2 = \frac{I_b}{(B - 1)C} \tag{6.33}$$

On the basis of 17.6 MeV per fusion event, $C = 1.54 \times 10^{-4}$ kg/MW$_{th}$ day. If an extra 2.5 MeV per lithium−neutron reaction is realized (i.e., about a net 20 MeV/fusion event), then $C = 1.38 \times 10^{-4}$ kg/MW$_{th}$ day. Thus, a 1-GW thermal fusion power plant will consume about 140 g of tritium per day. Doubling times as short as about 30 days have been calculated for some fusion tokamak reactor designs.

It has been shown that a tritium breeding ratio of about 1.02 tritons per D−T neutron is capable of producing a doubling time of between 2 and 7 years [60]. This means that a tritium breeding ratio of slightly greater than unity would meet the refueling requirement of a D−T fusion reactor for the whole plant lifetime of about 30 years. However, considering the uncertainties due to geometric assumptions in the neutron-transport calculation, and uncertainties in nuclear data, it is prudent that the tritium breeding ratio be greater than 1.10 to ensure adequate breeding capacity [61].

The breeding ratio calculated for the NUWMAK reactor [47], where the breeding material is lithium−lead eutectic $Li_{62}Pb_{38}$, chosen for thermal-design considerations, is about $B = 1.54$. More than 90% of the breeding is contributed from $^6Li(n,\alpha)T$ reactions. For pure lithium, $B \approx 1.46$. The result is nearly independent of the atom percentage of lithium above 20%. This nearly constant breeding of tritium is due to the competition between $^7Li(n,n'\alpha)T$ and $Pb(n,2n)$ reactions, while all the secondary reactions are absorbed by ^6Li and produce additional tritons. The NUWMAK reactor has an average thermal power of 2376 MW and a tritium inventory of 21.4 kg. Its doubling time is therefore about 117 days. NUWMAK burns 106 kg of tritium per year (0.225 gT/s) and keeps a single day's supply of 19.4 kg in storage. The STARFIRE reactor [30], on the other hand, is 4000 MW$_{th}$ and has only 11.6 kg of tritium because it was a design objective to achieve a low-T inventory.

There are three classes of tritium breeding materials: (1) liquid lithium, (2) lithium alloys and intermetallic compounds, and (3) lithium oxide and ternary lithium oxides. In choosing a breeding material, an important safety criterion is that it not have vigorous chemical reactions with air or water. This safety requirement excludes liquid lithium. Among the lithium compounds that are satisfactory for air/water acceptability and appear suitable for breeding are $Li_{17}Pb_{83}$, $LiPb_4Bi_5$, and Li_2O. If the breeding capability is to be maximized, the ternary oxides such as $LiAlO_2$ are usually eliminated. Tritium breeding calculations for one INTOR design blanket module gave the following results:

	Breeding Ratio (INTOR)
Li	1.23
Li_2O	1.28
$Li_{17}Pb_{83}$ (natural Li)	0.87
$Li_{17}Pb_{83}$(90% ^6Li)	1.40

In the NUWMAK design, $Li_{62}Pb_{38}$ was used because its eutectic temperature (464°C) was important in the design to permit the blanket to operate as a thermal flywheel, smoothing out variations in energy delivered from this pulsed reactor to the steam-generating system. On the other hand, the STAR-FIRE design *a priori* decided upon a solid lithium compound. After considering breeding performance, chemical stability with the coolant, compatibility with structural materials, and tritium release characteristics, the ceramic α-$LiAlO_2$ was chosen. To achieve adequate breeding for STARFIRE, the neutron multiplier Zr_5Pb_3 (or beryllium) is needed in the blanket.

The problems of reprocessing tritium from the blanket breeding material, from the fusion exhaust gas, from cryo-pumping panels, etc., require special technology. Development of tritium-handling equipment is underway in the United States at the Los Alamos National Laboratory in the Tritium Systems Test Assembly (TSTA). The U.S. program in tritium handling and development for fusion applications is described by Anderson [62]. The TFTR tokamak will have a tritium inventory of 5 g, but no on-site fuel reprocessing. Later tokamaks such as FED or INTOR will require a tritium inventory of a few kilograms.

An important design criterion in the use of tritium-handling equipment results from radiation degradation in materials like hydrocarbons and elastomers such as used in gaskets, valve seats, etc. Special problems occur because tritium undergoes exchange reactions with fluorine, and therefore Teflon and Viton should not be used. A very serious problem is the generation of extremely corrosive and radioactive tritium fluoride (TF). Fluorinated hydrocarbons must not be used in the design of tritium systems. Tritium will cause hydrogen embrittlement in metals susceptible to that problem, and tritium facilities must use materials known to be resistant to hydrogen embrittlement.

Vacuum pumps will have to be tritium compatible. Compound, two-stage pumps are being developed. The first stage is cryocondensation of hydrogen on a metal surface cooled to or near liquid-helium temperature. This will pump all the hydrogen isotopes, but not helium. The helium is pumped by cryosorption on a molecular-sieve surface, or a charcoal surface at liquid-helium temperature. Or helium may be pumped by argon cryotrapping. The separation of the hydrogen isotopes from each other and from helium and other species (fuel cleanup) is achieved by cryogenic fractional distillation, and by the use of molecular sieves. Metal-bellows pumps are used for gas transfer. An emergency tritium cleanup system, in the case of a gaseous tritium release, is described by Anderson [62].

Tritium can be conveniently stored as uranium tritide in a uranium bed. By heating the bed, the internal tritium gas pressure is raised to about 1 atm. The tritium gas can then be metered and pumped to the equipment used to inject it into the torus. All tritium-handling equipment requires the highest levels of reliability, safety, and tritium containment.

6.8. RADIOACTIVITY AND SHIELDING

A D−T fusion plasma is a powerful source of 14 MeV neutrons. Therefore, shielding is required to protect reactor components from excessive nuclear heating and radiation damage. Shielding also must provide protection for workers in the fusion plant. These neutronic aspects of D−T fusion play an extremely important role in the reactor power plant design and in the choice of materials throughout the system. For example, if the average energy of the D−T fusion plus breeding blanket reactions is taken as 20 MeV per fusion event, then there are 3×10^{11} primary neutrons per second per thermal watt. In a typical tokamak reactor design, the first wall has a 14 MeV neutron flux of about 3×10^8 n/m^2/watt-thermal. That is, the first wall has a neutron flux equivalent to about 2 MW/m^2. This latter value, and the choice of the first-wall material, have a great influence on the lifetime of the first wall because of material damage due to atomic displacements and swelling due to helium formation. Neutron wall loadings less than about 5 MW/m^2 are currently believed necessary to obtain adequate first-wall lifetimes. Future material testing in high-fluence 14 MeV-neutron environments may alter that estimate.

Shielding must protect the superconducting magnets from neutron-radiation heating and damage. Neutrons affect both the critical temperature T_c and critical current density j_c of superconductors. It has been reported [63] that there is only a very small reduction in T_c in NbTi with a neutron fluence up to about 10^{21} n/m^2. There is, however, substantial reduction in j_c at neutron fluences above about 10^{22} n/m^2. The neutrons will also damage the material used for stabilizing the superconductor (usually copper or aluminum).

Radiation damage to insulating materials is an important design criteria. Radiation damage is usually measured in terms of the dose, or the amount of energy, including neutrons and gamma rays, absorbed by the material. The unit employed is the rad or gray (G).* It has been reported [64] that nylon-based insulators suffer severe property changes if the dose exceeds 10^6 G (10^8 rad). For epoxy insulators, the endurable range is $1-5 \times 10^7$ G ($1-5 \times 10^9$ rad).

---

*1 rad = 100 ergs absorbed by 1 g of material. The SI unit is: 1 gray = 1 J absorbed/kg = 100 rad.

Neutronics calculations use one- and two-dimensional discrete, ordinate transport codes like ANISN [65] and DOT [66]. The geometry is sometimes simplified, particularly in doing preliminary optimization studies. Nuclear cross-section libraries, neutron group structure, displacement cross sections, etc., are available (see references in References 30 and 47). The model usually consists of the first wall, magnet, the shield, the blanket, the plasma, and the vacuum regions. The neutronics code provides detailed spatial neutron and gamma-ray fluxes, nuclear damage to components, and irradiation of components. An example of the blanket and shield, in the region of the NUWMAK reactor design, is illustrated in Figure 6.11A. In NUWMAK, the blanket and shield are made from the titanium alloy Ti-6242, and the piping from Ti-6A1-4V. The coolant is boiling water. The magnet system consists of NbTi superconductors, aluminum stabilizers and structure, superinsulation, cryogenically cooled by liquid helium to 1.8 K.

Tritium breeding, neutron multiplication, and nuclear heating for the final NUWMAK blanket and shield design are given in Table 6.1. The total nuclear heating in the blanket−shield system is about 17.15 MeV per D−T neutron, of which about 35% is gamma-ray heating. The total recoverable energy is found to be more than 99% of the total. The maximum atomic displacement rate in the aluminum stabilizer, the neutron fluence in the superconductors, and the

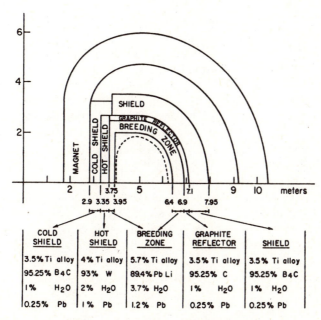

COLD SHIELD	HOT SHIELD	BREEDING ZONE	GRAPHITE REFLECTOR	SHIELD
3.5% Ti alloy	4% Ti alloy	5.7% Ti alloy	3.5% Ti alloy	3.5% Ti alloy
95.25% B₄C	93% W	89.4% Pb Li	95.25% C	95.25% B₄C
1% H₂O	2% H₂O	3.7% H₂O	1% H₂O	1% H₂O
0.25% Pb	1% Pb	1.2% Pb	0.25% Pb	0.25% Pb

FIGURE 6.11A.   The blanket and shield schematic of NUWMAK used for neutronic calculations.

TABLE 6.1.   Tritium Breeding, Neutron Multiplication, and Nuclear Heating in NUWMAK

	Inner	Outer	Total
Tritium Production (T/D−T neutron)			
$^6Li(n,\alpha)$ (T_6)	0.2604	1.1249	1.3853
$^7Li(n,n'\alpha)$ (T_7)	0.0375	0.1192	0.1567
$T_6 + T_7$	0.2979	1.2441	1.5420
Neutron Multiplication (Reaction/D−T Neutron)			
$Pb(n,2n)$	0.1339	0.4181	0.5520
$W(n,2n)$	0.0137	—	0.0137
Total	0.1476	0.4181	0.5657
Nuclear Heating (MeV/D−T Neutron)			
Neutron	2.2665	8.9298	11.1963
Gamma ray	2.4143	3.5422	5.9565
Neutron + Gamma ray	4.6808	12.4720	17.1528

dose rate in the insulators are 2×10^{-6} dpa/yr,* 7×10^{19} n/m^2 per year, and 3×10^5 G/yr, respectively. In the NUWMAK design, after 2 years operation the resistivity change of the magnets will necessitate plant shutdown for magnet annealing and for blanket replacement. Epoxy-based superinsulators must be used in regions of intense radiation dosage.

Using the flux spectrum from the neutronics calculations and nuclear data, codes have been developed to calculate specific activities. A result of such calculation is shown in Figure 6.11B where the radioactivity per thermal watt is given as a function of time after shutdown, when the plant had been operating 1 year. The activity changes very little for about the first month.

The radioactivity, afterheat, and biological hazard potential after 1 year of continuous operation are 0.8 Ci/W_{th}, 0.5% of operating power, and 2×10^2 km^3 air/kW_{th}, respectively. They drop by between four and five orders of magnitude in 10 years after shutdown. The post-shutdown dose rates outside the regular outer shield permit hands-on shift maintenance. They drop to less than 2.6×10^{-2} mS/hr,** 1 day after shutdown. The radioactivity present in the structure is very significantly influenced by the choice of materials in the shield, first wall, blanket structure and breeding zone. For an understanding of the tradeoffs used in the NUWMAK design, the interested reader should consult Reference 47. The STARFIRE reactor, described in Chapter 7 and in

---

*dpa = displacements per atom.
**1 rem = dose having the same biological effect as 1 rad of electrons or x-rays. SI unit: 1 Sievert (S) = 100 rem.

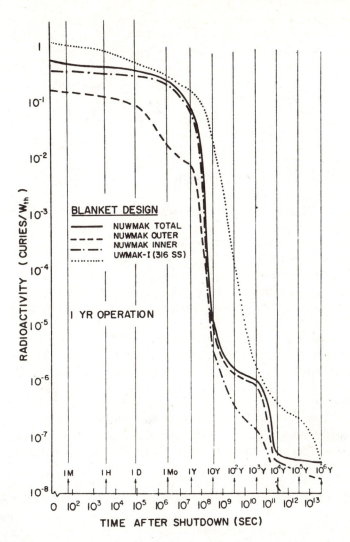

FIGURE 6.11B. Radioactivity of NUWMAK as a function of time after shutdown. A comparison is made with an early tokamak design, UWMAK-I, which employed 316-stainless-steel construction.

Reference 30, features a low-activation shield, and a design philosophy that all materials outside the blanket are recyclable within 30 years. This influenced the choice of materials, and the long-term radwaste from STARFIRE is dominated by the austenitic stainless steel used in the first wall and blanket structure.

An important shield consideration is radiation streaming through regions that penetrate the blanket and shield. The radiation problem associated with neutral-beam ports, divertors, etc., is one of the primary factors that influenced major changes from early to current tokamak reactor designs; that is, the more recent designs use RF heating rather than neutral beams, and eliminate any direct radiation streaming path from the plasma to the reactor exterior. Components inside neutral-beam injectors must use insulators, and there appears to be no way that they can be shielded during injection. A movable shield plug has been suggested for INTOR. Similar problems exist for cryo-absorption panels in divertors, etc.

6.9. MAGNETS AND CRYOGENICS

Large superconducting magnets are believed by many fusion engineers to be essential for economic fusion power by magnetic confinement. Because they are large, expensive, and often interlocked, fusion magnets are expected to have mean times between failures comparable to the lifetime of the reactor. Journal papers and proceedings of meetings devoted to large superconducting magnets are scattered through the engineering and fusion literature. A review paper by Montgomery [67] has a good and extensive list of references on this subject.

The magnetic mirror program had a small superconducting baseball coil in 1965, generating 3.8 T at the winding. In 1970, Baseball II, weighing 11 tonnes, was used in the Livermore mirror program. The superconducting tandem mirror MFTF-B, which will store 3 GJ of energy, is being constructed.* The USSR has a small superconducting tokamak T-7 which is operating and the next major Soviet tokamak T-15 is expected to utilize superconductors. The next French tokamak, TORE SUPRA, will also utilize superconductors. EBT, an alternate concept for magnetic fusion, proposes to use in the EBT-P device, 36 superconducting coils.

Magnet structural problems are similar for both copper and superconductors. Three large copper tokamak toroidal magnets are under construction: TFTR in the United States, JET in England, and JT-60 in Japan. A paper by File et al. [68] describes water-cooled copper magnets and their associated structural designs for the PLT, PDX, TFTR, and JET tokamaks. The TFTR magnet has $R_0 = 2.8$ m and $B_0 = 4.6$ T; it weighs 18 tonnes/coil, stores 4.7 GJ, and has the capacity of a 9-s pulse. The toroidal field varies as a function of the major radius [see Equation (5.12)], giving rise to asymmetric loading and a large net inward force. Circular-shaped coils have a large bending force, resulting in high shear stress between layers of the coil. If the tension is adjusted by changing the shape until the bending is eliminated, the resulting coil shape is a dee. The dee shape is particularly convenient for elongated plasmas. Other coil shapes can be achieved using wedges and other forms of mechanical support. For tokamaks, circle and dee shapes are typical, with a

*It operated successfully in early 1982.

strong center post taking up the net inward force. In tokamaks there are several intertwined magnetic field systems and they can give rise to complex mechanical forces which require careful analysis. The stress in the coil results from $\mathbf{j} \times \mathbf{B}$ force. For example, the poloidal equilibrium magnetic field interacting with the current in the toroidal field coil causes large forces out of the plane of the toroidal field magnet. These must be taken up by lateral supports between the coils. The lateral support structure must also be designed to allow for the failure of one coil, and this force associated with a coil failure is larger than the poloidal-field toroidal-current force.

Poloidal field coils in a tokamak are subject to high voltages during a plasma disruption. Peak values of 35 kV between terminals and 48 kV between ground and the highest terminal are the disruption voltages estimated for TFTR. Designers of superconducting poloidal systems must choose insulation and cooling concepts compatible with such voltages and which do not degrade under cyclic loading.

The two superconducting materials considered for fusion reactor magnets are niobium–titanium (NbTi) and niobium–tin (Nb_3Sn). NbTi has good ductile mechanical properties, but Nb_3Sn is very brittle. However Nb_3Sn can, at the same cryogenic temperature, produce a higher superconducting magnetic field than NbTi. Typically, NbTi can be used to produce $B_0 = 8$ T magnets with helium pool boiling, which implies a temperature of 4.2 K at 1 atm. At this same temperature, Nb_3Sn magnets can achieve $B_0 \approx 12$ T. Higher fields are possible with both materials by using lower cryogenic temperatures. The stabilizing material used for both superconductors is either copper or aluminum. Typical conductor current densities are about 2×10^8 A/m² for NbTi at 4.2 K and 8×10^8 A/m² for Nb_3Sn at 10 T. Values of critical current density for NbTi and Nb_3Sn as a function of temperature and magnetic field are given, for example, in the NUWMAK design report [47]. NbTi is relatively insensitive to irradiation, but Nb_3Sn degrades rapidly above a fluence of 3×10^{22} n/m² (E > 1 MeV) as a result of radiation-induced disorders. This effect is shown in the data of Figure 6.12. This radiation limit on the conductor is, however, above that usually set by the radiation limit on the insulation.

The cost of refrigeration increases rapidly with decreasing temperature. Power for refrigeration is approximately proportional to the Carnot factor $(T_w - T_c)/T_c$, where T_w is room temperature and T_c is the cryogenic environment temperature. The power requirement for 1.8 K is about three times that for 4.2 K. In the NUWMAK design, the heat loss in the 1.8 K coil is 1300 W, of which conduction is 500 W, nuclear heating is 500 W, and joint losses are 300 W. The refrigeration power required to remove this heat is 1.3 MW, which the designers consider small. One NUWMAK TF coil contains about 5000 liters of liquid helium.

Although it is anticipated that the development of the large superconducting coils needed for fusion reactors can be achieved by extrapolation from present experience using good engineering design, there is as yet no agreement on design details. Different conductor configurations, different winding configu-

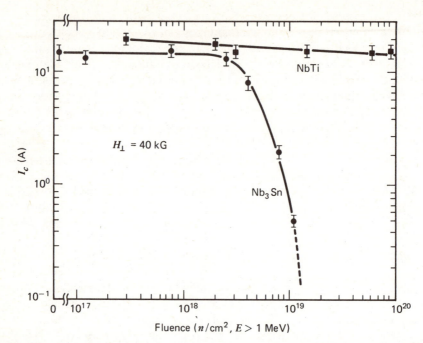

FIGURE 6.12. Effect of neutron irradiation on the critical current in NbTi and Nb_3Sn. Data from 1979 U.S. Contribution to the International Tokamak Reactor Workshop.

rations, different material choices, and different heat transfer and structural arrangements characterize the designs in the present U.S. large superconducting coil program.

As examples of some of the design possibilities, it is instructive to examine both the NUWMAK and the STARFIRE magnets. In Figure 6.13 is shown a cross section of NUWMAK illustrating its magnetic-coil configurations. A design philosophy for NUWMAK was to increase space for maintenance and other reactor components. To achieve this, the designers used only eight superconducting toroidal-field magnet coils. To correct for the resulting large field ripple, water-cooled-copper saddle coils are employed, and these are located at the back of the outer blanket. They consume about 25 MW. The field on axis was chosen to be 6 T, and the peak field at the coil is 12 T. The designers suggest either NbTi at 1.8 K or Nb_3Sn at 4.2 K, but there is as yet no experience to guide the choice between these two. There are four magnets linked inside the TF coils, and these are cryogenically cooled, normal Al coils. The system is designed so that a TF coil can be removed in case of failure.

The STARFIRE cross section illustrating the magnetic-coil configurations is shown in Figure 6.14. A relatively conservative design philosophy was used for STARFIRE's magnets with the intention that the large toroidal-field coils would last the reactor's lifetime. Twelve superconducting coils are used with a field on axis of 5.8 T and a peak field at the coils of 11.1. T. The super-

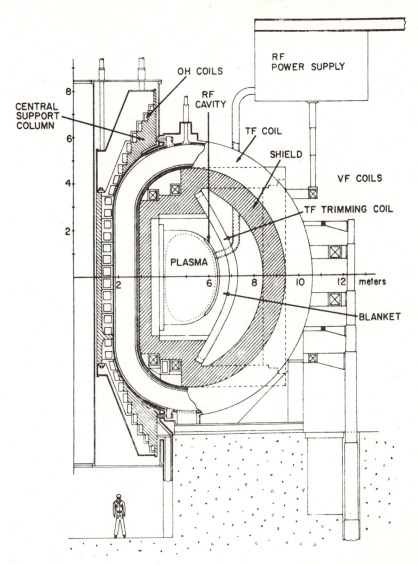

FIGURE 6.13.   NUWMAK cross section showing the magnetic-coil configuration.

conducting magnets are designed using NbTi at 4.2 K everywhere except in those regions where the magnetic field is in excess of 9 T. In these regions, a Nb_3Sn superconductor is used. No ohmic-heating (OH) or equilibrium-field coils (EF) link the toroidal-field coils. However, correction-field (CF) coils, needed to respond rapidly to plasma displacements and thus stabilize the plasma, do link the toroidal-field coils. These CF coils are constructed of watercooled copper and they are demountable. The NUWMAK and STAR-FIRE design reports [30, 47] contain many more engineering details of the magnet designs, which the interested reader should consult.

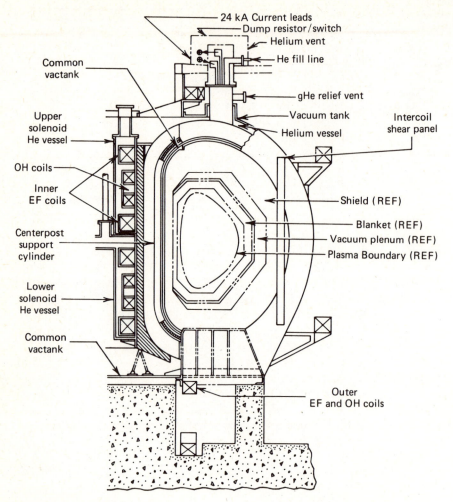

FIGURE 6.14.   STARFIRE cross section showing the magnetic-coil configuration.

6.10.   MATERIALS

Fusion power plants require a wide variety of materials, and when fusion becomes a major energy source, consideration will have to be given to the world resource base of some metals and the importance of being able, in a timely manner, to recycle critical materials. Fusion requires both metallic and insulating materials that can retain their desirable properties in the presence of radiation and corrosive substances. The choice of materials used in the limiter, the first wall, the blanket and cooling system, the superconducting magnets, and the radiation shields is particularly significant in influencing the character of the design of the reactor. Conferences completely devoted to fusion reactor materials have been held [69], and their proceedings provide a good source of references. An overview of the status of materials research and development related to the needs of fusion power reactors has been given by Gold et al. [70].

In early fusion-reactor design studies, done in the late 1960s and early 1970s, refractory metals (Nb, Mo) were favorite structural materials for the blanket. This choice was motivated by the desire to operate at high wall temperatures ($\sim 1000°C$) and at very high neutron-flux wall loadings (~ 10 MW/m^2). But the realities of cost, fabrication difficulties, radiation effects, and the present-day, small industrial base for refractory metals have lead more recent designs to choose austenitic stainless steels or titanium alloys for the higher-temperature structural materials. This shift in material choice required a drop in coolant temperatures to the regime of about 500°C and a drop in the neutron wall loading to about 2 MW/m^2 because of swelling and thermal stress problems. Structural materials considered for the first wall and the blanket in the near term are given in Table 6.2. Analysis of fusion materials is available in the 1979 U.S. INTOR Workshop Contribution [70] and in the NUWMAK reactor design report [47]. An assessment of titanium for use in the first wall and blanket structure of fusion plants has been given by Davis and Kulcinski [71]. It is stated that titanium will permit a higher neutron fluence on the first wall with fewer maintenance and long-term radiation-waste storage problems.

The major objectives in the choice of materials for a fusion power plant are availability, good mechanical and fabrication properties, reliability (long service life), no extremely long-lived isotopes (waste-storage problem), and the ability to withstand the neutron wall loadings associated with high-power-density operation used to obtain better economical performance. The choice of which alloy to use in a particular component of a fusion plant is still evolving, and there will continue to be changes in the materials choices, as the material data base for fusion increases and as further design-optimization studies are completed.

To help reduce the material problems caused by mechanical and thermal fatigue, the STARFIRE reactor-design philosophy stressed steady-state reac-

TABLE 6.2. Compositions of Structural Materials for First Wall and Blanket[a]

Designation	Nominal Composition (wt%)
316 SS	Fe-17Cr-13Ni-2.5Mo-0.05C
HT-9	Fe-12Cr-1Mo-0.5W-0.3V-0.20C
9Cr-1Mo	Fe-9Cr-1Mo-0.5W-0.15V-0.10Nb-0.08C
Ti 64	Ti-6Al-4V
Ti 6242	Ti-6Al-2Sn-4Zr-2Mo

[a]The industrial base for these alloys is adequate for fusion-reactor development. Their physical properties, fabrication characteristics, and behavior to radiation are reviewed in the 1979 U.S. INTOR Workshop Report [48].

tor operation. Material choices for STARFIRE are described in Chapter 7, and the materials data base is in Reference 30.

Titanium (Ti-6A1-4V) was chosen as the structural material in the high-temperature regions of NUWMAK. It has a high electrical resistivity, which reduces eddy currents as the plasma current is increased and decreased during each 245-s pulse. Titanium exhibits good fabricability, good specific strength, excellent creep resistance up to 535°C, and excellent fatigue and corrosion characteristics.

A major concern with the use of titanium in the blanket structure is diffusion of tritium through the titanium into the coolant and, subsequently, out of the reactor. Oxides and/or thin films are suggested for reducing the tritium loss and lowering the tritium inventory in the structure. A sample calculation of tritium permeability through Ti tubing in NUWMAK yields a maximum release of 1 kG/yr of tritium into NUWMAK's 2.5×10^5 kG of cooling water. These 30,000 Ci/day imply that a water cleanup system is needed.

There is a reasonable data base for stainless steels employed in a fission-reactor neutron environment, including information on displacements per atom, abbreviated as dpa, and gas production (H and He) in metals as a result of nuclear transmutations. Gas production in metals caused by transmutations and fission events are reported in terms of atomic parts per million, abbreviated appm. There is very little physical data on the properties of alloys that have been subjected to irradiation damage by a neutron spectrum characteristic of that which will be in a fusion reactor. A critical lack of data exists on insulator material response to fusion neutrons. The U.S. fusion materials test and development program, which will help provide this information, is described by Nygren [72]. It is impossible at present to calculate the working lifetime of a fusion-reactor first wall, but NUWMAK designers estimate for the Ti first wall, a lifetime of about 10 MW-yr/m^2 which, for that reactor, is about 2 years. Studies of the costs of replacement of the first wall suggest that significant economics are gained as the neutron flux increases from 1 to 3 MW/m^2.

The minerals-resource implications of a tokamak fusion reactor economy have been analyzed by Cameron et al. [73] and are discussed in the design reports for NUWMAK [47] and STARFIRE [30]. There are, of course, differences, depending on the designer's choices of materials. For STARFIRE, assuming 100 reactors (a total of 120 GW$_e$), the metals Be, B, Cu, Fe, Pb, Li, Mo, Si, and Zr should be available in the quantity needed from domestic U.S. mines during the early part of the 21st century. There is some controversy concerning beryllium assessments. For Al, Cr, Ti, Mn, Ni, Sn, V, and Nb conventional U.S. reserves will most probably be depleted by that time but new deposits, not now economically viable, could be developed. Niobium needed for the super-conducting magnets may be a problem. Each reactor requires about 95 tonnes of Nb, and the 100 reactor total would exceed U.S. reserves. But the world reserves of Nb are very large; for example, Canada and Brazil have large deposits. Tantalum (30 tonnes/reactor) and tungsten (842 tonnes/reactor) reserves are being steadily depleted in the United States and even today furnish only a small fraction of current total production needs. The United States imported 97% of its tantalum and 50% of its tungsten requirements in 1978. World resources appear adequate. Lithium, used for breeding tritium, will initially require about 115 tonnes per reactor and 1866 tonnes over the unit lifetime; its resource base is adequate.

6.11. DIAGNOSTICS, CONTROLS, AND COMPUTERS

Every electric power plant has an instrumentation and control (IC) system and very rapid changes are taking place in these areas of technology. For this reason, it is difficult to accurately forecast the instruments (diagnostics), the controls, and the computer systems that will be employed in a fusion reactor built a few decades in the future. A very interesting attempt at projection of IC technology for fusion is contained in the STARFIRE design report (Reference 30, Section 18).

The status of tokamak plasma diagnostics (instrumentation) has progressed enormously during the past decade, and even greater progress should be forthcoming. The books by Huddlestone and Leonard [74] and Lochte-Holtgreven [75] show the state of the art of plasma diagnostics in the mid 1960s. The review paper by the Fontenay-aux-Roses TFR group details tokamak diagnostics in 1978 [76]. Nearly all relevant plasma and machine variables can be measured, but the reliability of these measurements and frequency and spatial resolution will have to be improved to provide the data believed to be needed for good fusion power plant system control.

All large tokamaks and similar-sized fusion plasma experiments employ a variety of computers to rapidly access, process, and display the data obtained from the system diagnostics. There is, as yet, relatively little effort to control the plasma during a shot. On long-pulse experiments, plasma radial position is sensed, and feedback of this information into the equilibrium-field-coil supplies changes their current to help maintain the plasma in the desired location. The large amount of data collected is usually examined by the machine operators and the plasma scientists after the shot is completed. Some of the current-day problems of instrumentation and controls for an ignited tokamak are described in a report by Becraft et al. [77]. They conclude that a great gulf exists between present capability and requirements for control of a large, burning tokamak. The biggest aspect of control problems is the development of adequate mathematical models of plasma behavior. Assuming that these will be forthcoming from research, reactor control design should be a reasonable task.

In the attempt to forecast the instrumentation and control aspects of STARFIRE, it is estimated that about 50,000 signals will be required. This is an increase of one order of magnitude above that needed for modern electric power plants. To accommodate performance compatible with both plasma control requirements and electromechanical components, a system update rate of once per millisecond is suggested. This produces an overall system bandwidth of 50×10^6 signals/s. This is well beyond the capabilities of present-day process control systems, but is predicted to be comfortably within the capabilities of the technology in the early 21st century. The STARFIRE IC system is designed as a network, with functions decentralized and performed at an appropriate hierarchical level, but with centralized control of data and communications. There will be about 200 controllers throughout the STARFIRE facility, each with a maximum bandwidth of 5×10^5 transfers per second, performing 10^3 instructions on each of 500 signals every millisecond or 5×10^8 instructions per second. This requires a 20 ns instruction time, well within microprocessor technology. Assuming 1% of all signals generate message requirements for supervisory level results in a 2.5 MHz communication bandwidth.

At the next level are communication concentration systems. They concentrate communications and distribute the information. Certain local controllers will have significant processing and/or control capability, for example, computing a plasma temperature from Thomson—laser scattering data, and controlling the fuel injection rate or the impurity control system. At the supervisory level the displays/controls system operates the human interfaces in the plant control room, and responds to operator requests, A random-access main memory storage capability of about 50×10^6 bytes and an on-line archive of about 500×10^6 bytes are estimated to be needed. This too is somewhat larger than present capability, but it is believed to be within practical reach by the time they will be needed. At the highest level of network hierarchy is a computer system capable of simulating operation of the entire power plant at 100 times real time. Thus, projection of consequences of a particular operating decision some 2 hours in the future would be available within a few minutes of elapsed time.

Modular instrumentation of high reliability will be required, together with wireless data links, and the ability to perform remote maintenance. The demands of speed, quantity, and diversity for information handling are expected to evolve, driven by the needs of other nonfusion technologies. The special problems for IC of fusion power plants are associated with the radiation environment, and the need to develop standard and reliable plasma measuring systems.

6.12.   VACUUM SYSTEM

The primary purpose of the vacuum system is to remove helium ash and impurities from the fusion reactor. Early designs of tokamak reactors (for example, References 4 and 7) employed divertors to guide the alpha particles and edge impurities into a chamber where they are neutralized on collector plates and subsequently pumped out of the system (see Sections 6.3 and 6.6). The employment of divertors, however, seriously complicate the reactor design by making the system large, costly, and difficult to maintain. A divertor-less tokamak is attractive because it can be a small reactor system, and it is easier to assemble and to maintain. Whether long-burn tokamak reactors can satisfactorily remove the ash and impurities without a divertor, that is, by pumping the primary chamber, has not yet been established. The more recent STARFIRE [30] and NUWMAK [47] reactor designs, however, both employ divertorless tokamak configurations.

When divertors are used in the vacuum system, the openings are a small percentage of the first-wall area. There is a practical limit, somewhere between 5 and 10% of the first-wall area, which can be used for pumping because of other considerations such as neutron streaming, space limitations, and inter-ference with other reactor components. Also, highly efficient removal of helium ash will remove large amounts of unburnt fuel with the result that tritium inventories in the vacuum pumps and fuel recycle systems will be large. Tradeoff studies between conflicting objections of the size of first-wall penetra-tions, neutron streaming, and tritium inventory should be made relatively early in a fusion reactor design.

There are at least three choices for the location of a tokamak reactor vacuum wall. They are: (1) at the torus first wall, (2) at the front or the back of the shield, and (3) at the building structural wall. Early designs used the first wall as the vacuum surface, but recent designs locate the vacuum seals just behind the shield. This results in a larger vacuum volume than the first-wall choice, and a larger surface to be outgassed, but it removes the seals from regions of severe radiation damage and it makes maintenance and repair easier.

The principal parameters that affect the cost and complexity of the vacuum system are the required particle pumping rates and the pressure permitted in the torus (or divertor) chamber. These parameters set the size, and sometimes the feasibility, of the pumping system. The degree of burn-up and the reactor duty cycle in pulsed reactors are also important factors influencing the vacuum system.

An assessment of the INTOR vacuum system together with the state of present and near-future vacuum technology are given in Reference 48. Very large vacuum systems have been built previously for simulation of space vacuum conditions, but fusion-reactor vacuum systems have special and rela-tively unique requirements. In particular, it must pump large quantities of helium and all the pumped gases must be processed to recover at least the tritium. Condensation and sorption cryopumps are the vacuum sources em-ployed in recent reactor designs. Cryocondensation and cryosorption pump speeds up to 1×10^5 liter/m^{2}s for D$-$T and 3×10^4 liter/m^{2}s for helium are available. The major radiation effect on liquid-helium-cooled cryopanels is the additional heat load due to neutrons and gamma rays. However, the estimated refrigeration load ($\sim$20 W/m^2) will be well below the critical film boiling flux for liquid helium at 1 atm ($\sim$10^4 W/m^2).

Turbomolecular pumps, whose bearings are altered for fusion use, have also been considered. Turbomolecular pumps are currently available in sizes up to

3500 liter/s. Vacuum seals and valves do not appear to present any unique problems other than being functional in a radiation environment, and there is considerable experience for these components from fission reactors.

. The major components of the STARFIRE fusion reactor vacuum system are shown in Figure 6.15. The designers chose a toroidal-limiter−vacuum system (i.e., no divertor) which is described in Chapter 7; further design details are given in Reference 30. STARFIRE has 703 MW of alpha power and the helium production rate is 1.24×10^{21} particles per second. The vacuum system consists of the limiter slots, limiter ducts, plenum region, vacuum ducts, and

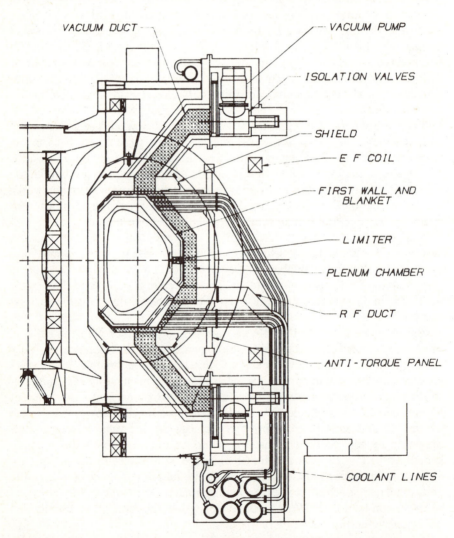

FIGURE 6.15. A cross section of the STARFIRE reactor showing the components of the vacuum system [30].

vacuum pumps. There are 48 compound cryopumps operating on 24 vacuum ducts. Two pumps are provided on each duct so that regeneration can be accomplished during plasma operation. Each pump has a rated helium pumping speed of 120 m^3/s. The vacuum system is designed to produce a base pressure of about 1.3×10^{-6} Pa (10^{-8} torr).* Tritium inventories in the pumps are minimized by high tritium fractional burnup (42%) and by minimizing the pump regeneration time (2 h). The maximum tritium inventory in a single pump is 2.6 g.

The STARFIRE limiter/vacuum system configuration is perhaps the most detailed tokamak reactor–vacuum system yet designed. The vacuum vessel is calculated to reach 1×10^{-6} Pa in about 60 min, at which time the cryopumps are turned on and the vessel pressure reaches less than 1×10^{-9} Pa in another 10 min. The STARFIRE vacuum duct and pump configuration are illustrated in Figure 6.16a and a cross section of the liquid-helium cryopump is shown in Figure 6.16b. The cryocondensation element will pump D_2 and T_2 and most impurity gases. This element also provides an optically tight shield for the cryosorption surface so that only noncondensable gases such as helium are pumped by the adsorbent. The designers chose 3700 g of Mole Sieve 5A (a zeolight) distributed over 2.7 m^2. Roughing pumps will be used in regeneration, and they will also interface with the fuel-processing system.

6.13. BALANCE OF PLANT

The balance of plant facilities for a tokamak fusion electric power plant usually concern features common to any large power plant, for example: steam generators, turbine generators, condensors, cooling towers for heat rejection (if the site is not close to a large body of water), buildings, building services, electrical and mechanical systems, switchyard, etc. However, the balance of plant may also include some systems particular to fusion such as a hot cell for remote repair and maintenance functions, and perhaps the RF or neutral-beam power-supply building. Although engineering of the balance of plant is a well-developed subject, the design does have a very significant impact on the economics of the power plant and on construction time. The balance of plant design will follow industry codes and standards consistent with technical and licensing requirements. The facility must be designed to accept appropriate seismic loads, storm conditions, etc. The architectural and structural features and other balance of plant items of STARFIRE are described in Chapter 7 and further details are available in Reference 30. It is an important subject, but because there is a well-developed engineering base for design and construction of the balance of plant, we do not pursue this subject further in this book.

---

*1 atm = 1.013×10^5 Pa; 1 torr = 1.33×10^2 Pa; 1 lb/in.^{2} = 6.90×10^3 Pa.

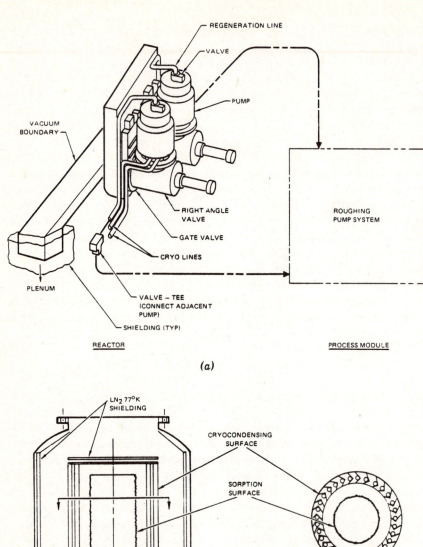

FIGURE 6.16. *(a)* STARFIRE vacuum duct and pump configuration. *(b)* Section view of liquid helium cryopump from Reference 30.

6.14. MAINTENANCE

Design of fusion reactors must include practical techniques for servicing, maintenance, and repair of all components of the power plant throughout its working life. The intense neutron flux from the D−T fusion reactions cause high activation and often shorten the service life of some of the structure. Remotely controlled machines or robots must be used for servicing operations on the first wall, blanket, shield, and many components such as neutral-beam or RF heating systems, cryopumps, valves, and instruments near the plasma.

Fusion-power systems are characterized by high capital costs and rather low fuel and operating costs. Consequently, short downtime and high availability are very important in minimizing total fusion-energy costs. Early tokamak reactor designs [4, 7] were very large and little attention was given to maintenance. Widespread criticism from the utility industry plus the realization that the cost of energy was only marginally dependent on the power output of a tokamak reactor between 3000 and 7000 MW_{th} [78] drove the designers toward smaller and more serviceable fusion designs. In particular, for smaller toroids, the outer, return-leg of the toroidal-field coils could be placed far enough back so as to allow unimpeded radial extraction of first-wall, blanket, and shield segments.

Further maintenance and economic studies [79] have shown that short downtimes and lower costs are achieved in designs based on replacement of larger assemblies rather than many smaller subassemblies. After the reactor is returned to operational status, further disassembly and repair may be carried out in a hot cell adjacent to the reactor room. The cost of replacement energy is so high that the substantial investment needed for large, complex, remote-operating machines to replace large module sections is justified.

Philosophy of maintenance must be incorporated in the reactor design from the very beginning. In the STARFIRE [30], NUWMAK [47], and Culham tokamak Mark II [80, 81] tokamak reactor designs, major components are designed to move linearly (i.e., radially) so that dismantling and reassembly can be more easily done by remotely controlled machines. This is illustrated in Figure 6.17. There is engineering experience for remote maintenance devices in diverse fields such as high-energy experimental physics, fission-power reactors, and the space and underseas development programs.

The nuclear-fission-power-plant average availability* has varied between 68% and 74%, and individual plants have achieved availabilities as high as 85%. Complexities of a fusion system will reduce availability as compared to light-water reactors, if the same technology is assumed. However,

---

*Availability $\equiv \dfrac{\text{time available for operation}}{\text{total time}}$

$\approx \dfrac{\text{mean time between repairs}}{\text{mean time between repairs} + \text{mean time for repairs}}$

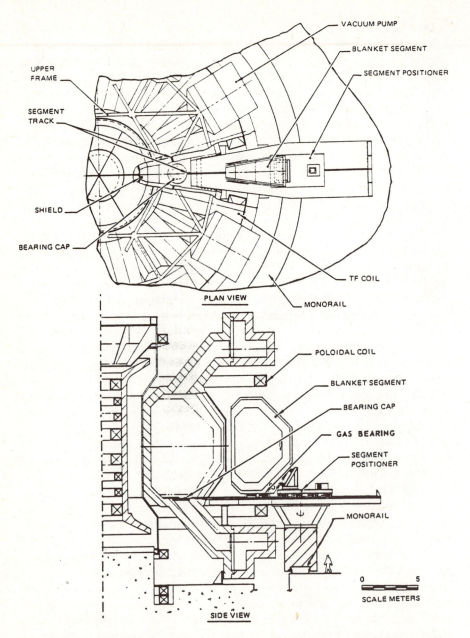

FIGURE 6.17. STARFIRE reactor maintenance showing the blanket-sector handling machine (from Reference 30). The blanket sector can be transported to a hot cell via the monorail.

176

advances in robotics, computers, and maintenance technology should help offset the burden of increased complexity. There is, of course, no substitute for good and simple design. Because of a lack of any experience and actual history, fusion-reactor components have unknown reliability. However, drawing on experience with similar equipment used in fission-power plants, etc., estimates have been made [82] of the expected reliability and operating life expectancy of fusion-power plant components. Some components can be designed for the lifetime of the plant, for example, shield, structure, and perhaps the toroidal-field coils. But other components must be periodically replaced; for example, blanket modules, limiter, first wall, cryogenic system components, pumps, etc. Estimates of downtime for the STARFIRE power plant result in 37 days scheduled for maintenance, 34 days of unscheduled downtime due to reactor component difficulties, and 20 days for the balance of plant. The expected STARFIRE plant availability is therefore 75% (91 days total outage per year).

Even with good design, careful attention in manufacture, and excellent quality control, things can and do fail or require replacement. Think, for example, about remote vacuum leak detection, location, and repair. Remote maintenance will have to be performed and remote maintenance vehicles or robots suitable for operation in the reactor room must be developed. Examples of such devices, designed for the STARFIRE power plant, are shown in Figure 6.18.

The maintenance aspect of fusion reactors is an area of fusion engineering which is in its early stages, and inventive design, together with developments in robotics, offer important areas for development. Design criteria for fusion components are: they must be accessible for inspection and disassembly; they should be modularized to ease disassembly; and the simplest of designs is usually the best design. The STARFIRE design study gave serious attention to maintenance and the interested reader should consult the full design report [30].

6.15. SAFETY

Safety characteristics of a fusion-power plant depend on many design details such as choices of materials, coolants, and tritium inventory. Fusion power is generally perceived as being relatively safe for the general public, but it does present hazards and it is very important to design fusion-power plants in ways which maximize their safety and minimize their environmental impact. Safety analysis of fusion-power plants began in the latter part of the 1970s as more detailed reactor designs became available. For example, Hafele et al. [83] explored the relative safety aspects of fusion and fast breeder reactors. The U.S. Electric Power Research Institute held a workshop on fusion safety in 1978 [84]. There is a U.S. DOE fusion-reactor safety program [85] and Kulcinski has summarized the environmental features of fusion and fission after a

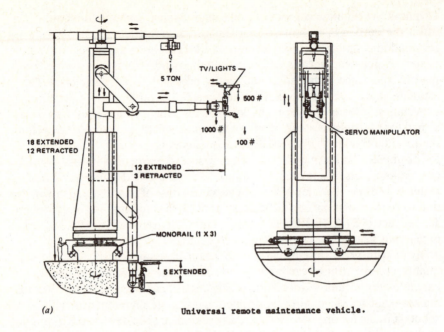

(a)   Universal remote maintenance vehicle.

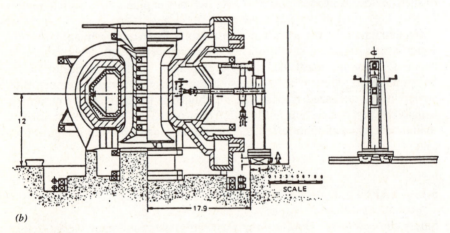

(b)

FIGURE 6.18. (a) STARFIRE remote maintenance vehicle. (b) Illustration of remote maintenance vehicle working on STARFIRE.

decade of study [86]. The STARFIRE power plant described in Chapter 7 and assessed in greater detail in Reference 30 includes its safety and environmental features.

The major sources of hazards in a tokamak electric-power plant are associated with the tritium inventory, the structure which is radioactive, the stored energy in the magnets, and the possibility of a hydrogen fire or detonation. The STARFIRE steady-state tritium inventory is 11.6 kg or 2.8×10^4 Ci/MW$_{th}$.

Tritium has a high permeability in materials, especially at high temperature, and is therefore difficult to contain. It is estimated that about 10 Ci/day will accumulate in the primary cooling water. Relatively large releases of tritium are possible through system failures, accidents, or fires. The use of redundant systems and triple barriers, where practical, reduces the possibility of such releases.

Although the reactor structure within the shield becomes radioactive, the associated afterheat is small, being only from 1 to 2.2% of operating thermal power at shutdown. There is no credible scenario known today that could produce a nuclear runaway or meltdown in a fusion reactor. No mechanism that would cause melting of the STARFIRE structural material has been identified as being credible. The selection of materials was such as to minimize long-term radioactivity and to minimize rad-waste storage. The reactor is designed to be accessible within 24 hr after shutdown, even though remote operations are planned for all maintenance. Loss-of-coolant accident, resulting from a pipe break or mistaken valve closure, within the reactor could be serious and would force shutdown of the reactor for repair. It should not pose a hazard to the public. A helium leak in the superconducting magnets, which store about 50 GJ of energy, might drive the conductor normal with the possibility of melting in the coil. Plasma disruption could deposit up to 900 MJ of plasma energy onto the first wall, causing ablation, loss of ignition, and, at the worst, a failure in the first wall. Hydrogen is flammable in air (between 4 and 75 vol%) and can detonate (18 to 59 vol%). Detection and correction of any deuterium or tritium leaks should be prompt. The health hazards associated with fringing magnetic fields are negligible.

Potential hazards such as storms, earthquakes, or floods are common to all power plants, and a fusion plant must be designed to be safe in the event of such natural phenomena. The environmental impact of a fusion-power plant is similar, but in many aspects more benign, than coal or fission electric plants. The biological hazard potential for fusion systems is considerably less, by factors of at least 10 to 100, than for nuclear fission plants [86]. It is possible to use the excess fusion neutrons to make nuclear weapons materials, but that is easily detected and a relatively poor way to do it.

In summary, fusion has the potential, with good design, to be among the most safe of all large electric-power-plant systems and to be devoid of serious potential for the spread of nuclear weapons materials. It is relatively benign to the environment, and the safety risk to the public should be very small. Safety concern must be an integral part of the fusion-reactor design from its very inception.

6.16. ECONOMICS

To be an attractive energy source, fusion-plant performance and reliability must equal that of other major commercial energy systems. It also must be economically competitive with alternate energy systems. Although the absolute cost of any future power system can not be forecast precisely, relative costs have been studied. STARFIRE, NUWMAK, and other fusion-tokamak power plants have undergone cost analysis according to standard methods set by the Department of Energy, which have been used to evaluate different central power plants options [87]. The results indicate that fusion's cost of energy should be competitive with that from fission or coal power plants.

REFERENCES

1. L. Spitzer, Jr., D. J. Grove, W. E. Johnson, L. Tonks and W. F. Westendorp, "Problems of the Stellerator as a Useful Power Source," USAEX, Report N.Y.O.-6047, 1954 Project Matterhorn, Princeton, N.J., 1954.

2. R. F. Post, "Some Aspects of the Economics of Fusion Reactors," UCRL-6077, Office of Technical Services, U.S. Dept. of Commerce, 1960.

3. D. J. Rose, *Nuclear Fusion* **9**, 183 (1969).

4. R. G. Mills et al., "A Fusion Power Plant," MATT-1050, Aug. 1974, Princeton Plasma Laboratory.

5. F. Ribe, "Fusion Reactor Systems," *Rev. Mod. Phys.* **47**, 7 (1975).

6. R. Carruthers et al., "The Economic Generation of Power from Thermonuclear Fusion," Culham Laboratory Report, CLM-R-85, 1967.

7. G. Emmert et al. UWMAK-I, Wisconsin Tokamak Reactor Design UWFDM-68, 20 Nov. 1973.

8. D. Steiner, "The Technology Requirements for Power by Fusion," *Proc. IEEE* **63**, 1568 (1975).

9. *Proceedings of Topical Meetings on the Technology of Controlled Nuclear Fusion:* 1st Meeting, San Diego, CONF-740402-P1 (1974); 2nd Meeting, Richland, Wash. CONF 760935-P1 (1976); 3rd Meeting, Santa Fe, N.M. CONF 780508 (1978), 4th Meeting, King of Prussia, Pa. CONF-801011 (1980).

10. *IEEE Proceedings of Symposia on Engineering of Fusion Research:* 5th Symp., Princeton, NJ., IEEE Pub. No. 73CH0843-3-NPS (1973); 6th Symp., San Diego, CA, IEEE Pub. No. 75CH1097-5-NPS (1975); 7th Symp., Knoxville, Tenn., IEEE Pub. No. 77CH1267-4-NPS (1977); 8th Symp., San Francisco, CA, IEEE Pub. No. 79CH1441-5-NPS (1979).

11. European Proceedings of Symposia on Fusion Technology: 8th Symp., Noordwijkerhout, Netherlands, EUR 5182e (1974); 9th Symp., Garmisch-Partenkirchen, Germany, EUR 5602 (1976); 10th Symp., Padova, Italy, EUR 6215 (1978).

12. *Nuclear Technology/Fusion.* A journal of the American Nuclear Society and the European Nuclear Society, Vol. 1, 1981.

13. *Journal of Fusion Energy*, Plenum Press Vol. 1, Jan. 1981.

14. *Nuclear Engineering and Design*, Vol. 63, 1981.

15. M. Nozawa and D. Steiner, *An Assessment of the Power Balance in Fusion Reactors*, ORNL-TM-4221, 1974.

16. T. Kammash, *Fusion Reactor Physics: Principles and Technology*, Ann Arbor Science, Ann Arbor MI, 1975.

17. M. Abdou et al., *ANL Parametric System Studies*, ANL/FPP/TM-100, Nov. 1977.

18. E. Teller (ed.), *Fusion*, Academic, New York, 1981, see R. W. Conn, Chapter 14, p. 193, Vol. 1, Part B.

19. G. M. McCracken and P. E. Stott, "Plasma-Surface Interactions in Tokamaks," *Nuclear Fusion* **19**, 889 (1979).

20. "Proceedings of the Conference on Surface Effects in Controlled Thermonuclear Fusion Devices and Reactors," *J. Nucl. Mater.* **53** (1974); "Proceedings of the 2nd Conference on Surface Effects in Controlled Thermonuclear Fusion Devices," *J. Nucl. Mater.* **63** (1976); *Proceedings of the International Symposium on Plasma-Wall Interaction*, Julich, Pergamon, New York, 1977; "Proceedings of the 4th International Conference on Plasma-Surface Interactions in Controlled Fusion Devices," Garmish-Partenkirchen, *J. Nucl. Mater.* **93/94** (1980).

21. H. Niedermeyer et al., "Conditioning of Discharges in ASDEX," *J. Nucl. Mater.* **93/94**, 286 (1980).

22. I. Langmuir and L. Tonks, *Phys. Rev.* **34**, 876 (1929).

23. G. D. Hobbs and J. A. Wesson, *Plasma Phys.* **9**, 85 (1967).

24. A. E. Robson and P. C. Thonemann, *Proc. Phys. Soc.* **73**, 508 (1959).

25. J. Bohdansky, *J. Nucl. Mater.* **93/94**, 44 (1980); J. Roth et al., *Garching Report IPP 9/26*, May 1979.

26. H. Knoepfel and D. A. Spong, *Nuclear Fusion* **19**, 785 (1979).

27. R. W. Conn, *J. Nucl. Mater.* **76/77**, 103 (1978).

28. J. W. M. Paul et al., "Review of Results from DITE Tokamak," *Proceedings of the 8th European Conference on Controlled Fusion and Plasma Physics*, Prague, 1977, Vol. 2, p. 49. See also *Review of DITE Work*, Culham Report CLM-P597, 1979.

29. *Proceedings of the 8th International Conference on Plasma Physics and Controlled Nuclear Fusion Research*, IAEA, Brussels, 1980: D. Meade et al., PDX Experimental Results, Paper CN-38/X-1; M. Keilhacker, Impurity Control Experiments in the ASEEX Divertor Tokamak, Paper CN38/0-1.

30. *Starfire—A Commercial Tokamak Fusion Power Plant Study*, Argonne National Laboratory ANL/FPP-80-1, Sept. 1980, pp. 2–23.

31. W. B. Kunkel, "Neutral-Beam Injection," *Fusion*, Academic, New York, (1981); Vol. 1, Part B; Chap. 12, p. 103. E. Teller (ed.)

32. M. M. Menon, "Neutral Beam Heating Applications and Development." *Proc. IEEE* **69**, 1012 (1981); see also, ORNL preprint TM #209, Jan. 1981.

33. D. R. Sweetman, *Nuclear Fusion* **13**, 157 (1973).

34. J. A. Rome, D. G. McAlees, J. D. Callen, and R. H. Fowler, *Nuclear Fusion* **16**, 55 (1976).

35. G. Schilling et al., *Neutral Beam Injection Performance on the PLT and PDX Tokamaks*, PPPL-1728, Feb. 1981.

36. T. Ohkawa, Nuclear Fusion **10**, 185 (1975).

37. D. L. Jassby, *Nuclear Fusion* **17**, 309 (1977).

38. D. Q. Hwang and J. R. Wilson, "Radio Frequency Wave Applications in Magnetic Fusion Devices." *Proc. IEEE*, **69**, 1030 (1981).

39. M. Porkolab, in Reference 18, Vol. 1, Part B, Chap. 13, p. 151.

40. "Heating in Toroidal Plasmas," Proceedings of the 3rd Joint Grenoble–Varenna International Symposium, C. Gormezano, G. G. Leotta, and E. Sindoni (eds.), Pergamon Press, New York, 1982.

41. T. H. Stix. *The Theory of Plasma Waves*, McGraw-Hill, New York, 1962.

42. T. H. Stix. *Nuclear Fusion* **15**, 737 (1975).

43. J. C. Hosea et al., "Fast Wave Ion Cyclotron Heating in the Princeton Large Torus," *8th IAEA International Conference on Plasma Physics and Controlled Nuclear Fusion Research*, Brussels, Paper IAEA-Cn-38/D-5-1, July, 1980.

44. J. C. Hosea et al., "Fast Wave Heating in the Princeton Large Torus," *Proceedings Course and Workshop on Physics of Plasma Close to Thermonuclear Conditions*, Varenna, 1979 (to be published).

45. D. Q. Hwang et al., *Fast Wave Heating in the Two-Ion Hybrid Regime on PLT*, PPPL-1676, Jan. 1981.

46. Equipe TFR, "ICRF Heating in TFR 600," *8th IAEA International Conference on Plasma Physics and Controlled Nuclear Fusion Research*, Brussels, Paper IAEA-CN-38/D-3, July, 1980.

47. B. Badger et al., *NUWMAK, A Tokamak Reactor Design Study*, Nuclear Engineering Dept., University of Wisconsin, UWFDM-330, March 1979.

48. W. M. Stacey et al., *U.S. INTOR Contributions to the International Tokamak Reactor Workshop*, Nov. 1979.

49. J. J. Schuss et al., "Lower Hybrid Heating in the Alcator A Tokamak," *Nuclear Fusion* **21**, 427 (1981).

50. M. Brambilla, *Nuclear Fusion* **16**, 47 (1976).

51. L. Dupas, P. Grelot, F. Parlange, and J. Weisse, "Lower Hybrid Wave Penetration and Effects on Electron Population," *8th IAEA International Conference on Plasma Physics and Controlled Nuclear Fusion Research,* Brussels, Paper IAEA-CN-38/T-1-1, July 1980.

52. R. McWilliams and R. W. Motley, *Currents Generated by Lower-Hybrid Waves*, PPPL 1774, April 1981.

53. R. Cano, A. Cavallo, and H. Capes, "Electron Cyclotron Heating Versus Neutral-Beam Heating: A Comparison on a Tokamak Ignition Experiment," *Nuclear Fusion* **21**, 481 (1981).

54. R. M. Gilgenbach et al., "Heating at the Electron-Cyclotron Frequency in the ISX-B Tokamak," *Phys. Rev. Lett.* **44**, 647 (1980); see also ORNL/TM-7368.

55. J. M. Rawls (ed.), *Status of Tokamak Research*, DOE/ER-0034, October 1979.

56. C. T. Chang, L. W. Jørgensen, P. Nielsen, and L. L. Lengyel, "The Feasibility of Pellet Re-fueling of a Fusion Reactor," *Nuclear Fusion* **20**, 859 (1980).

57. S. L. Milora, "Review of Pellet Fueling," *J. Fusion Energy* **1**, 15 (1981).

58. Ya. I. Kolesnichenko, "The Role of Alpha Particles in Tokamak Reactors," *Nuclear Fusion* **20**, 727 (1980).

59. D. G. Jacobs, *Sources of Tritium and its Behavior upon Release to the Environment*, USAEC Critical Review Series, 1968, TID-24635.

60. W. F. Vogelsang, "Breeding Ratio, Inventory, and Doubling Time in a D-T Fusion Reactor," *Nucl. Tech.* **15**, 470 (1972).

61. M. A. Abdou and R. W. Conn, "A Comparative Study of Several Fusion Reactor Blanket Studies." *Nucl. Sci. Eng.* **55** 256 (1974).

62. J. L. Anderson, "Tritium Handling Requirements and Development for Fusion," *Proc. IEEE* **69**, 1069 (1981).

63. M. A. Abdou, *Radiation Considerations for Superconducting Magnets in a Fusion Reactor*, ANL/FPP/TM-92, Argonne National Laboratory Report, 1977.

64. R. D. Hay, "Superconducting Magnetic Electrical Insulations," *Proceedings of the Meeting on CTR Electrical Insulators*, May 17–19, 1976. LASL, CONF-760558, Feb. 1978.

65. W. W. Engle, Jr., *A Users Manual for ANISN*, K-1693, Oak Ridge Gaseous Diffusion Plant, 1967.

66. F. R. Mynatt et al., *The DOT-III Two-Dimensional Discrete Ordinates Transport Code*, Oak Ridge National Laboratory, ORNL-TM-4280 (1973).

67. D. B. Montgomery, "Magnet Development," *Proc. IEEE*, 69, 977 (1981).

68. J. File, D. S. Knutson, R. E. Marino, and G. H. Rappe, PPPL-1698, October 1980.

69. "Proceedings of the First Topical Conference on Fusion Reactor Materials," Miami Beach, Florida, F. W. Wiffen et al. (eds.), *J. Nuclear Materials*, Jan 29–31 (1979).

70. R. E. Gold, E. E. Bloom, F. W. Clinard, Jr., D. L. Smith, R. D. Stevenson, and W. G. Wolfer, "Materials Technology for Fusion; Current Status and Future Requirements," *Nucl. Tech/Fusion* **1**, 169 (1981).

71. J. Davis and G. L. Kulcinski. *Assessment of Titanium for Use in the 1st Wall/Blanket Structure of Fusion Power Plants*, EPRI—ER 386, 1977.

72. R. E. Nygren, "Materials Requirements and Development." *Proc. IEEE* **69**, 1056 (1981).

73. E. N. Cameron et al., *Minerals Resource Implications of a Tokamak Fusion Reactor Economy*, UWFDM-3/3, 1979.

74. R. H. Huddlestone and S. Leonard (eds.), *Plasma Diagnostic Techniques*, Academic Press, New York, 1965.

75. W. Lochte-Holtgreven (ed.), *Plasma Diagnostics*, North Holland Press, Amsterdam, 1968.

76. C. DeMichelis (ed.), Equipe TFR, "Tokamak Plasma Diagnostics," *Nuclear Fusion* **18**, 647 (1978).

77. W. R. Becraft et al., *The Instrumentation and Controls of an Ignited Tokamak*. ORNL/TM-7153, Oct. 1980.

78. D. A. DeFreece, G. M. Fuller, and L. M. Waganer, "The Impact of Neutron Wall Loading on the Cost of Tokamak Fusion Power Systems," *Proceedings of the 2nd Topical Meeting on the Technology of Controlled Nuclear Fusion*, ANS. Richland, Washington (Sept. 1976) CONF 760935-P1.

79. G. M. Fuller et al. *Developing Manualability for Tokamak Fusion Power Systems*, McDonnell Douglas Rept. C00-4184-6, Nov. 1978.

80. J. T. D. Mitchell, "Remote Operations and Fusion Reactor Design." Preprint of lecture, Ispra course on Engineering Aspects of Thermonuclear Fusion Reactors, Ispra, Varese, Italy. 23–27 June 1980.

81. J. T. D. Mitchell. "Blanket Replacement in Toroidal Fusion Reactors," *ANS 3rd Topical Meeting on Technology of Controlled Nuclear Fusion*, Santa Fe, New Mexico, 1978, p. 954.

82. *Fusion Reactor Remote Maintenance Study*, EPRI-ER-1046. Westinghouse Electric Co., April 1979.

83. W. Hafele, J. P. Holdren, G. Kessler, and G. L. Kulcinski, *Fusion and Fast Breeder Reactors*. International Institute for Applied Systems Analysis, A-2361 Laxenburg, Austria. Rev. July 197.

84. H. J. Willenberg (ed.), "Workshop Proceedings," *Safety and Environmental Aspects of D-T Fusion Power Plants*, EPRI-ER-821-WS, June 1978.

85. J. G. Crocker and S. Cohen, *Fusion Reactor Safety Research Program Annual Report FY-79*, U.S. DOE, Idaho National Engineering Lab, EG&G-2018, August 1980.

86. G. L. Kulcinski, *The Current Perception of the Environmental Features of Fusion Versus Fission after a Decade of Study*, UWFDM-338, December 1980.

87. S. C. Schulte et al., *Fusion Reactor Design Studies—Standard Accounts for Cost Estimates*, PNL 2648, May 1978; also PNL 2987, September 1979.

7 | STARFIRE: A COMMERCIAL TOKAMAK FUSION-POWER PLANT

7.1. INTRODUCTION

A comprehensive conceptual design of the STARFIRE reactor and balance-of-plant has been developed. The detailed analyses and description of the reference design for the power plant systems are given in Reference 1. The purpose of this chapter is to present an overview of the STARFIRE study and a brief description of the reference design.

The primary criteria for commercial attractiveness emphasized in the STARFIRE study are economics, safety, and environmental impact. The approach to meeting these criteria involved building upon experience from previous studies, developing additional innovative design concepts, and selecting features that simplify the engineering design and enhance reactor maintainability. Table 7.1 shows the key features of STARFIRE. The reactor is operated steady state with the plasma current maintained by lower-hybrid waves. This mode of operation results in a reduction in the plant capital cost and an increase in the plant availability. The capital cost savings are due to the

This chapter is taken directly from Reference 1, STARFIRE Design Report, and is reprinted here by the courtesy and permission of the Argonne National Laboratory. It is the most detailed and extensive study of what a tokamak commercial power plant may be like in the early 21st century, as perceived by the U.S. fusion community in 1980.

TABLE 7.1.   Key Features of STARFIRE

Steady-state plasma operation.
Lower-hybrid rf for plasma heating and current drive.
ECRH-assisted startup.
Limiter/vacuum system for plasma purity control and exhaust.
All superconducting EF coils outside TF coils.
Vacuum boundary at the shield; mechanical seals.
Total remote maintenance with modular design.
Water-cooled, solid tritium breeder blanket with stainless-steel structure.
All materials outside the blanket are recyclable within 30 yr.
Less than 0.5 kg of vulnerable tritium inventory.
Minimum radiation exposure to personnel.
Conventional water/steam power cycle with no intermediate coolant loop and no thermal-energy storage.

elimination of electrical and thermal energy storage, derating of power supplies, and the reduction in the reactor size, which is made possible by the increase in the permissible wall loading. The improvement in reactor availability is brought about by the increase in component reliability, elimination of material fatigue as a life-limiting effect in the first wall, and the reduction in the probability of plasma disruption occurrence. The reactor design is simplified by utilizing the lower-hybrid rf system, with its attractive engineering features, for the dual purpose of plasma heating and current drive. The problems associated with plasma initiation and startup have been eased by the use of electron-cyclotron resonance heating to reduce the OH voltage.

The limiter/vacuum system concept has been selected for the plasma impurity control and exhaust system. Compared to divertors, the limiter/vacuum system greatly simplifies the reactor design and improves its reliability and accessibility. Detailed analysis showed that the system can be designed to credible engineering standards.

The characteristics of the plasma operating point and the plasma support systems in STARFIRE are different from those in previous conceptual designs. The major differences are due to the choice of the steady-state operation and the limiter/vacuum system. These choices were motivated by the desire to simplify the engineering design. It was assumed in the early stages of the design that these options could be developed in the STARFIRE time frame. Fortunately, results from recent plasma physics experiments on noninductive current drive and on limiters are very encouraging, and they suggest that these options can be developed in the next few years.

A major effort has been devoted in STARFIRE to enhancing reactor maintainability and improving plant availability. The approach was to select design features and develop a design configuration that reduced the frequency of failure and shortened the replacement time. Relevant examples are: (1) steady-state operation with current drive; (2) limiter/vacuum system for impurity control and exhaust; (3) vacuum boundary located at the shield with all mechanical seals (no welds); (4) all service connections (e.g., for high-pressure coolant) are located outside the vacuum boundary (shield); (5) optimized modular design; (6) all superconducting EF coils are outside the TF coils; (7) conservative TF-coil design; (8) fully remote maintenance permitting some repairs during reactor operation; (9) "remove and replace" maintenance approach (failed parts are replaced with spare parts and the reactor is operated while repairs are made in the hot cell) that minimizes downtime; (10) combining components for simplicity (e.g., TF-coil room-temperature Dewar provides support for the EF coils and shield); and (11) providing redundancy of strategic components (e.g., for the EF coils trapped below the reactor). These features as well as potential future improvements in component reliability provide optimism that the plant availability goal of 75% can be achieved.

The safety and environmental considerations have played a major role in the STARFIRE design effort. A solid-tritium breeder was selected in preference to liquid lithium in order to minimize the stored chemical energy. The impurity control and exhaust system was selected and designed so that the tritium fractional burnup is maximized and the vulnerable tritium inventory in the fueling and vacuum pumping systems is minimized. Furthermore, the reactor design was developed to contain the tritium with multiple barriers and minimize the size of potential tritium releases. The shield was designed and all reactor materials selected to permit recycling of all materials outside the blanket in less than 30 yr. Radiation exposure of personnel has been minimized by the use of extensive remote maintenance operations and by providing adequate shielding. The use of resource-limited materials was avoided. Mechanisms for fast reactor shutdown and auxiliary cooling systems have been incorporated into the design. The beryllium coating on the first wall and limiter provide an inherent safety feature that terminates the plasma burn if the metal temperature reaches $\sim 900°C$. Calculations show that the reactor will be automatically shut down in less than 1 s if a hot spot forms on 10% of the first wall, without the need for any active control system. No major damage, other than some first-wall-coating ablation, will occur.

The use of water coolant, steam cycle, and conventional materials in STARFIRE makes the heat-transport and energy-conversion system a state-of-the-art technology. The balance of plant has been designed to maximize the utilization of current power-plant features. However, the reactor hall, hot cell, and tritium facility are unique to fusion reactors. The tritium facility utilizes current-day design practices of the Tritium Systems Test Assembly (TSTA). The reactor building houses the reactor and modules for auxiliary systems that may become contaminated.

The STARFIRE power plant design, shown in Figure 7.1, represents a single 1200-MW_e generating unit. The plant is part of a utility grid that normally uses the STARFIRE plant as a base-load unit; however, the plant can load follow at 5% of full load per minute.

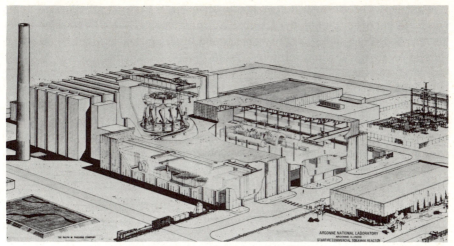

FIGURE 7.1.   STARFIRE plant site.

7.2.   KEY ASSUMPTIONS

The STARFIRE reactor design was developed using the assumption that it was the tenth commercial plant constructed from a standarized design. This assumption provides a highly predictable machine that will have resulted from an extensive R & D program and utility operating experience from earlier plants. The first year of operation is assumed to be approximately 2020. Although this date does not directly enter in the design considerations, it is useful in considering the availability of certain advanced technologies.

Remote maintenance of all equipment within the reactor building has been the design basis; however, provision for personnel entry into the reactor building, on a contingency basis, has also been included in the design. A goal of an overall plant availability of 75% is justified on the basis of a maintainable design, a first wall life of greater than 6 yr, and the assumption that STARFIRE is the tenth-of-a-kind power plant.

The major plasma physics assumptions and guidelines are discussed in detail in Reference 1 Chapters 6–8. Key assumptions include ion heat transport that is three times faster than the neoclassical value, electron heat transport that is given by the empirical Alcator scaling, and particle transport that is taken as twice as fast as the empirical energy-confinement time. Plasma cross-field transport at the edge is assumed to be Bohm diffusion with ion-sound-speed flow parallel to the field lines. The particle density and energy are assumed to fall exponentially in the scrape-off region. Neutral recycling at the plasma edge with some gas puffing is assumed to be an effective fueling mechanism. The plasma equilibrium with a somewhat hollow current profile is assumed to be stabilized by a conducting first wall for β's of at least 6.7%.

Lower-hybrid-wave-driven currents are calculated from standard quasilinear and Fokker–Planck theory. Linear theory is assumed when computing the transmission characteristics of the waveguide antenna and when performing the ray-tracing calculation in the plasma.

The economic analysis assumes private utility ownership and a single generating unit at the site. Certain economic advantages would result from locating multiple units at one site. The construction plan assumes siting near navigable water for transportation of large assemblies to the site. Otherwise, siting would be nearly universal except that minor modifications would be required for seismic zone 3. The plant lifetime for economic analyses was taken to be 30 yr; the engineering design lifetime was taken to be 40 yr.

7.3. REACTOR OVERVIEW

7.3.1. Reactor Configuration

The major reactor parameters for STARFIRE are listed in Table 7.2. These parameters were derived based on results from system analyses to minimize the cost of energy subject to constraints of physics, engineering, and technology. A discussion of these trade-off studies is given in Chapter 5 of Reference 1.

The reactor design has a major radius of 7.0 m and operates at a first-wall average neutron loading of 3.6 MW/m^2. The reactor delivers 1200 MW_e to the grid in addition to providing 240 MW_e for recirculating power requirements. The reactor operates with a continuous plasma burn and develops 4000 MW of useful thermal power. Approximately 3800 MW is provided to the main heat-transport system and 200 MW is collected from the active limiter for use in feed water heating. An isometric view of the reactor is shown in Figure 7.2. The reactor cross section is shown in Figure 7.3 and a top view is shown in Figure 7.4.

The reactor configuration utilizes 12 toroidal-field (TF) coils and 12 superconducting poloidal coils (EF and OH) located external to the TF coils. Additionally, four small normal conducting control coils (CF) are located inside the TF coils and outside the bulk shield to provide the necessary response time for plasma control while permitting good access for reactor maintenance. The magnet systems and shield are expected to last the full 40-yr design life under normal operating conditions; however, provisions are incorporated for their replacement. Blanket sectors, including the limiters and rf ducts, will require replacement every six years. Vacuum pumps and the isolation vacuum valves will require replacement every two years.

The reactor configuration was developed to permit each component of the reactor to be replaced in a time consistent with its anticipated life (i.e., rapid replacement for components with short life) using remote-maintenance techniques. Emphasis has been placed on overall reactor simplicity. This has led to combining components, where possible, to minimize the number of replace-

TABLE 7.2.   STARFIRE Major Design Parameters

Net electrical power, MW	1200
Gross electrical power, MW	1440
Fusion power, MW	3510
Thermal power, MW	4000
Gross turbine cycle efficiency, %	35.7
Overall availability, %	75
Average neutron wall load, MW/m^2	3.6
Major radius, m	7.0
Plasma half-width, m	1.94
Plasma elongation, b/a	1.6
Plasma current, MA	10.1
Average toroidal beta	0.067
Toroidal field on axis, T	5.8
Maximum toroidal field, T	11.1
Number of TF coils	12
Plasma burn mode	Continuous
Current-drive method	rf (lower hybrid)
Plasma-heating method	rf (lower hybrid)
Plasma startup	ECRH-assist
TF coil material	$Nb_3Sn/NbTi/Cu/SS$
Blanket structural material	PCA[a]
Tritium breeding medium	Solid breeder (α-LiAlO$_2$)
Wall/blanket coolant	Pressurized water (H$_2$O)
Plasma impurity control	Limiter and vacuum system supplemented by low-Z coating, enhanced radiation, and field margin
Primary vacuum boundary	Inner edge of shield

[a]Primary Candidate Alloy (PCA), an advanced austenitic stainless steel.

able parts and to efficiently use the materials and space of various components. Modularization has also been emphasized so that all reactor components can be removed and replaced in a simple and practical manner. The remove and replace philosophy permits the reactor to resume operation while time-consuming repairs are made in the hot cell where more time and equipment are available.

The TF coil, and hence the reactor configuration, was developed primarily with the desire to keep the superconducting EF coils external to the superconducting TF coils so that their replacement is possible without fabrication of a new coil on the reactor. External placement of the EF coils increases the incentive to reduce the TF-coil size to minimize the stored energy of the EF system. The TF-coil outer radius is constrained to 13 m by the clearance required for shield installation.

The TF coils are joined into a common vacuum tank at the center of the reactor. The common vacuum tank was chosen to minimize the heat leak to and

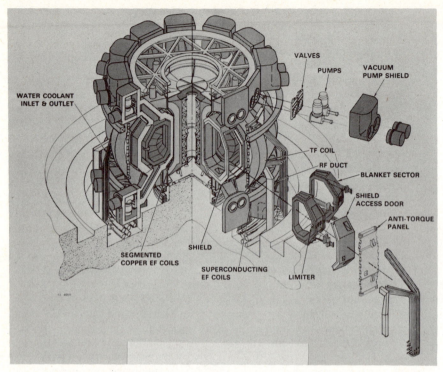

FIGURE 7.2.   STARFIRE reactor design.

the shield thickness at the magnet inner leg. The common TF-coil vacuum tank also provides the vacuum boundary for the center post and EF/OH coils in the center region. Since the common vacuum tank does not contain a dielectric break, care was taken to minimize its thickness and hence the image currents in the vacuum tank so that the CF coils could properly control the plasma.

The EF/OH coils inside the center post are grouped in two modules to simplify their removal from the top of the reactor without significantly increasing the overhead crane or building height. The outer EF coils and upper EF/OH coil can be removed vertically. Spares have been provided for the lower EF/OH coils that are trapped under the reactor because the inherent complications of replacing a failed coil, even if only once in every few plant lifetimes, make it cost effective.

The shield is assembled as 24 sectors to permit its installation between TF coils. The 12 sectors that fit under the TF coils also incorporate dielectric breaks in every other sector. The other sectors incorporate an access door and two vacuum ducts. The sectors are joined together by a welded vacuum seal and are not expected to require frequent replacement (i.e., they are life-of-plant components).

The vacuum-boundary location was selected at the shield interior with access door seals located at the outer surface in order to (1) provide a conve-

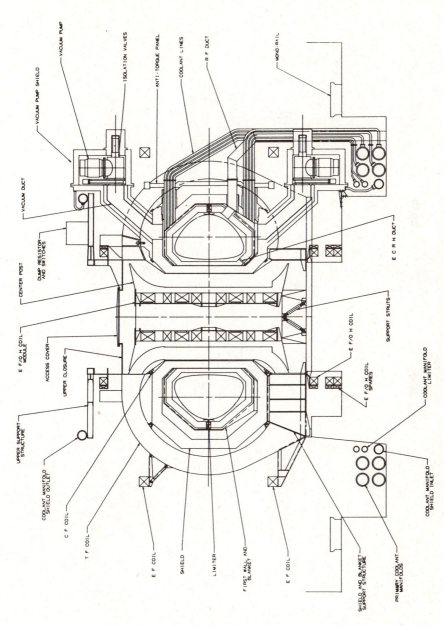

VACUUM PUMP

VACUUM PUMP SHIELD

ISOLATION VALVES

ANTI-TORQUE PANEL

COOLANT LINES

R F DUCT

MONO-RAIL

VACUUM DUCT

CENTER POST

DUMP RESISTOR AND SWITCHES

E F/O H COIL MODULE

ACCESS COVER

UPPER CLOSURE

UPPER SUPPORT STRUCTURE

COOLANT MANIFOLD SHIELD OUTLET

C F COIL

T F COIL

E F COIL

SHIELD

LIMITER

FIRST WALL AND BLANKET

E F COIL

SHIELD AND BLANKET SUPPORT STRUCTURE

PRIMARY COOLANT MANIFOLDS

COOLANT MANIFOLD SHIELD INLET

COOLANT MANIFOLD LIMITER

E F/O H COIL SPARES

E F/O H COIL

SUPPORT STRUTS

E C R H DUCT

FIGURE 7.3. STARFIRE commercial tokamak—cross section.

191

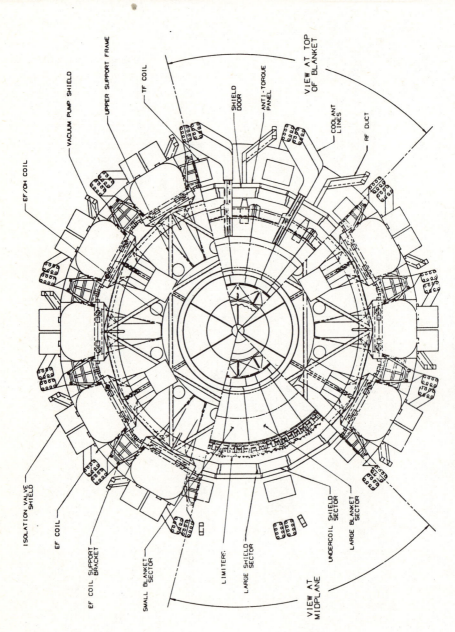

EF/OH COIL

VACUUM PUMP SHIELD

UPPER SUPPORT FRAME

TF COIL

SHIELD DOOR

ANTI-TORQUE PANEL

VIEW AT TOP OF BLANKET

COOLANT LINES

RF DUCT

ISOLATION VALVE SHIELD

EF COIL

EF COIL SUPPORT BRACKET

SMALL BLANKET SECTOR

LIMITERS

LARGE SHIELD SECTOR

UNDERCOIL SHIELD SECTOR

LARGE BLANKET SECTOR

VIEW AT MIDPLANE

FIGURE 7.4. STARFIRE commercial tokamak—top view.

192

nient way of pumping for the limiter slot system, (2) minimize the complexities of providing a vacuum boundary at the blanket/first wall and (3) permit the inboard vacuum seals, which have limited access, to remain intact during maintenance. The vacuum seals that must be opened for maintenance were located at the outer shield surface to provide access for maintenance and to reduce the damage to seal materials by radiation exposure. The shielding is effective enough to permit use of elastomer seals, which can be sealed repeatably and easily. The vacuum pumps were located at the top and bottom of the reactor to minimize the neutron heating on the cryopanels and to permit the pumps to remain in-place during blanket replacement.

The blanket was divided into large sectors to permit replacement with a minimum number of in-reactor maintenance actions. Twenty-four torodial sectors of two different sector sizes are used to permit installation in the space between adjacent TF coils. The overall blanket installation was simplified by mounting the limiter, rf duct, and ECRH duct to the sector for removal as a unit. Coolant connections to the blanket sector were located outside of the vacuum boundary to minimize the effects of irradiation on the joint and to permit use of less than high-integrity "leak-tight" mechanical seals. The penetration through the vacuum boundary is sealed with elastomer seals located at the external shield surface.

The limiter consists of 96 elements that form a near continuous toroidal ring at the outer midplane of the blanket. Four limiters are mounted on each blanket sector in front of a slot through the blanket that provides a path for particles to a plenum. Particles are then pumped by 24 vacuum pumps at the top and bottom of the reactor. An additional 24 vacuum pumps are provided to permit pump rejuvenation every 2 h.

Twelve rf ducts provide for heating and current drive of the plasma. These ducts are mounted in the blanket sector located between TF coils. An rf window and phase monitor are located in the duct near the shield while phase shifters, circulators, and crossed-field amplifiers are located in the reactor building basement where personnel access during operation is possible. Twenty-four ECRH ducts are provided for initial plasma breakdown and wall cleaning. Two vacuum ducts are located on each blanket sector between TF coils.

Fuel is provided to the reactor by extracting bred tritium from a solid breeding blanket and injecting it into the plasma via gas puffing. Two gas ports are provided. Gas enters the plasma through the limiter, which incorporates a drilled passage to the innermost protrusion of the plasma at the outer blanket midplane.

7.3.2. Plasma Engineering

STARFIRE employs a D–T burning, dee-shaped plasma to produce 3510 MW of fusion power. The plasma is operated at a moderate β of 6.7% and is moderately elongated, with a height to width ratio of 1.6. The major plasma parameters and plasma engineering features of STARFIRE are listed in Tables 7.3 and 7.4, respectively. The plasma current is driven in steady state with 90 MW of lower-hybrid rf power. The first wall and all other components in the vacuum chamber are coated with Be. The impurity-control system maintains a steady-state concentration of 14% He and 4% Be in the plasma. The fairly low D–T removal efficiency (15%) of the impurity-control system permits a high fractional burnup of tritium. For the same reason, most of the plasma fueling is done automatically by D–T neutrals recycling from the first wall and limiter. Additional fueling is done by gas puffing.

In order to minimize the heat-transport load on the limiter, as well as to establish a thermal equilibrium, the plasma is operated in an "enhanced-radiation" mode, whereby a small amount of high-Z material, nominally iodine, is added along with the fuel stream. This serves to radiate most of the heating energy and stabilizes the thermal operating point.

The plasma magnetohydrodynamic(mhd) equilibrium is of the low-current, hollow profile type. The plasma position is controlled with two sets of coils, a main equilibrium-field (EF) coil set and a control-field (CF) coil set. The main EF coils are superconducting and are located outside of the TF coils. They are

TABLE 7.3.   STARFIRE Plasma Parameters

Parameter	Unit	Value
Major radius, R	m	7.0
Aspect ratio, A	—	3.6
Elongation, $\varkappa$	—	1.6
Triangularity, d	—	0.5
Safety factor at limiter	—	5.1
Average beta, β		0.067
Maximum toroidal field at coil, B_M	T	11.1
Toroidal field at plasma center, B_0	T	5.8
Plasma current, I_p	MA	10.1
Plasma volume, V_p	m^3	781
Average electron temperature, T_e	keV	17.3
Centerline electron temperature, T_{e0}	keV	22.5
Average ion temperature, T_i	keV	24.1
Centerline ion temperature, T_{i0}	keV	31.3
Average fuel density, N_{DT}	m^{-3}	0.806×10^{20}
Center fuel density, N_{DT}	m^{-3}	1.69×10^{20}
Electron-energy-confinement time, τ_E	s	3.6
Ion-energy-confinement time, τ_I	s	10
Particle confinement time, τ_p	s	1.8
Fractional helium concentration, N_α/N_{DT}	—	0.14
Fractional beryllium concentration, N_{Be}/N_{DT}	—	0.04
Fractional iodine concentration N_I/N_{DT}	—	0.001
Fusion power, P_F	MW	3510
Lower-hybrid rf power to plasma, P_{rf}	MW	90
Average neutron wall load, P_{WN}	MW/m^2	3.6

TABLE 7.4. Plasma Engineering Features of STARFIRE

Operating Point

Equilibrium type:	Elongated, dee-shape, moderate β, hollow current profile.
Equilibrium generation method:	Outside superconducting-equilibrium-field-coil system.
Position stabilization method:	Inside control-field coils and conducting first wall with 300-ms time constant.

Burn Cycle

Startup time:	~ 24 min.
Method:	Tritium lean startup; vary rf power, D−T density, T fraction; 5%/min fusion-power ramp
Normal shutdown time:	~ 24 min.
Emergency shutdown:	Induced disruption method, time < 3 s.
Plasma initiation method:	5-MW electron-cyclotron-resonance heating.
Burn method:	Steady-state, lower-hybrid current drive.
Thermal stabilization:	Enhanced-radiation-mode operation by iodine injection.

Fueling

Fueling method:	Recycling D−T plus gas puffing.

used to provide the basic positional equilibria. The CF coils consist of small copper coils inside the TF coils and are used to control position and to stabilize against plasma disruptions. To further aid in the latter task, the first wall is designed with a time constant of 300 ms to stabilize against rapid vertical instabilities.

Because of the need to minimize the lower-hybrid rf current-drive power, the plasma density is lower and the plasma ion temperature is higher than most previous tokamak reactor designs. The plasma is operated with $T_i > T_e$, which makes better use of the available β. The tradeoffs, between the operating point and rf power, and the selection of density and temperature parameters, are discussed in detail in Chapter 7 of Reference 1.

Most of the STARFIRE burn cycle is substantially different from pulsed-reactor burn cycles. Plasma breakdown is done with 5 MW of electron-cyclotron-resonance heating (ECRH) and does not require a high-voltage OH coil. The startup period takes 24 min and conforms to the requirement that the fusion power should be ramped at a 5%/min rate, to minimize thermal problems in the energy recovery and conversion systems. The OH coil as well as the OH and EF power supplies have modest requirements compared to pulsed-reactor requirements. The steady-state burn phase of the burn cycle has a thermal equilibrium maintained by the addition of iodine. The equilibrium and stability of the enhanced-radiation mode of operation has been studied with a global code and with the one-dimensional WHIST code. These studies indicate that this mode of operation is feasible.

Several types of shutdown scenarios have been developed for STARFIRE. The normal shutdown is basically the reverse of the startup period, whereby the fusion power is reduced at a 5%/min rate by reducing the tritium fraction in the plasma. There are three types of emergency shutdowns. The fastest is an "abrupt" shutdown whereby a plasma disruption is induced by injecting excess high-Z material. There is a more orderly "rapid" shutdown which also uses a disruption, but where most of the plasma energy is radiated away prior to the disruption. Finally, a naturally occurring "ablative-induced shutdown" has been identified, which occurs as result of a hot-spot formation on the first wall or limiter.

Various fueling options for STARFIRE were studied. The high fractional burnup rate of 42% in STARFIRE permits a fairly low fueling rate from an external source. Gas puffing is the most desirable engineering option and has been adopted as the STARFIRE fueling method.

We have considered dee-shaped plasma equilibria in toroidal geometry with pressure profiles characterized by a width parameter α. Using a diamagnetism function $F^2(\psi) = F_0(1 - \delta\hat{\psi}^\gamma)$ and a pressure function $p(\psi) = \hat{\psi}^\alpha$, it is possible to generate a broad range of low- and high-β equilibria with various axis and limiter values of the safety factor, q_a and q_b, respectively. The highest stable β is, in general, a function of A, $\varkappa$, d, α, q_a, and q_b, where A is the aspect ratio, $\varkappa$ is the elongation, and d determines the triangularity. This functional dependence is under active investigation in the physics community, and the operating point for STARFIRE is based on a survey of equilibria and subsequent stability analysis.

7.3.3. Plasma Heating and Current Drive

The design of a tokamak reactor that can run in a steady-state mode is basically different from the design of a pulsed tokamak, because the circulating power required to sustain the torodial current against collisional dissipation may be a substantial fraction of the power plant's electric output. Consequently, the STARFIRE design focused on efforts to minimize the circulating electric power for steady-state operation, and the resulting lower-hybrid rf system was optimized with this goal. In addition, the same system appears adequate to provide auxiliary heating during the startup phase to bring the plasma to ignition temperatures.

One obvious means of reducing rf power to the reactor is the selection of operating regimes with the lowest plasma currents. Accordingly, a large variety of mhd equilibria (solutions to the Grad–Shafranov pressure-balance equation) were surveyed in order to identify the most desirable candidate. The aspect ratio A was selected after studying the power requirements for generating equilibria with $\beta_\phi = 0.25/A$. For a fixed reactor power, wall loading, and toroidal field strength, the plasma current and electron density increase as A is reduced. This leads to increasing rf power at lower A. However, a larger A requires a larger major radius. This study concluded that $A = 3.6$ is best, with $R = 7.0$ m.

The selection of the plasma beta was another crucial decision. Generally, the higher the plasma current, the higher is the stable β_ϕ that can be achieved. The increasing rf power required at higher β_ϕ motivated the selection of a comparatively modest design value—$\beta_\phi = 6.7\%$. It was shown that hollow current density profiles can have favorable stability while requiring less total current than more conventional centrally peaked profiles.

An elongated ($\varkappa = 1.6$), highly dee-shaped plasma cross section was chosen, which requires $I = 10.1$ MA with $B_0 = 5.8$ T on axis. The plasma cross section and lower-hybrid system are shown in Fig. 7.5. The design of the equilibrium-field (EF) coil system required to position and shape the poloidal flux surfaces is eased by the ability to locate large coils in the central hole. The safety-factor profile does not allow double-tearing modes, although other resistive modes have not been ruled out. Analysis with the BLOON and ERATO codes shows the equilibrium is stable to local interchange and ballooning modes but requires a close-fitting, conducting first wall/blanket to stabilize low-n kinks.

Lower-hybrid current-drive theory shows that the rf driving power is proportional to the local electron density where the current is generated, which

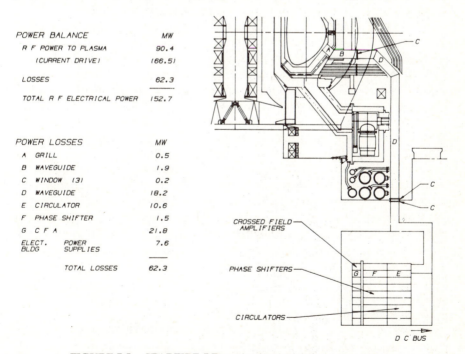

POWER BALANCE	MW
R F POWER TO PLASMA	90.4
(CURRENT DRIVE)	(66.5)
LOSSES	62.3
TOTAL R F ELECTRICAL POWER	152.7

POWER LOSSES		MW
A	GRILL	0.5
B	WAVEGUIDE	1.9
C	WINDOW (3)	0.2
D	WAVEGUIDE	18.2
E	CIRCULATOR	10.6
F	PHASE SHIFTER	1.5
G	C F A	21.8
ELECT. BLDG	POWER SUPPLIES	7.6
	TOTAL LOSSES	62.3

FIGURE 7.5. STARFIRE RF system showing plasma cross section.

makes the hollow current profiles especially attractive. In addition, for a fixed β, the average electron density $(\bar{n}_e)$ may be reduced by operating the plasma at a higher temperature. Above 20 keV, the decreasing fusion reactivity of D–T tends to offset these reductions in the rf power at low n_e. (Maximum Q occurs in the range 20–30 keV.) Despite low-Q operation, the net electric output peaks for $T_e \simeq T_i \approx 11$ keV. However, the capital outlay for rf power supplies at 11 keV far exceeds that needed for auxiliary heating to ignition. The minimum cost of electricity appears when $T_e = 17$ keV, $T_i = 24$ keV, and $n_e = 1.2 \times 10^{20}$ m^{-3}, which results in a fusion power of 3510 MW.

Using the WKB form of the electromagnetic dispersion relation, the Landau damping of externally launched lower-hybrid waves is computed in cylindrical geometry, and current generation is calculated from standard quasilinear/ Fokker–Planck formulas. By minimizing the parallel wave index of refraction, $n_\parallel$, and adjusting the spectrum width, $n_{\parallel 1} - n_{\parallel 2}$, the equilibrium current density may be generated with a minimum amount of power. Wave accessibility (to avoid reflection) limits the lower bound on $n_{\parallel 2}$ to 1.40 in STARFIRE; setting $n_{\parallel i} - n_{\parallel 2} = 0.46$ yields a hollow profile with a total parallel ("force-free") current of $I_\parallel = 9.1$ MA. In equilibrium, owing to the field-line rotational transform, diamagnetic currents add another toroidal current-density increment, with the result that 66.5 MW of driving power creates 10.1 MA of toroidal current. In these calculations, we have set the wave frequency at $\nu = 1.677$ GHz, which exceeds the lower-hybrid frequency at the point of deepest wave penetration and thus avoids parametric instabilities.

The Brambilla grill, a horizontal array of phased waveguides, is the ideal wave launcher in the rf frequency range. The waveguides have a vertical opening of 17.0 cm and transmit only the TE$_{10}$ mode. The horizontal opening is 2.95 cm, and the guides are separated by 0.70-cm metal partitions. Phasing adjacent guides by $2\pi/3$ results in an asymmetric spectrum of toroidally traveling waves with average $n_\parallel = 1.63$. We find that a large number ($N = 18$) of guides is necessary to concentrate the spectral power into the range $n_{\parallel 1} - n_{\parallel 2} = 0.46$. The grill performance has been studied for a variety of electron-edge-density profiles, since this determines how well the grill couples to plasma waves. Two figures of merit appear. One is the average grill reflection coefficient, $\bar{R}$; the other is the fraction of transmitted power ε, which is contained in the region $n_\parallel = 1.40–1.86$ and which is thus useful for driving toroidal current. Wave interference results in some power transmission at $n_\parallel = -3.2$, but, if reasonable control of the electron-edge density is possible, ε will be in the range of 0.6–0.8. The overall $\bar{R}$ has been dramatically reduced in the STARFIRE design by replacing those waveguides with high individual reflection coefficients with passive reflectors. The final design has nine active and nine passive guides, which results in $\bar{R} = 0.44$ and $\varepsilon = 0.74$. The total transmitted power to the plasma is thus 66.5 MW/ε = 90 MW.

In view of the large $\bar{R}$ of the grill, it is necessary to circulate the reflected power in a given waveguide directly into its neighboring guide. Thus a klystron drives the first of the nine active elements in an array, and circulators direct returning power through the other eight. Crossed-field amplifiers (CFAs) are used to boost the reflected power to the required output level. Owing to high reflection coefficients, the required amplifier gains are low, in the range 3–7 db. Present-day CFAs have achieved impressive efficiency in CW operation at low gains in this frequency range ($\eta_{CFA} = 78\%$ at 400 kW). We are confident that $\eta_{CFA} = 85\%$ is achievable with a modest R and D effort, and we have assumed this value in designing the STARFIRE system. Each CFA operates at 420 kW of rf output and requires a 20–30 kV power supply at 10–15 A, depending on the gain. Unlike klystrons, the CFA does not require a highly regulated power supply. The CFA may be operated with a cold cathode, eliminating costly heater power supplies and producing long-life operation (20,000 h or 3 calendar years of STARFIRE operation).

Each waveguide facing the plasma is plated with Cu and Be to present a low-sputtering, high-conductivity surface. The wave intensity is low—1.4 kW/cm^2 assuming 90 MW broadcast by a total launcher area of 6.4 m^2—to avoid nonlinear spectral modification; the electric fields are consequently well below the multipactor limit. A BeO window and dc break are located between the outboard legs of the toroidal-field coils, and the electron-cyclotron-resonance region beyond is pressurized. Directional couplers monitor the phase on the vacuum side of each window to ensure proper grill operation. Neutron damage to the window is negligible over the reactor lifetime, the fluence being less than 8×10^{11} n/cm^2. The pressurized waveguide is routed through the building floor to a basement area (see Fig. 7.5) where the circulators, phase shifters, and CFAs are located. Power supplies are in an auxiliary building. Inside the reactor hall, the rf equipment is passive, requiring little maintenance, while the CFAs and phase shifters are accessible during reactor operation if repair should prove necessary. By stacking four grills poloidally in each of the 12 reactor toroidal sectors, a system with 432 transmission lines and CFAs is achieved, with an input electric power of 153 MW (compared with 1200 MW net electric supplied to the grid). The redundancy in this system ensures continued reactor operation in the event of single component failures. The total system cost, assuming prior amortization of an R&D program for CFAs and assuming mass production of the tubes, is $33.5 million for the hardware, exclusive of power supplies. Power supply costs are $14 million. This study concluded that a CFA system can provide lower-hybrid power for current-drive applications with much less circulating power than klystrons (which would require over 200 MW electric input) and at a fraction of the cost of a klystron system. The rf system is summarized in Table 7.5.

With minor modifications from this basic design, 3 of the 12 rf ducts can serve to heat and drive current during the startup phase of the tokamak discharge. These ducts contain grills with 36 waveguide elements with narrower openings. By appropriate phase control, these grills can heat the plasma at temperatures as low as $T_e = 0.5$ keV and can drive currents once T_e reaches ~ 1.5 keV. A small OH coil and ECR heating serve to bring the discharge up to the temperatures at which the lower-hybrid system functions. Beyond this

TABLE 7.5. Lower-Hybrid System Parameters[a]

Parameter	Symbol	Value
Wave frequency	ν	1.677 GHz
Spectrum required	$n_{\parallel 1} - n_{\parallel 2}$	1.40–1.86
Toroidal wavelength	$\lambda_{\parallel}$	10.94 cm
Parallel phase speed	ω/k	1.83×10^8 m/s
Parallel current	$I_{\parallel}$	9.11 MA
Toroidal current	I_p	10.1 MA
Wave intensity at antenna	I	1.6 kW/cm^2
Nine active/nine passive guides	N	$9 + 9 = 18$
Phase difference	$\Delta\phi$	$2\pi/3$
Narrow guide opening	b	2.95 cm
Septum	d	0.70 cm
Vertical guide opening	a	17.0 cm
Edge-density gradient	dn_e/dx	5.0×10^{10}cm^{-4}
Vacuum distance	x_0	1.0 cm
Spectral fraction driving current	ε	0.736
Antenna reflection coefficient	$\overline{R'}$	0.443
Power required to drive current	P_{LH}	66 MW
Net heating power to plasma	P_h	90 MW
Net electric power to rf system	P_e	152.7 MW
CFA efficiency	η_{CFA}	85%

[a]The rf system requires 432 CFA tubes operating at 420 kW with gains from 4.0 to 6.6 db.

point, we have demonstrated that there is a smooth evolution of the discharge toward the steady state.

7.3.4.   Plasma Impurity Control and Exhaust

A plasma impurity-control and exhaust system was developed for STARFIRE to satisfy the following goals: (1) engineering simplicity compatible with ease of assembly/disassembly and maintenance; (2) a high tritium burnup to minimize the tritium inventory in the fuel cycle; (3) a reasonable and reliable vacuum system that minimizes the number and size of vacuum ducts; and (4) manageable heat loads in the medium where the alpha and impurity particles are collected.

These goals are found to be best satisfied by a limiter/vacuum system together with a beryllium coating on the first wall, limiter, and all other surfaces exposed to the plasma. In order to minimize the heat load to the limiter, most of the alpha-heating power to the plasma is radiated to the first wall, by injecting a small amount of high-Z material, for example, iodine, along with the D–T fuel stream. The iodine atoms enhance the line-and-recombination radiation over most of the plasma volume. The helium removal efficiency of the limiter/vacuum system is intentionally kept low for three reasons: (1) to reduce the heat load on the limiter; (2) to simplify the vacuum system and reduce radiation streaming; and (3) to minimize the tritium inventory tied up in the vacuum and tritium processing systems. The major features of the STAR-FIRE impurity control and exhaust system are summarized in Table 7.6.

Figure 7.6 shows a cross section through the limiter, the limiter slot, the limiter duct, the plenum region. The limiter consists of 96 segments that form one toroidal ring centered at the midplane and positioned at the outer side of the plasma chamber. This location was selected because (1) it is the least likely place for a thermal energy dump from a plasma disruption and (2) it ensures symmetry in particle and heat load on the upper and lower branches of the limiter. Each of the limiter segments is 1 m high and 0.6 m wide. The physical dimensions of the system are shown in Fig. 7.6. The limiter slot, which is the region between the limiter and first wall, leads to a 0.4 m high limiter duct that penetrates the 0.7 m thick blanket. The limiter duct opens into a plenum region that is located between the blanket and shield and extends all the way around the torus. This plenum region is large enough so that it spreads the radiation leakage from the limiter duct into a larger surface area of the bulk shield. The conductance of the plenum region is large enough to permit locating the vacuum ducts in the bulk shield sufficiently removed from the midplane so that radiation streaming from the limiter duct in the blanket to the

TABLE 7.6.   Major Features of STARFIRE Impurity Control/Exhaust

A limiter/vacuum system
    One toroidal belt-type limiter centered around midplane
    Simplified, credible engineering
Low-Z coating (beryllium) on all surfaces exposed to plasma
Enhance plasma radiation
    To reduce heat load at collection plate
    By injecting small amount of iodine
A low helium removal efficiency (25%)
    Much simpler vacuum system
    Less radiation streaming
    High tritium burnup, low tritium inventory
    Penalty: Modest increase in toroidal field (0.85 T on axis)
Simple vacuum system
    Limiter duct penetrates blanket leading to a plenum region between blanket and
      bulk shield
    Significantly reduced radiation streaming; less shielding and lower nuclear heat load
      in cryopanels

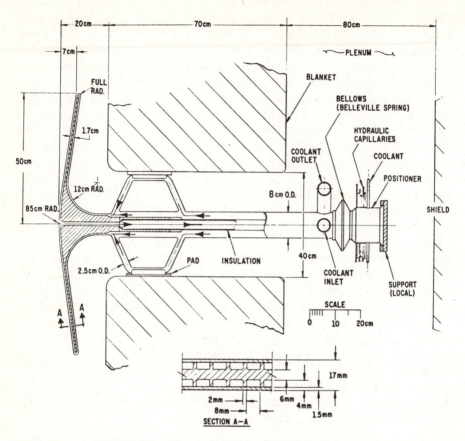

FIGURE 7.6. Cross section of the STARFIRE limiter design.

vacuum pumps is acceptable. There are 12 vacuum ducts at the top and another 12 at the bottom of the reactor. Each of these vacuum ducts has an equivalent diameter of 1 m and penetrates the bulk shield leading to the vacuum pumps. Fig. 7.7 shows an isometric view of the limiter system.

The basic principles of how the limiter works are rather simple. Ions that hit the front face of the limiter will be neutralized and reflected back into the plasma. Ions that enter into the limiter slot hit the back surface and are neutralized. Some of the scattered neutrals will directly reach the limiter duct and follow a multiple-scattering path into the plenum region and into the vacuum ducts where they are pumped out by the vacuum pumps. Other particles neutralized at the back surface of the limiter will scatter back in the direction of the plasma. These neutrals have a high probability of being ionized and returned back to the limiter surface. Calculations show that this trapping or "inversion" effect is so large for helium that ~ 90% of the helium entering the

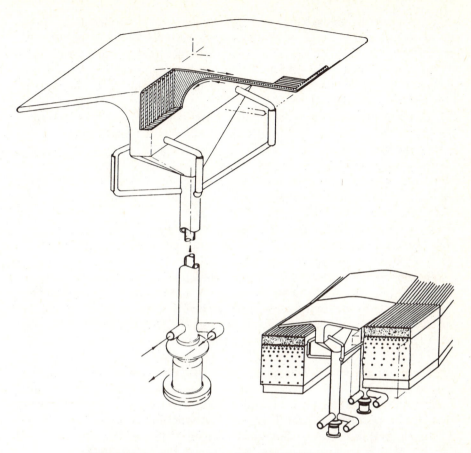

FIGURE 7.7.   Isometric view of the STARFIRE limiter design.

limiter slot will be pumped. This inversion effect greatly simplifies the limiter/vacuum system design in at least two ways:

1. *Location of the Leading Edge.*   Since the inversion probability is very high, the fraction of particles that enter the limiter slot need to be only slightly greater than the helium removal efficiency. This permits locating the two leading edges at the top and bottom of the limiter sufficiently away from the plasma edge and outward into the scrape-off region so that the peak heat flux at the leading edge is reasonably low.

2. *Neutral Pressure.*   This inversion effect causes the neutral-gas pressure at the limiter duct to be considerably higher than the neutral pressure around the plasma. Such high pressure greatly simplifies the vacuum system design.

Hydrogen can charge-exchange as well as be ionized. These charge-exchange events significantly reduce the inversion probability for hydrogen because the resulting neutral will tend to make its way out of the slot region into the plasma. Therefore, the beneficial effect of higher helium pumping probability and enhanced hydrogen recycling into the plasma is obtainable in the limiter/ vacuum system.

The plasma parameters related to the impurity control and exhaust system are shown in Table 7.7. The design parameters for the limiter and vacuum system are shown in Tables 7.8 and 7.9, respectively.

The charged particle flux in the scrape-off region falls off exponentially as e^{-x/δ_p} with $\delta_p = 10$ cm and x being the distance into the scrape-off region. About 28% of the helium particles diffusing out of the plasma will enter into the limiter slot, that is, between $x = 8.7$ cm and $x = 20$ cm. The transmission (pumping) probability for these particles is 0.9 giving an overall helium reflection coefficient $R_\alpha = 0.75$ and helium removal efficiency $(1 - R_\alpha)$ of 0.25. The reflection coefficient for deuterium/tritium is 0.9.

The transport heat flux on the limiter (due to charged particles) varies as $e^{-x/\delta_E} \sin \theta$ in units of MW/m with x in cm, $\delta_E = 5$ cm, and θ being the angle between the field lines in the poloidal plane (nearly vertical in Fig. 7.6) and the surface of the limiter. The limiter surface from the tip (at plasma edge) to the top (or bottom) leading edge is slanted to spread the heat load. The leading edge (region where $\theta = 0$ occurs) forms approximately one-half a cylinder shell with a diameter of 1.7 cm and extends from $x \sim 7$ cm to $x \sim 8.7$ cm. The location of the leading edge was determined from tradeoffs between the helium removal efficiency (and the associated toroidal-field margin) and the peak heat flux. The peak transport heat flux is ~ 3.4 MW/m^2 and occurs at $x = 7.8$ cm. The average transport heat flux on the region of the leading edge is ~ 2.2 MW/m^2. The front surface of the limiter (from the tip of the leading edge) receives an additional surface heat load of 0.9 MW/m^2 due to plasma radiation

TABLE 7.7.   Plasma-Related Parameters for the Limiter/Vacuum System

Fusion alpha power (P_α), MW	703
Lower-hybrid power to plasma (P_{LH}), MW	90
Transport power to the limiter, MW	90
Helium production rate, s^{-1}	1.24×10^{21}
Alpha-particle concentration (n_α/n_{DT})	0.14
Beryllium (low-Z coating) concentration ($n_{\text{Be}}/n_{\text{DT}}$)	0.04
Iodine (radiation-enhancement) concentration ($n_{\text{I}}/n_{\text{DT}}$)	1.0×10^{-3}
Helium reflection coefficient, R_α	0.75
Toroidal-field margin at plasma center, T	0.85
Scrape-off region thickness, m	0.2
Particle-confinement time (τ_p), s	1.8
Particle e-folding distance in scrape-off zone (δ_p), cm	10
Energy e-folding distance in scrape-off zone (δ_E), cm	5
Plasma-edge temperature (T_{edge}), keV	1.2

TABLE 7.8. Limiter Design Parameters

Coolant		Water		
Reference structural materials		Ta-5W, AMAX-MZC, FS-85, or V-20Ti		
Low-Z coating material		Beryllium		
Total heat removed from limiter, MW (90 MW transport, 56 MW radiation plus neutrals, and 54 MW nuclear)		200		
Average surface heat load, MW/m^2		2.3		
Peak surface heat load,[a] MW/m^2		4		
Coolant inlet temperature, °C		115		
Coolant outlet temperature (two-pass), °C		145		
Coolant pressure, MPa (psia)		4.2 (600)		
Coolant channel size, mm × mm		8 × 4		
Wall thickness, mm		1.5		
Maximum temperature, °C	Ta-5W	AMAX-MZC	FS-85	V-20Ti
Water side	193	182	192	191
Coating side	290	196	404	449

[a]Includes transport load (3.4 MW/m^2) plus load from radiation and charge-exchange neutrals.

and charge-exchange. The magnitude of the volumetric nuclear heating depends on the specific limiter materials and is in the range of 30–80 MW/m^3 for the materials to be discussed shortly.

Water is selected as the limiter coolant because of its good heat transfer characteristics. This choice is consistent with the use of water cooling in the first wall and blanket. The limiter segments are connected so that the coolant passes through two segments (two passes). The water inlet temperature to the first pass is 115°C, and with a temperature rise of 15°C per segment, the outlet temperature is 145°C for the second pass. The coolant pressure is 4.2 MPa (600 psia). The water temperature is kept low to minimize pressure stresses. Since the 200 MW of heat removed from the limiter represents only 5% of the reactor thermal power, this heat is used effectively for feedwater heating in the steam cycle without significant loss in thermal efficiency.

A large number of materials were evaluated as to their suitability for the limiter structure. The evaluation included the capability of withstanding high heat fluxes, resistance to radiation damage, fabricability, and compatibility with the surrounding environment. This resulted in identifying four reference alloys as the primary candidate materials. These included a copper alloy AMAX-MZC, and the refractory metal alloys of vanadium (V-20Ti), niobium (FS-85), and tantalum (Ta-5W). Three-dimensional thermal-hydraulic and stress analyses were carried out for these four materials. A summary of the results is shown in Table 7.10.

The limiter-wall temperature at the coolant side is essentially the same, <200°C, for all materials with small differences due to axial conduction. At this

TABLE 7.9. Vacuum System Parameters

Component	Dimensions (cm)	Conductance (m³/s)
Limiter slots (2)	5650 × 10 × 50	4,300
Limiter ducts (2)	3170 × 16 × 70	4,100
Plenum	6000 × 67 × 600	13,700
Vacuum ducts (24)	100 × 640	730
	120 × 560	
Vacuum pumps (24)	—	2,900

Rated helium speed per pump, m³/s	120
Rated D–T speed per pump, m³/s	200
Total helium pumping speed, m³/s	490
Transmission probability[a] (helium)	0.9
Reflection coefficient (helium), R_α	0.75
Maximum helium pressure, Pa	0.016
Total D–T pumping speed, m³/s	480
Transmission probability[a] (D–T)	0.40
Reflection coefficient (D–T), R_{DT}	0.9
Maximum D–T pressure, Pa	0.024
Tritium fractional burnup	0.42
Total gas load, Pa-m³/s	18.7
D–T-gas load, Pa-m³/s	10.85
Helium-gas load, Pa-m³/s	7.85
Temperature, K	573
Number of vacuum pumps, on-line/total	24/48
Regeneration time, h	2
Maximum tritium inventory per pump, g	2.6

[a]Transmission probability per particle entering the limiter slot.

TABLE 7.10. Thermal/Stress Analysis of Candidate Limiter Materials[a,b]

	Temperature (°C)		Maximum Effective Stress (MPa)	Yield Stress (MPa)	Effective Yield
	Outer	Inner			
Tantalum, Ta-5W	290	193	249	342	0.7
Niobium, FS-85	404	192	370	370	1.0
Vanadium, V-20Ti	449	191	537	452	1.2
Copper, AMX-MZC	196	182	178	431	0.4

[a]Coolant: Pressure = 600 psi, T_{in} = 115°C, T_{out} = 145°C; channels = 4 × 8 mm, 1.5 mm thick at outer side.
[b]Peak heat load = 4 MW/m².

low temperature, the corrosion rate of these materials in water should be acceptable. The maximum temperature in the structure (coating side) varies from 196°C in copper to 449°C in vanadium reflecting the large difference in the thermophysical properties. The ratio of the effective stress to the yield stress is also shown in Table 7.10. These results indicate that under normal operating conditions, all of the materials meet the allowable stress criteria of Code Case 1592. However, only AMAX-MZC and Ta-5W can meet the more restrictive criterion of 0.75 of the yield strength. Since the thermal-stress component dominates the total stress in the limiter, the materials with the highest thermal conductivity and lowest thermal expansion will experience the lowest stress. It should be noted, however, that the results in Table 7.10 are based on conservative assumptions. Furthermore, several modifications in the reference limiter design that can significantly reduce the thermal stress have been identified and are discussed in Section 8.4 of Reference 1. Therefore, all the four alloys in Table 7.10 are considered viable candidates, and the selection of one of them must be made based on additional data from future experimental results in areas such as resistance to radiation damage. For the purposes of this Chapter, the alloy Ta-5W is identified as the reference structural material whenever the need arises to provide only one set of parameters.

The limiter and the first wall are coated with beryllium to eliminate sputtering of the underlying high-Z structural materials. Beryllium is selected as the low-Z coating because its properties make it superior to other candidates. Estimates of the erosion of the beryllium coating were made. The coating on the first wall will erode at a rate of 0.14 mm/yr; therefore, a 1.2-mm coating is adequate for a 7-yr life. The limiter coating will sputter by all ion species with a spatially varying rate. Redeposition of beryllium from the plasma and first wall will also occur. The net effect is that the coating will erode on the wall while it grows on the limiter. The STARFIRE design is developed such that there is no net erosion or growth on the leading edge. This is accomplished by maintaining a beryllium density in the plasma of $\sim 4\%$ of the hydrogen ion density. There will be a net growth of beryllium on the rest of the limiter averaging ~ 0.6 mm/yr. A simple grinding process in place can be performed if necessary to restore the beryllium coating to its original thickness.

The response of the limiter to off-normal conditions was considered as an integral part of the design. The important off-normal events are (1) plasma disruptions and (2) loss-of-coolant flow. The concerns with plasma disruptions are the thermal-energy dump and the induced electromagnetic forces. The limiter is intentionally located at the outer side of the plasma and centered around the midplane, where a thermal-energy dump from a plasma disruption is unlikely. However, when a plasma thermal-energy dump on the limiter occurs, only the coating will be affected. The rate of ablation of beryllium is small enough that several disruptions per year with thermal-energy dumps on the limiter can be tolerated.

The electromagnetic forces will always be induced in the limiter in the case of a plasma disruption regardless of where the plasma energy dump occurs. Three electromagnetic effects are produced, with the magnitude strongly dependent on the plasma disruption (current-decay) time. The first is a uniform pressure, acting on the outside panels of the limiter. For a plasma disruption time of >10 ms, the maximum induced stress due to this uniform pressure is 0.6 MPa (90 psi), which is a small fraction of the yield stress for the copper, tantalum, niobium, and vanadium alloys. The second effect is a force tending to bend the limiter arm about a toroidal axis. Accommodating this force required an iterative process in the limiter design. In particular, providing a thick root for the limiter (see Fig. 7.6) was found necessary to reduce the moment's arm and the magnitude of the force. With the present reference design, the maximum bending stress is 154 MPa (22,000 psi), which is < 40% of the yield stress for the reference structural materials when the plasma disruption time is >10 ms. The third electromagnetic effect is a torque that tends to twist the limiter about a radial axis. For a plasma disruption time of 10 ms, the maximum torque is 46 kN-m resulting in an effective stress that is < 60% of the yield stress for all of the four primary structural materials. The magnitude of these forces and torques is reduced substantially at longer, and perhaps more realistic, plasma-disruption (current-decay) times. The reference limiter design can withstand the electromagnetic effects without any permanent deformation for an unlimited number of plasma disruptions.

7.3.5. Magnets

The magnet systems provide the plasma with confinement and a stable equilibrium configuration as well as some current initiation. The magnets must be superconducting, except for a few control coils carrying relatively low current; otherwise they would consume unacceptable amounts of electrical power. The magnets are large components that experience large forces, which must be resisted with structural material in a manner that minimizes heat conduction to the magnets operating at liquid-helium temperature. Most importantly, they must be extremely reliable, as a magnet replacement is a laborious and time-consuming process.

The superconducting toroidal-field (TF) and poloidal-field (PF) coils have been designed with a cabled conductor consisting of a copper stabilizer and NbTi superconductor, except for the inner turns of the TF coils, where field requirements in excess of 9 T have led to the choice of Nb_3Sn superconductor. In both the TF and PF coils, each cable conductor is contained in its own structure, which bears against the structure of neighboring conductors to transmit radial and axial forces. All coils are bath cooled by pool boiling liquid helium at 4.2K. The structure around the conductor contains transverse and longitudinal channels, to carry liquid helium to where cooling is needed and to carry helium vapor away.

The superconducting toroidal-field coil system for STARFIRE is a logical, straight forward extension of present superconducting magnet technology. Table 7.11 is a summary of selected TF coil features and parameters.

TABLE 7.11.   STARFIRE Toroidal-Field-Coil Parameters

Number of coils	12
Total ampere turns	2×10^8 A-turns
Total stored energy	50 GJ
Total inductance	174 H
Peak field	11.1 T
Current	24 kA
Total weight	6×10^6 kg
Coil straight-section height	8 m
Mean radius of outer coil leg	13 m
Conductor	
Superconductor	$Nb_3Sn/NbTi$
Stabilizer	Copper
Configuration	Cable
Coil cooling	He Bath, 4.2 K
Structural material	Austenitic, high Mn stainless steel

The TF coils bear radially inward against the G-10 fiberglass–epoxy center-post support cylinder, within whose bore is located the inner ohmic-heating (OH) and equilibrium-field (EF) coils. All of these elements share a common vacuum volume. The centerpost region is surrounded by a common vacuum tank section with individual vacuum tanks surrounding each TF coil outer leg.

The 24 kA conductor for the TF coils is a three-level, unsoldered, unin-sulated "Rutherford cable," consisting of sixteen 1500 A cables, each of which is a six-around-one bundle of similarly configured subcables. Inherent in a cabled conductor design, particularly when using Nb_3Sn, is its limited ability to support hoop and transverse bearing loads (the latter occurring in the center-post region of a TF coil). Therefore, in the selected design, the conductor is sandwiched between two pretensioned stainless-steel strips for hoop-load sup-port, and flanked by two bearing-load support strips. Stainless steel is used for these support elements owing to its high elastic modulus and strength.

Nb_3Sn is employed only in the high-field (9–11 T) region. A bronze diffu-sion geometry is envisioned, with a tantalum barrier to shield the surrounding stabilizer material of the composite strand. The Nb is reacted with the Sn of the bronze matrix after the cabling process is complete, but before coil winding.

The NbTi alloy is specified for the three 0–9 T field grades. Grading is based on three centerpost region parameters: amount and type of superconductor required (determined by magnetic field); amount of copper stabilizer required (determined by magnetoresistance, radiation degradation, cryostability, and protection criterion limit); and required bearing-load support (determined by cumulative radial bearing load).

The STARFIRE reactor has three sets of poloidal-field (PF) coils. These are the ohmic-heating (OH) coils, the equilibrium-field (EF) coils, and correction-field (CF) coils. Even though STARFIRE operates in a steady-state mode, it incorporates a small OH coil to provide an inductive voltage over several

TABLE 7.12.   Ohmic-Heating and Equilibrium-Field Coil Parameters

	OH Coils	EF Coils
Superconductor/stabilizer	NbTi/Cu	NbTi/Cu
Conductor configuration	Cable	Cable
Stability	Cryostable	Cryostable
Cooling	Bath cooled	Bath cooled
Operating temperature	4.2 K	4.2 K
Operating current	100 kA	100 kA
Average current density	1400 A/cm	1400 A/cm
Total amp-turns	51 MA-turns	86 MA-turns
Total amp-meters	600 MA-m	2900 MA-m
Peak field	8.0 T	4.5 T
Maximum dB/dt (normal operation)	0.6 T/s	0.2 T/s
Stored energy (self)	1.1 GJ	10.0 GJ
Self inductance [a]	55 mH	500 mH

[a]Based on equivalent parallel current of 200 kA.

seconds to initiate a plasma current. The EF coils provide the field that maintains the plasma at equilibrium with the desired position, shape, and current profile. The correction-field coils, which link the TF coils, respond to displacements of the plasma to correct those displacements and thus stabilize the plasma. They can respond to plasma motion more quickly without excessive power demands than can the EF coils that are located outside the TF coils. Parameters for the OH and EF coils are given in Table 7.12. The CF coils, because they link the TF coils and must be demountable, are constructed of water-cooled copper.

The EF and OH coils must be superconducting; normal conducting coils would consume an unacceptable amount of power. Being superconducting, these coils must be outside the TF-coil system to facilitate maintenance and possible coil replacement. External location of the EF coils exposes the outer TF-coil region to large fields, which interact with the TF-coil current to generate large out-of-plane (overturning) loads. The magnitude of the overturning moment on each coil is about 1.5×10^9 N-m. The centerpost region of the TF coil reacts a small portion of this load. The major portion of the load is reacted in the outer curved coil region, where the distributed out-of-plane load is transmitted from the helium vessel to the surrounding vacuum tank by closely packed pairs of cold-to-warm tiebars. The individual coil vacuum tanks are in turn supported by substantial intercoil shear panels.

7.3.6.   First-Wall/Blanket

The primary functions of the first-wall/blanket of a commercial tokamak reactor are to provide the first physical barrier for the plasma, to convert the fusion energy into sensible heat and provide for the heat removal, to breed tritium and provide for tritium recovery, and to provide some shielding for the magnet system. The first wall must withstand high particle and energy fluxes from the plasma, high thermal and mechanical stresses, and elevated-temperature operation. Also, the wall must not be a source of excessive plasma contamination. The first wall may or may not be integral with the blanket. The blanket must withstand high neutron fluences, elevated-temperature operation, and thermal and mechanical stresses, and must be compatible with the chemical environment, the plasma, and the vacuum.

In the present study, the technological and design aspects of various first-wall/blanket concepts have been considered in the selection of potentially viable designs for STARFIRE. The objectives of the present study involve identification of key technological constraints of candidate tritium-breeding-blanket design concepts, establishment of a basis for assessment and comparison of the critical problem areas and design features of each concept, and development of optimized first-wall/blanket designs for STARFIRE. The major emphasis has been placed on the development of a blanket design that is safe and environmentally acceptable. The primary guidelines established to meet these criteria are low tritium inventory in the blanket, minimal long-lived activation products, and minimal stored energy.

Since breeding of tritium is considered essential, and since lithium is the only viable tritium-breeding medium, lithium in some form is required in the blanket. On the basis of engineering and design considerations, liquid lithium provides many advantages for the tritium breeder; however, because of perceived safety problems associated with the liquid-lithium system, an *a priori* decision was made to focus the present study on the use of solid-lithium compounds for breeding. Although previous studies have assessed the viability of alternate blanket options, a technical evaluation of the design and safety problems associated with the liquid lithium, liquid-lithium alloys, and molten-salt breeding materials was not performed in the present study. The primary objective was to assess the design and performance characteristics of a blanket concept based on solid tritium breeding materials.

The development of the reference STARFIRE first-wall/blanket design involved numerous tradeoffs in the materials selection process for the breeding material, coolant, structure, low-Z coating, neutron multiplier, and reflector. The coolant and structural material selections were greatly influenced by the choice of the solid-breeder concept, which was used as a basis for the STARFIRE design. The most important criteria considered in the selection of potentially viable solid breeding materials include breeding performance, chemical stability, compatibility, and tritium-release characteristics. Of the two types of solid breeding materials considered as primary candidates, namely, intermetallic compounds and oxide ceramics, only selected ceramics appear to have satisfactory tritium-release characteristics. The α-$LiAlO_2$ is selected for the reference design on the basis of the best combination of these critical materials requirements. It is one of the most stable compounds considered and compatibility should not be a major problem; however, adequate tritium breeding is attainable only with the aid of a neutron multiplier. The high tritium solubility and greater reactivity with the structural materials were primary factors in the elimination of Li_2O as the reference breeding material.

Pressurized water, both H_2O and D_2O, and helium were considered for the coolant. Major concerns regarding the use of helium relate to difficult neutron-shielding problems, large manifold requirements, leakage into plasma chamber, lower tritium breeding because of the large structure requirements, and the high temperatures required for the energy-conversion system. An acceptable structural material for use with high-temperature helium in a radiation environment has not been identified. Also, design constraints associated with the use of helium as a first-wall coolant appear to be prohibitive. Major advantages of the water coolant are its characteristically low operating temperature and its excellent heat-transfer characteristics. However, the use of water with the intermetallic-compound breeder materials is probably not acceptable because of the high reactivity, and hence, safety concern. Although D_2O has several neutronic advantages compared to H_2O, the cost is considered prohibitive.

The choices of breeding materials and coolant limit the number of viable candidate structural materials. Key factors in the selection of the advanced austenitic stainless steel relate to the steady-state reactor operation and the low operating temperatures characteristic of a water-cooled system. Because of the high thermal stress factor associated with austenitic stainless steel, acceptable first-wall lifetimes could not be attained with a cyclic burn. Also, radiation-damage effects are less severe at the proposed operating temperatures than at temperatures above 500°C.

The low-Z coating concept for the first wall is incorporated as part of the plasma-impurity-control system. The low-Z coating concept provides flexibility in that the structural material can be selected primarily on the basis of structural requirements and the coating can be selected primarily on the basis of surface-related properties. Favorable properties such as high thermal conductivity, high heat capacity, and compatibility with hydrogen were important considerations in the selection of beryllium as the first-wall coating/cladding material. A primary consideration in the selection of the candidate coating/cladding is that it can be used on all components exposed to the plasma. This is important because considerable redistribution of the material throughout the chamber is expected as a result of sputtering and ablation.

An effective neutron multiplier is required to obtain adequate tritium breeding with the $LiAlO_2$. Two candidate materials are proposed. Beryllium provides good neutronics performance and can be easily incorporated into the blanket design since it has low density, high thermal conductivity, and high heat capacity. Because of the concern regarding limited resources of beryllium, an alternate neutron multiplier, Zr_5Pb_3, is also proposed. This compound retains some of the beneficial neutron characteristics of lead but remains solid at the operating temperatures.

Low activation, low cost, and inherent safety characteristics were key factors in the selection of graphite over water and stainless steel as the reflector.

A schematic diagram of the reference STARFIRE blanket concept is given in Fig. 7.8 and the key reference parameters are summarized in Table 7.13. The water-cooled blanket module, with a thickness of 68 cm, consists of 1-cm-thick

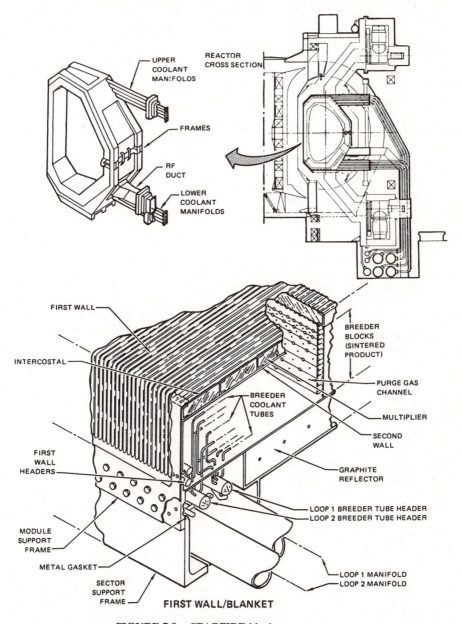

FIGURE 7.8.   STARFIRE blanket concept.

TABLE 7.13. Summary of First-Wall/Blanket Design Parameters

First-Wall

Form	Be-coated panel
Structural material	Austenitic stainless steel[a]
Outer wall structural thickness, mm	1.5
Maximum structural temperature, °C	<423
Coating/cladding	Beryllium
Coating/cladding thickness, mm	1.0
Coolant	Pressurized water, H_2O
Coolant outlet temperature, °C	320
Coolant inlet temperature, °C	280
Coolant nominal pressure, MPa	15.2
Coolant velocity, m/s	6.1

Neutron Multiplier

Material Options	Be	Zr_5Pb_3
Maximum temperature, °C	490	840
Thickness, m	0.05	0.05
Theoretical density, g/cm^3	1.8	8.9
Effective density, %	70	100
Total mass, kg	51,800	356,000

Breeding Region

Structural material	Austenitic stainless steel[a]
Maximum structural temperature, °C	425
Breeder material	α-$LiAlO_2$ (natural Li with Be) (60% ^{6}Li with Zr_5Pb_3)
Theoretical density, g/cm^3	3.4
Effective density, %	60
Grain size, 10^{-6} m	0.1
Maximum/minimum temperature, °C	850/500
Region thickness, m	0.46
Coolant	Pressurized water, H_2O
Coolant outlet temperature, °C	320
Coolant inlet temperature, °C	280
Coolant nominal pressure, MPa	15.2
Tritium processing fluid	He (0.05 MPa)

Reflector

Material	Graphite
Thickness, m	0.15
Maximum temperature, °C	<800
Structure	Austenitic stainless steel (low Mo)
Structure temperature, °C	300–400

[a]Prime Candidate Alloy, an advanced titanium-modified Type-316 austenitic stainless steel.

first wall, a 5-cm-thick neutron multiplier, a 1-cm-thick second wall, a 46-cm-thick breeding zone, and a 15-cm-thick reflector zone that contains the blanket support structure and the manifolding. The modules are 2–3 m wide by ~ 3 m high depending on the location within the reactor. The module walls and all support structures in the high-radiation zone are fabricated from an advanced low-swelling austenitic stainless steel. The internal structure is integrally cooled to remove the nuclear heating and maintain the structure below 400°C.

The first wall, which is a water-cooled austenitic stainless-steel panel coil, is an integral part of the blanket module. The corrugated plasma side of the first-wall panel is constructed of 1.5-mm-thick advanced austenitic stainless steel. The 3.5-mm-thick back plate is formed from the same material. The pressurized-water coolant is maintained between 280 and 320°C throughout the first wall and blanket. For the average neutron wall loading of 3.6 MW/m^2, the average surface heat flux on the first wall is 0.92 MW/m^2 with a peak-to-average value of ~ 1.2. The maximum structural temperature in the stainless-steel wall is ~ 450°C for the reference conditions. For steady-state operation at these relatively low temperatures, an estimated wall design life of six years is considered reasonable for the advanced austenitic stainless steel. The proposed panel-type construction provides integral cooling of the blanket wall and avoids the necessity for a large number of pressure-boundary-tube welds in the high-radiation zone. Also, the panel-type structure is perceived to have less vibration problems than an unsupported tube bank.

A ~ 1-mm-thick beryllium coating or cladding on the first wall serves to protect the plasma from the high-Z wall material. This thickness will provide sufficient material to withstand the predicted surface erosion for the reference blanket lifetime of six years. The beryllium coating/cladding on the inboard wall will also accommodate the projected number (~ 10 per wall lifetime) of plasma disruptions for the assumed conditions.

A 50-mm-thick neutron multiplier is placed directly behind the first wall to permit adequate breeding with the LiAlO$_2$ breeding material. Two neutron multiplier options, beryllium and Zr$_5$Pb$_3$, are carried in the reference design. Beryllium is generally considered to be the most favorable neutron multiplier; however, resource limitations are a concern. The present analysis indicates that beryllium requirements for several hundred reactors are only a few percent of the estimated U.S. beryllium reserves. An important part of the present study was to provide an alternative to beryllium. The Zr$_5$Pb$_3$ with an estimated melting temperature of 1400°C provides some of the benefits of a lead multiplier while maintaining the design simplicity of the solid materials. Approximately 30% of the neutron heating is deposited in the multiplier zone. The back side of the first-wall panel and a water-cooled panel (second wall) between the multiplier and breeder region provides cooling that maintains the maximum multiplier temperature at 480°C for beryllium and 850°C for Zr$_5$Pb$_3$. Structural webs between the first and second walls provide support for the first wall. The overall reactor analyses of the energy-conversion system, the shield, the remote maintenance and repair, and safety are based on the blanket design option with the Zr$_5$Pb$_3$ neutron multiplier.

The 46-cm tritium-breeding zone consists of a packed bed of α-LiAlO$_2$ with 1.25-cm-diameter stainless-steel coolant tubes spaced appropriately throughout the zone to maintain a maximum breeder temperature of 850°C (Fig. 7.9). The spacing of the horizontal tubes increased from ~ 2 cm at the front of the breeder zone to ~ 10 cm at the back. There are approximately 60,000 coolant

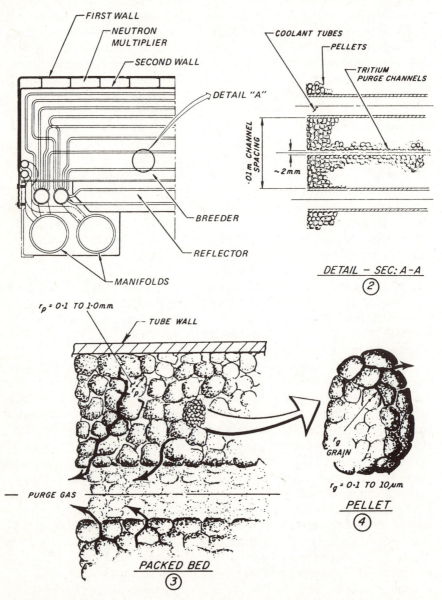

FIGURE 7.9. Schematic diagram of STARFIRE blanket concept showing solid breeder microstructure and bimodal pore distribution and tritium-removal scheme.

tubes in the blanket. The nominal coolant pressure is 15.2 MPa (2200 psi) with a coolant inlet temperature of 280°C and an outlet temperature of 320°C. The relatively low temperature of the austenitic stainless-steel tubes ($< 400°C$) and the oxide film on the water side of the tubes provide an adequate tritium barrier for in-leakage into the coolant. Natural lithium is used for the beryllium neutron-multiplier option; however, 60% enriched Li is required to achieve adequate tritium breeding with the Zr_5Pb_3 neutron-multiplier option. The $LiAlO_2$ is in the form of low-density (60%) sintered product with a tailored bimodal pore distribution, that is, a small grain size (< 1 μm) and a fine porosity within particles that are fairly coarse (~ 1 mm) with a much coarser porosity between particles. The sintered $LiAlO_2$ is perforated with $\sim$ 2-mm-diameter holes through which low-pressure (0.5 atm) helium passes to recover the tritium from the breeder. The low-density ceramic with a tailored microstructure is proposed to facilitate percolation of tritium (as T_2O) to the helium purge channels. A breeder lifetime of six years before lithium burnup becomes excessive is considered feasible.

The reflector consists primarily of $\sim$ 15 cm of graphite. The support structure to which the blanket modules are attached also serves as the containment for the graphite reflector. In order to conserve space and improve the vacuum characteristics of the blanket, the manifolds and headers for the blanket are embedded in the reflector region. The large number of coolant tubes are joined to the headers in a vented chamber that is isolated from both the breeder region and the vacuum chamber. This concept provides both safety and reliability benefits since the most probable coolant leakage problem is at the coolant tube-to-header welds. A coolant-tube weld failure in this chamber would not likely lead to coolant ingress into the breeder region. Also, a small leak at a weld would not destroy the vacuum, and therefore, might not require immediate reactor shutdown. Isolating the geometrically complex manifolds and headers from the plenum region with a relatively smooth rear-blanket surface should substantially improve the vacuum characteristics of the blanket. A modified austenitic steel with low molybdenum contents is used in this low-flux region to reduce the long-term activation.

A two-loop coolant system is provided in the blanket to reduce the consequences in the event of a loss-of-flow or loss-of-coolant accident. One loop provides coolant for the first wall and alternate tube banks in the breeder region beginning with the first row of tubes. The second loop provides coolant for the second wall and the remaining coolant tubes in the blanket. Under the reference plasma shutdown conditions, cooling provided by either loop is sufficient to prevent excessive temperatures in all regions of the blanket. The two-loop concept will also reduce the pressure release and activation release in the event of a coolant-tube failure.

For plasma stability, an electrical conducting path equivalent to 2 cm of stainless steel is required near the first wall. The conductivity of the first wall and the neutron multiplier meets this requirement in the modules. Bimetallic contacts between the modules are provided adjacent to the multiplier region to complete the current path. Upon cooling, these contacts recess into the module wall to allow for sector removal.

The inner blanket is similar in most respects to the outer blanket just described. The major differences, which relate to the breeder-zone thickness, the reflector, and the coolant flow direction are designed to minimize the inner blanket/shield thickness. The breeding zone thickness is reduced from 46 to 35 cm and the graphite reflector is eliminated with a modest penalty on tritium breeding capability. Vertical coolant flow in the inner blanket eliminates the need for manifolds in the back of the blanket module, thereby improving the effective shielding capability of the inner blanket/shield.

7.3.7. Radiation-Shielding System

The shield design in STARFIRE has evolved from a comprehensive approach that involved the following considerations: (1) recognition of the importance of the shield system and its impact on reactor component reliability, simplicity, maintainability, and economics; (2) a full accounting of shielding considerations in the selection process of key reactor subsystems from the early stages of the reactor design—examples of this are the choices of the limiter instead of divertor for plasma-impurity control and exhaust and the selection of lower-hybrid rf system for plasma auxiliary heating in preference to neutral beams; and (3) comprehensive trade-off analyses for determining the material composition and dimensions of the shield components.

The design criteria for the shield included (1) protection of all reactor components from excessive nuclear heating and radiation damage; (2) the biological dose rate outside the shield at 24 h after shutdown should be sufficiently low, $\sim 1-2$ mrem/h, to facilitate personnel access into the reactor building; and (c) material composition and dimensions of the shields were selected so that all reactor components, including the shields, outside the blanket are recyclable within 50 yr or less.

The shield system in STARFIRE consists of the blanket, primary bulk shield, penetration shield, component shield, and biological shield. Table 7.14 shows the material compositions and dimensions of the bulk shield (inboard and outboard regions) and penetration shields. State-of-the-art analyses, including multidimensional Monte-Carlo calculations, were performed for the reference design.

The space problems in the inboard region have been resolved by trade-off studies as discussed in Chapter 5 of Reference 1. The inboard blanket/shield thickness, Δ_{BS}^i, is 1.2 m. This includes space for 9-cm vacuum gaps between the blanket and shield, shield and TF coils and thermal insulation inside the TF vacuum tank; 3-cm vacuum tank (alloy Fe14Mn2Ni2Cr, referred to as Fe-1422) and 7-cm helium vessel (stainless steel). The inner blanket is 37 cm thick and must breed tritium as the breeding margin with the solid breeder is small. The inboard shield is 54 cm thick and consists of alternating layers of tungsten and boron carbide with water for cooling and Fe-1422 for structure as shown in Table 7.14.

TABLE 7.14. Material Composition and Dimensions of Bulk and Penetration Shields

Component	Major Radius (m)	Thickness (m)	Composition
A. Inboard Shield[a]			
Shield jacket	4.47	0.02	Fe-1422
Shield 1	4.45	0.15	10% Fe-1422 + 10% H_2O + 80% W
Shield 2	4.30	0.075	10% Fe-1422 + 10% H_2O + 80% B_4C
Shield 3	4.225	0.15	10% Fe-1422 + 10% H_2O + 80% W
Shield 4	4.075	0.075	10% Fe-1422 + 10% H_2O + 80% B_4C
Shield 5	4.00	0.075	10% Fe-1422 + 10% H_2O + 80% W
Shield 6	3.925	0.075	10% Fe-1422 + 10% H_2O + 80% B_4C
Shield jacket	3.85	0.02	Fe-1422
B. Outboard Shield[b]			
Shield jacket	10.54	0.02	Fe-1422
HFS shield	10.56	0.50	5% Ti_6Al_4V + 65% TiH_2 + 15% B_4C + 15% H_2O
MFS shield	11.06	0.40	70% Fe-1422 + 15% B_4C + 15% H_2O
LFS shield	11.46	0.18	Fe-1422

(continued)

219

TABLE 7.14. *(Continued)*

Component	Major Radius (m)	Thickness (m)	Composition
C. Vacuum-System Shield			
Shield jacket	—	0.02	Fe-1422
C-1. Upper Duct Shield			
Shield 1	—	0.30	5% Ti_6Al_4V + 65% TiH_2 + 15% B_4C + 15% H_2O
Shield 2	—	0.28	70% Fe-1422 + 15% B_4C + 15% H_2O
C-2. Lower Duct Shield			
Shield 1	—	0.30	5% Ti_6Al_4V + 65% TiH_2 + 15% B_4C + 15% H_2O
Shield 2	—	0.18	70% Fe-1422 + 15% B_4C + 15% H_2O
C-3. Pump Pod Shield	—	0.48	70% Fe-1422 + 15% B_4C + 15% H_2O
D. rf-System Shield			
In the Plenum	—	0.10	Fe-1422
Outside the bulk shield	—	0.15 max.	Fe-1422

[a]The major radius is for outer surface at the reactor midplane.
[b]The major radius is for the inner surface at the reactor midplane.

Table 7.15 shows the maximum radiation effects in the inboard section of the TF coils. The maximum radiation-induced resistivity in the copper stabilizer after a 40-yr operation is $2.2 \times 10^{-10} \, \Omega \cdot m$, assuming a magnet anneal every 10 yr with 83% recovery. The maximum radiation dose in the shield dielectric break is 7.4×10 Gy after a 40-yr operation.

The outboard bulk shield is 1.1 m thick. It includes 2-cm shield jacket at the plenum region with the rest divided into three regions. The first region, in the high flux zone, is 0.5 m thick and has a material composition of 5% Ti alloy + 65% TiH_2 + 15% B_4C + 15% H_2O. The second region, middle zone, is 0.40 m thick with the material composition as 70% Fe-1422 + 15% B_4C + 15% H_2O. The third region, outer zone, is 0.18 m thick of Fe-1422. At reactor shutdown, the biological dose rate in the reactor building is ~ 130 mrem/h and is dominated by the contribution from the decay of ^{56}Mn and ^{54}Mn. Owing to the short half-life of ^{56}Mn, the biological dose rate in the reactor building decays very rapidly and reaches ~ 1.5 mrem/h at 24 h after shutdown. These dose rates are calculated with all shielding in place. Although the STARFIRE plans call for fully remote maintenance, the dose rate of 1.5 mrem/h shows that personnel access into the reactor building with all shielding in place is permissible within 1 day after shutdown. This provides a degree of confidence in improving the plant availability factor, if desired, by allowing some maintenance tasks to be carried out in contact or semiremote mode.

One of the important shield considerations is radiation streaming through void regions that penetrate the blanket and bulk shield regions. In general, the direct radiation flow in neutral-beam ports, divertors, etc., has been one of the primary sources of design complexity and shielding difficulties in previous tokamak designs. In the STARFIRE design, a serious effort has been devoted to minimizing possible design difficulties associated with these penetrations.

TABLE 7.15.   Maximum Radiation Effects in the TF Coil (in the Inboard Region)

Maximum nuclear heating in the super-conductor (MW/m^3)	1.54×10^{-5}
Maximum nuclear heating in the helium tank (MW/m^3)	3.18×10^{-5}
Maximum nuclear heating in the vacuum tank (MW/m^3)	4.76×10^{-5}
Maximum dose in the electrical insulator[a] (Gy)	$1.22 \times 10^{+7}$
Maximum dose in the thermal insulator[a] (Gy)	$2.39 \times 10^{+7}$
Maximum dose in the dielectric break[a] (Gy)	$7.41 \times 10^{+7}$
Maximum fast neutron fluence ($E > 0.1$ MeV) in the superconductor [a] (n/m^2)	$1.87 \times 10^{+21}$
Maximum radiation induced resistivity in the Cu stabilizer[a] ($\Omega - m$)	2.17×10^{-10}

[a]After 40 yr of operation.

The STARFIRE design features the selection of a lower-hybrid rf system in preference to a neutral-beam heating system and a limiter-impurity-control concept rather than divertors. A great advantage of the rf and limiter/vacuum systems is the elimination of any direct radiation streaming path from the plasma to the reactor exterior. These design features have helped reduce the shielding problems to a manageable level and brought about overall simplicity in the shield design.

Radiation streaming through the limiter duct increases the nuclear heating rate by about a factor of 10 in a relatively small region of the bulk shield (centered around the midplane). This presents no difficulty in the shield design. The maximum nuclear heating in the cryopanels of the vacuum pumps is ~ 0.3 kW/m^3, which poses no difficult heat removal problems.

The STARFIRE design employs an rf-wave launching method for the plasma heating as well as for the plasma current drive. From a radiation shielding standpoint, the rf system is very attractive in that bends can be tolerated in the rf ducts and also that the waveguide structural materials (PCA stainless steel in the reference design) along with the water coolant flowing through the structure prevent, to a substantial degree, the direct radiation flow from coming out through the waveguides.

In order to hinder radiation streaming into the plenum region, the duct portion inside the plenum is completely surrounded by a 0.1-m-thick Fe-1422 shield. In addition, the duct portion in the reactor room is fully shielded by the same material of 0.15m, which is aimed primarily at reducing the impact of the rf penetration on the potential increase in the post-shutdown biological dose inside the reactor building.

7.3.8. Radioactivity

The most important concern with regard to induced activation is the production of radioisotopes with very long half-lives ($> 50-100$ yr) in relatively large volumes of materials; this results in (1) requirements for permanent radwaste storage, and (2) depletion of some resource-limited materials. An important strategy for fusion-reactor development, therefore, is to avoid generating any large inventories of high-level, long-term activation products so that a majority of the reactor construction materials could be recycled on a reasonably short time-scale, for example, a human generation of ~ 30 yr after component replacement or reactor decommissioning. This strategy has been adopted in STARFIRE and has considerably affected the material selection.

Component Activation and Material Recycling. The importance of the major radioisotopes has been examined in terms of radioactivity and radioactivity-related parameters such as biological hazard potential (BHP) in air and BHP in water. With regard to material recycling, an effort has been devoted to establishing a criterion for potential material recycling categorization. In addition to the conventional waste-level classification by specific radioactivity concentration (Ci/unit volume), a criterion based on the contact biological dose has been suggested and used for the recycling analysis.

It is found that the magnitude of high-level, long-term radwaste from STARFIRE is dominated by the PCA first-wall/blanket-structural material. It is shown, however, that the average annual discharge rate of PCA from STARFIRE is only about 9.5 m^3 (~ 75 metric tons in weight), which is considerably less than a typical annual discharge of high-level waste from a liquid-metal fast breeder reactor (LMFBR) of the same power. Note that because the power density is lower in fusion than in fission systems, the reactor material volume that is potentially activated is larger. However, the degree of biological hazard potential associated with STARFIRE is significantly lower. For example, the BHP-air of STARFIRE varies as air volumes of 57 km^3, 0.03 km^3, and 0.004 km^3 per kW$_{th}$ reactor power at post-shutdown times of 1 yr, 100 yr, and 1000 yr, respectively. These values are compared to the corresponding BHP-air's of ~ 550 km^3, ~ 100 km^3, and 80 km^3 in a typical LMFBR system.

Figure 7.10 shows the time-dependent radioactivity concentration contributed by each constituent element of the PCA first-wall structure. In addition to the PCA activation, the radioactive products generated in the blanket come from the Zr$_5$Pb$_3$ neutron multiplier and the LiAlO$_2$ tritium breeder. The long-term activation of the LiAlO$_2$ breeder is governed solely by Al (half-life of 7.2 × 10^5 yr) and its concentration is ~ 2 Ci/m^3 at times greater than a few

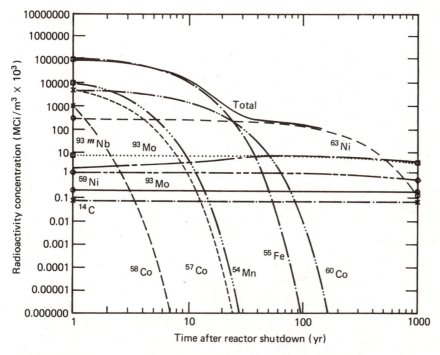

FIGURE 7.10. Isotopic radioactivity contribution for STARFIRE—PCA first wall.

months after shutdown. Recycling of $LiAlO_2$ or Li is desirable and appears to be technically feasible but needs to be addressed in future studies in detail.

The activation level of most reactor components external to the blanket, including the major penetration subsystems, decays to a category of low-level waste in 30 yr at most. Thus, the potential for recycling of materials from most reactor components is excellent. Table 7.16 classifies major reactor components according to high (R) and low (N) potential for recycling the material.

Decay Afterheat.   The total decay heat in the Zr_5Pb_3 multiplier design is ~ 88 MW at shutdown, which is $\sim 2.2\%$ of the total reaction thermal power. This decay heat is reduced to ~ 40 MW within 24 h, followed by a rapid decrease beyond that time period, reflecting the decay of the most dominant radioactive isotope of ^{89}Zr (78 h). Approximately half of the decay power comes solely from the Zr_5Pb_3 activation decay. The decay heat from the alternate neutron multiplier of beryllium is about an order of magnitude lower than that from Zr_5Pb_3.

Reactor-Room Activation.   Three different gases—air, nitrogen (N_2), and carbon dioxide (CO_2)—were considered for the reactor building. The magnitude of induced activation was compared for the three gases. It was found that the residual radioactivity in both air and N atmospheres is only about a factor of 4 lower than the current maximum permissible concentration (MPC) for ^{14}C.

TABLE 7.16.   A Classification of STARFIRE Component Radioactivities

Component	Mass (Mg)	Class[a]
First-wall/blanket		
PCA	450	N
Zr_5Pb_3	329	R
$LiAlO_2$	605	$N–R$[b]
C	163	R
Bulk shield	8,770	R
Vacuum shield	6,688	R
TF-magnets	6,240	R
OH/EF-magnets	1,931	R
CF-magnets	140	R
Support structure	1,852	R
Reactor building	175,400	R

[a]R: The contact biological dose rate at the surface of 1-m-diameter sphere is below 2.5 mrem/h within 30–50 yr after component replacement or reactor decommissioning; the potential for recycling is high. N: The biological dose is greater than 2.5 mrem/h; the potential for recycling is low.
[b]The dose is high but since it comes from Al, which can be chemically separated, recycling potential for lithium is high.

The isotope ^{14}C is produced in both air and nitrogen via the (n, p) reaction. In addition, the activation of argon in the air makes a large contribution during reactor operation and for a short time after shutdown. The CO_2 activity is due largely to the $^{16}O(n, p)$ ^{16}N reaction. The ^{16}N isotope has a 7.1 s decay half-life. Therefore, its activity decays very quickly (in about 10 min). Beyond this time period, the CO_2 activation is determined by the radioactivity of ^{14}C which comes in this case from the $^{13}C(n, \gamma)$ reaction.

7.3.9. Heat Transport and Energy Conversion

The thermal energy deposited in the blanket, first wall, and limiter is delivered via the heat-transport system to the steam-power-conversion system, where it is converted to electricity. Two separate heat-removal systems are utilized, a dual-loop circuit for the blanket/first wall and a single loop for the limiter.

The dual-loop system was chosen for the blanket/first wall in order to virtually eliminate the possibility of a complete loss of coolant in the blanket and to reduce the building over-pressure occurring as a consequence of certain loss of coolant events. The primary coolant system consists of piping and valves, pumps, pressurizers, steam generators, water makeup and conditioning equipment, and instrumentation and controls. Inlet and outlet ring manifolds are located beneath the reactor. Each circuit incorporates two vertical straight-tube-and-shell steam generators, two vertical single-stage shaft seal pumps, and a single pressurizer, all located within the reactor building. The main piping and manifolds are about 1 m diameter, and piping and valving are incorporated such that individual blanket segments, steam generators, or pumps can be isolated from the rest of the system. Because of the steady-state operating mode, a thermal-energy-storage system is not required, and tritium concentrations in the primary coolant are maintained low enough so that an intermediate loop is not needed.

Primary coolant at 15.2 MPa (2200 psi) leaves the blanket at 320°C and is returned at 280°C. The pressure drop through the system is about 1.0 MPa (150 psi), and the electric power required to pump the primary coolant is about 30 MW.

The thermal energy deposited in the limiter, about 5% of the total power, is transported via the single-loop limiter feedwater circuit and utilized for feedwater heating in the power-conversion system. This system incorporates piping and valves, pumps, a pressurizer, feedwater heaters, a water-conditioning system and the appropriate controls and instrumentation. Water coolant at an operating pressure of 700 psi leaves the limiter at 145°C and is cooled to 115°C in the feedwater heat exchangers. Coolant is transported via ring manifolds beneath the reactor and piping beneath the building floor. The pumping power is less than 3 MW. As in the primary loop, system components are located within the reactor building and are mounted in standard modules. This modularity approach is consistent with the total remote, remove-and-replace maintenance philosophy for those systems which are potentially contaminated with tritium or activated particulates.

When the reactor is shut down, initial cooling of the blanket and first wall is provided by one of the primary loops. However, as the heat load decreases and the coolant temperature drops, the residual heat-removal system takes over this task. This lower-pressure system is plumbed into both of the primary loops at the inlet and outlet ring manifolds so that cooling may be provided through either of the primary loop blanket circuits. The system is sized for a maximum heat load of 60 MW, the blanket and first-wall afterheat generation rate approximately 12 h after shutdown and the earliest time at which blanket maintenance would begin.

7.3.10. Tritium Systems

The fuel cycle for STARFIRE is shown in Fig. 7.11. The system is designed to reprocess tritium for fueling, process the tritium produced in the blanket, control the amount and location of tritium in the plant, and process tritiated wastes. Fully redundant and modular units, multiple processing paths, and location in a separate tritium facility provide maximum reliability, availability, and maintainability. The existence of multiple, isolated pathways reduces both the impact of a single unit's failure and the magnitude of an associated-tritium release.

The high fractional burnup (0.42) for STARFIRE results in a minimized tritium inventory in all fuel-processing systems. This reduces the magnitude of a possible tritium release in the reactor building to approximately 10 g (as T_2O). In the tritium facility as much as 50 g of T_2 could be released if multiple failures occur in an isotope separation unit.

The inventory in the STARFIRE plant is designated "vulnerable" or "nonvulnerable" (Table 7.17) depending on the degree of control that can be enforced on a system and also the physical state of the tritium in that system. The tritium within the blanket (10 kg) is considered "nonvulnerable" since it is tenaciously retained by the solid breeding material and thus is relatively immobile. The tritium in the pump and fuelers (in the reactor building) is considered "mobile." The total "vulnerable" inventory for STARFIRE is less than 400 g.

The design goal for tritium relases from the STARFIRE fusion plant is less than 5000 Ci/yr in all forms (gas, liquid, and solid waste) averaged over operating, standing, and maintenance phases and including in-plant releases. The sources and amounts of tritium release are shown in Table 7.17. The primary release pathway is leakage in the coolant system.

7.3.11. Electrical Storage and Power Supplies

The STARFIRE electrical system is somewhat simpler than that for a pulsed device due to the steady-state operation. There is no requirement for an energy-storage system, and the electrical energy required for startup of the power system is taken directly from the local power grid. Some capacitive energy storage is provided for use in conjunction with the correction-field (CF)

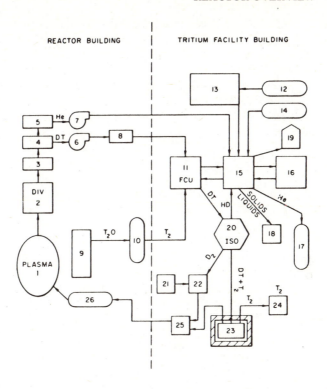

REACTOR BUILDING | TRITIUM FACILITY BUILDING

FIGURE 7.11. Fuel-cycle scenario for STARFIRE.

1. Plasma Chamber
2. Limiter Plates
3. Debris Separator
4. D-T Cryocondensation Pump
5. Helium Pump
6,7. Regeneration Pumps
8. Metal Bellows Pumps
9. Breeder Blanket
10. Electrolysis Unit
11. Fuel Cleanup Unit
12. Tertiary Enclosures
13. Atmospheric Tritium Recovery System
14. Secondary Enclosures, Purge Streams

15. Tritiated Waste Treatment
16. Tritiated Water Recovery Unit
17. Helium (tritium-free)
18. Tritiated Waste -- Liquids and Solids
19. Detritiated Gases: N_2, O_2, CO_2, Ar
20. Isotopic Separation Unit
21. D_2 Supply
22. D_2 Storage
23. DT and T_2 Storage
24. T_2 Shipment/Receiving
25. Fuel Blender
26. Gas Fueling

coil power supply. Major parameters of the electrical power system are shown in Table 7.18.

The TF coil power supply charges the TF coil and compensates for the power losses in the normal-temperature current-carrying devices, which make up the interconnections between superconducting coils. Because the coils are deenergized infrequently, the charging voltage (100 V) and the charging time

TABLE 7.17. Reference STARFIRE Tritium Parameters

Plasma Parameters	
Plant availability	0.75
Thermal power (MW)	4000
Energy per fusion (pJ)	3.22
Ion density (ions/m^3)	8.06×10^{19}
Plasma volume (m^3)	783
Particle-confinement time(s)	1.8
Reflection coefficient	0.9
Fractional burnup	0.42
Tritium Mass Flow Rates (g/d)	
Tritium burnup	536
Tritium fueled	1296
Tritium exhausted	760
Tritium bred	562
Deuterium Mass Flow Rates (g/d)	
Deuterium burned	360
Deuterium fueled	865
Impurity Mass Flow Rates (g/d)	
Helium exhaust	712
Iodine exhaust	~ 50
Protium Exhaust	~ 10
Carbon Exhaust	~ 10
Nitrogen Exhaust	~ 10
Oxygen exhaust	~ 10

(12 h) are moderate. Dump switches that open in the event of a TF coil fault are provided so that coil energy can be dissipated in dump resistors.

The EF coil power supply is a 1417 V system. A free-wheeling diode is connected across the power supply output terminals. In order to protect against loss of phase, failed thyristor, or over-current conditions, standard protection schemes and equipment are provided. A dump circuit is incorporated to limit the voltage across the coil to 85 kV during emergency shutdown. A form of liquid rheostat is utilized to increase the dump resistance from 0 to 0.8 Ω during the 2.5 s over which the energy is removed from the coil.

The CF coil power system provides the energy dissipated in the normal coils, bus work, thyristors, and so on. Energy is transferred between each of the four CF coils and a common energy-storage device (capacitors) via a dc-ac-dc convertor system. Energy is delivered to the coils through a force-commutated chopper switch and returned to storage from the coils by way of a force-commutated inverter, transformer, and rectifier.

The 12 V ohmic-heating-coil power supply with series-blocking diodes in the bus from the power supply is connected across the ohmic-heating coil. A series

TABLE 7.18. Electrical Power System Parameters

Toroidal-Field-Coil Power Supply	
Primary voltage	13.8 kV
Input power	3.2 MVA
TF-system charging voltage	100 V
TF-system charging time	12 h
Equilibrium-Field-Coil Power Supply	
Input voltage	13.8 kV
Input Power	290 MVA
Power supply voltage	1417 V
Ohmic-Heating-Coil Power Supply	
Primary voltage	13.8 kV
Input power	3.2 MVA
Charging voltage	12 V
Correction-Field-Coil Power Supply	
Primary voltage	13.8 kV
Power supply	3 MVA
Power supply voltage	5000 V
Energy storage capacity	10 MJ

combination of an adjustable resistor and a diode is in parallel with this superconducting coil. The series combination is connected so as to block current during the coil-charging period. The use of blocking diodes in series with the 12 V, 200 kA power supply protects the power supply from the coil reverse voltage and eliminates the necessity of mechanical switches to perform this function.

7.3.12. Support Structure

The primary requirement in the design of the reactor support structure system was to safely react all loads acting on reactor components resulting from magnetic, thermal, pressure, gravity, and seismic forces. Additional key considerations were the simplification of routine maintenance tasks (e.g., blanket sectors changeout) and the design of structural components to serve multiple functions where practical.

The design of the support structure system was strongly influenced by the large out-of-plane magnetic forces acting on the TF coils, which result from the requirement that all superconducting EF coils be located outside the TF coils. This requirement, which precludes "interlocking" of superconducting coils, was considered fundamental to achieving a maintanable reactor design for which a failed superconducting TF or PF coil could be replaced within acceptable downtimes.

The method selected to react the forces was to surround the helium vessel of each TF coil with a room-temperature vacuum tank and to rigidly connect the outer legs of the tanks of adjacent TF coils with thick vertical steel panels. The forces acting on each helium vessel are transmitted to the surrounding vacuum tank through struts fabricated from a high-strength fiberglass having low thermal conductivity. The tank beams the loads to the pinned joints, which attach the panel corners to the Dewar. The loads for each coil are equal in magnitude and opposite in direction on either side of the midplane; thus the loads in each panel are balanced out by attaching its vertical sides to the vacuum tank surface to effect a shear tie along the sides.

Other major components of the system are the centerpost, thermal isolation struts, blanket/shield support pedestals, upper and lower vacuum pump support assemblies, and EF coil vacuum tanks and supports. The centerpost, constructed of fiberglass laminate, reacts the centering forces from all 12 TF coils, and reacts a small part of the TF coil out-of-plane forces as two torques acting in opposite directions on the centerpost top and bottom. The centerpost loads are reacted to the building floor through a set of fiberglass thermal isolation struts.

The blanket/shield support pedestal is a beam-and-web structure, which attaches to the bottom of each TF coil vacuum tank to react loads from the blanket and shield sectors to the building floor. The overhead supports vacuum pump assembly and duct shielding. The two large-diameter EF coils are supported from the TF coil vacuum tanks by tripod-shaped beam assemblies.

All metallic components of the system are constructed of Nonmagne 30, a nonmagnetic austenitic manganese steel with low nickel and chrome content, selected to reduce activation of the structure and to minimize the use of resource-limited elements.

7.3.13. Cryogenic Systems

The cryogenic system for STARFIRE is a single central system that supplies the required quantities of liquid helium and liquid nitrogen to the user locations near the reactor and in the fueling facility. By far the largest cryogenic requirement is for the superconducting toroidal-field coils, with lesser requirements for the superconducting EF/OH coils, the fueling and tritium-processing system, and the vacuum-system cryopumps. The cryogenic system is a relatively straightforward system requiring no new technology, but should benefit from future improvements in reliability, economy, and operating efficiency.

Cryogenic refrigeration is supplied at two temperature levels. Vaporization of liquid nitrogen at 80 K is used for thermal shielding of the liquid-helium-cooled components and precooling of warm helium gas in the helium refrigerator−liquefier. This liquid nitrogen is supplied to the user systems from a central pressurized storage Dewar. The nitrogen gas is returned to the closed-cycle nitrogen liquefaction plant where it is condensed and returned to the

TABLE 7.19.   Cryogenic-System Parameters

Liquid helium supply rate	
TF coils	15,000 liter/h
EF/OH coils	6,600 liter/h
Other	4,900 liter/h
Helium refrigeration of 4.2–4.5 K	20 kW
Liquid helium storage	100,000 std. m^3
Gaseous helium storage	70,000 std. m^3
Liquid-nitrogen supply rate	
Helium liquefier–refrigerator	900 liter/h
TF coils	100 liter/h
EF/OH coils	50 liter/h
Other	450 liter/h
Liquid-nitrogen storage (at 0.3 Mpa)	40,000 liter
Total system electrical power	7 MW

supply Dewar. The nominal liquefaction capability of the nitrogen plant is 1,500 liter/h.

Refrigeration at 4.2 K is supplied by pumping liquid helium (LHe) from the central LHe storage Dewar through liquid-nitrogen (LN_2)-shielded vacuum-jacketed transfer lines to the location of the component requiring cryogenic refrigeration. The liquid helium is vaporized and returned to the central helium refrigerator–liquefier as either cold or warm gas depending on the user requirements. The return helium gas is liquefied and returned to the central supply Dewar. The return helium gas is liquefied and returned to the central supply Dewar. The nominal liquefaction capability of the liquid helium system is 26.5 kl/h. A large medium-pressure helium-gas storage facility supplies makeup gas to the system and is used to store the helium when the cryogenic system is warmed up.

The major parameters of the cryogenic system are listed in Table 7.19.

7.3.14. Instrumentation and Control

The conceptual design of STARFIRE has not included the detail required to completely specify and design the IC system. The system description anticipates future developments in IC technology.

It is estimated that up to 50,000 signals will require processing and that a 1-ms update rate will be required. Other key features of the system are:

Distributed network architecture

Integrated safety system

Multiple redundant elements for improved reliability

Modular instrumentation—integral with subsytems

Wireless data links—2.5 and 50 MHz

Architecture.   A distributed network should be a good match for the available computer and instrumentation technology, taking advantage of what appears to be a natural evolution toward distributed intelligence in control systems. In addition, it provides improved response for direct-digital-control-loop computation times compatible with plasma and electromechanical requirements. It further provides considerable reduction of network communications traffic to the control room through exchange of only summary and setpoint control information with local controllers.

Central-control-room displays are provided in the form of color graphic diagrams on flat panels of convenient size, which are capable of providing either summary information or any desired depth of detail to suit operational requirements at any particular time. Commands will be given by voice, leaving the operators free to concentrate on plant operations, except where hand/ eye or simulated touch are appropriate, as in controlling remote-handling equipment.

A central data base is maintained at the supervisory level, which centralizes information requests and updates. This central data base is also used by other supervisory-level computer systems. These service the control-room displays; provide historical records of normal and alarm conditions; provide sequencing information for startup, shutdown, and response to unusual conditions; and provide archive access to outside users, such as power dispatch controllers, utility management, and regulatory agencies.

Access is also provided to a plant simulation computer, for use in evaluating unusual situations and predicting plant responses to various commands. Since this system is capable of 100 times concurrent real-time simulations, it is assumed to be a large supercomputer located remotely and shared by many plants of the STARFIRE design.

Integrated Safety System.   Safety functions have been integrated into the design of the STARFIRE IC System. It is anticipated that this will be possible because of hardware advances that will provide electrical isolation between various sensors and measuring subsystems, distribution of intelligence to multifold dedicated subsystem controllers, and economical multiply-redundant elements at all levels of the IC System. This will provide operators with all information relevant to any potentially hazardous situations.

Redundancy.   Redundancy will be employed extensively to permit reliable operation with a minimum of system downtime, particularly in the area of the reactor itself.

Modularity.   It is intended that instrumentation be integrally designed with subsystems and components. All plasma instrumentation is intended to be similarly carried with and installed as part of the blanket modules and be modular with the annual blanket replacement cycle, so that a complete new redundant set of all required instrumentation will be installed each year. This means that such instrumentation will have sixfold redundancy and need only be designed for a 6-yr service life in the reactor environment.

Wireless Data Links.   Use of wireless transmission is designed to achieve electrical noise and fault isolation, reduction of labor and containment-wall penetrations, and elimination of radiation effects on cables and connectors. In addition, maintainability and replacement are simplified. These links may use light, microwaves, or both.

Remote Maintenance.   Requirements for remote maintenance will dictate many features of the reactor subsystems designs, but systems using industrial robot technology will be available for performance of the required functions remotely. It is expected that by the time of STARFIRE, such systems will be capable of performing such tasks as visual recognition of objects, grasping, feeling, hearing, and connecting basic operations and movements to perform complex operations with a minimum of human supervision.

Development Areas.   It is anticipated that much of the technology required for STARFIRE IC will be developed on natural evolutionary lines, independent of the fusion-energy program. Three areas requiring specific development have been identified, however. These include the development of radiation-resistant sensors, electronics, and optical components; provision for a plasma access as a deliberate part of reactor design; and development of fully engineered plasma instrumentation modules and systems.

7.3.15.   Material Inventory

The material requirements for a single STARFIRE reactor have been tabulated and are presented in Table 7.20. The material requirements for the remainder of the plant would typically be common with current generating systems.

7.4.   BALANCE OF PLANT

The balance of plant facilities for STARFIRE are a combination of features common to any large power plant, and elements peculiar to the fusion technology. For example, the steam generators, turbine–generator, and main condenser components of the power conversion system are generic to power plants. The tritium-reprocessing facilities, the electrical and rf power supply building, and the hot cell, in which fully remote repair and maintenance functions are performed, are unique to a fusion-power plant.

In the overall plan of this fusion-power plant, closely related facilities are combined into the same buildings that are then located to achieve a functional

TABLE 7.20. Materials Inventory for Each STARFIRE Reactor Categorized by Material

Material	Initial Requirements		Life of Unit Requirements[a]	
	Volume (m^3)	Mass (tonnes)	Volume (m^3)	Mass (tonnes)
PCA				
FW	3.6	28.6	30.6	214.5
SW	2.6	20.5	22.1	153.8
Breeder	22.2	174.7	188.7	1310.3
Reflector	5.7	44.7	48.4	335.3
Jacket	19.0	149.3	161.5	1119.8
Headers and manifolds	4.1	32.2	34.9	241.5
		450.0		3375.2
H$_2$O (Primary)				
FW	2.0	2.0		
SW	1.3	1.3		
Breeder	11.3	11.3		
Reflector	5.7	5.7		
Headers and manifolds	29.5	29.5		
H$_2$O (Shield)	9.3	9.3		
Inner shield				
Outer shield	144.8	144.8		
Vacuum duct shield	175.0	175.0		
Zr$_5$Pb$_3$[b]				
Multiplier	36.8	328.0	73.6	656.0
LiAlO$_2$ (60% enriched)[b]				
Breeder	178.1	606.5	356.2	1213.0
C[b]				
Reflector	102.0	164.0	204.0	328.0
W[d]				
Inner shield	44.0	840.0		
Outer shield	0.0	0.0		
Vacuum duct shield	0.0	0.0		
		840.0		
B$_4$C				
Inner shield	26.3	66.0		
Outer shield	144.8	362.0		
Vacuum duct shield	174.8	437.0		
		865.0		
Ti-64				
Inner shield	0.0	0.0		
Outer shield	25.9	117.0		
Vacuum duct shield	6.6	30.0		
		147.0		

TABLE 7.20. *(Continued)*

Material	Initial Requirements		Life of Unit Requirements[a]	
	Volume (m³)	Mass (tonnes)	Volume (m³)	Mass (tonnes)
TiH₂				
Inner shield	0.0	0.0		
Outer shield	336.0	1310.0		
Vacuum duct shield	86.1	336.0		
		1646.0		
Fe-1422 (low-Mo steel)				
Inner shield	12.1	95.9		
Outer shield	487.0	3858.0		
Vacuum duct shield	736.0	5830.0		
rf duct shield	6.9	55.0		
TF magnet vacuum tank	123.0	974.0		
Common Dewar	8.8	69.9		
Antitorque panel	132.4	1050.9		
Blanket/shield pedestal	52.8	418.0		
Equipment/coil suports	4.1	32.2		
OH magnet vacuum tank	6.9	54.9		
EF magnet vacuum tank	31.4	249.1		
		12,687.9		
Cu				
TF coil stabilizer	179.3	1598.0		
OH coil stabilizer	8.6	77.1		
CF coil conductor	15.7	140.0		
EF coil stabilizer	41.9	372.9		
		2188.0		
C-10 insulator				
TF coils	32.1	61.0		
OH/EF coils	45.8	87.0		
Centerpost supports	0.7	1.4		
Centerpost	147.0	279.3		
		428.7		
Nb₃Sn				
TF magnet	6.4	51.0		
NbTi				
TF magnet	8.9	57.0		
OH, EF magnets	5.1	33.0		
Ta5W[b,c]				
Limiter	1.9	32.0	3.8	64.0

(continued)

TABLE 7.20. *(Continued)*

	Initial Requirements		Life of Unit Requirements[a]	
Material	Volume (m³)	Mass (tonnes)	Volume (m³)	Mass (tonnes)
Be				
FW coating	0.5	1.0	4.3	8.5
Limiter coating	0.1	0.2	0.9	1.7
		1.2		10.2
304 Stainless Steel				
TF magnet helium tank	143.0	1124.0		
TF coil	293.1	2304.0		
OH magnet helium tank	12.8	101.0		
OH magnet helium tank	10.2	80.2		
EF magnet helium tank	62.1	488.0		
EF coil	49.4	388.0		
		4,485.2		

[a]For reactor only life-of-unit requirements are same as initial requirements except where otherwise indicated.
[b]Life-of-unit requirements for these materials are twice the initial requirements, to account for anticipated recycling within a short period of time following removal from the reactor.
[c]Tantalum (as Ta-5W) is only one of several candidates for the reference design limiter.
[d]Tungsten can be replaced by another shielding material.

and economical layout. The principal buildings and their spatial relationships are summarized below:

1. The reactor building contains the tokamak fusion reactor and supporting systems.
2. To the south of the reactor building is the turbine and support building that contains the energy-conversion equipment, a reactor service area in which blanket sections and other new reactor subsystems are prepared, a plant auxiliary area that houses the closed-loop cooling water system, and the hot cell.
3. The electrical and rf power supply building is placed to the east of the reactor building and the tritium-reprocessing and cryogenics buildings is located to the north of the reactor building.
4. The administration, facility control, and site service complex is located to the south of the turbine and support building.

The balance of the plant facilities are shown on the site plan of Figure 7.1 and in layouts of Chapter 20 of Reference 1.

7.4.1. Site Plan

Principal site elements include buildings, roads, walks, fencing, and surface and subsurface mechanical and electrical equipment and utilities, most notably the large natural draft hyperbolic cooling towers. The site elements are located and arranged to minimize piping, electrical, and utility runs and at the same time provide adequate separation between buildings and other elements.

7.4.2. Facility Buildings

The reactor building is a steel-lined, hardened ribbed box structure constructed of reinforced concrete designed for DBE seismic loading and 0.16-MPa (24-psig) internal pressure. Modifications of the primary coolant loop have resulted in decreasing the maximum anticipated over-pressure to 100 kPa (15 psi). The 1.5 m thick building outside walls and roof prevent penetration by tornado missiles, withstand a tornado-induced differential pressure, and provide adequate shielding.

The reactor building houses the reactor and related systems that can potentially become contaminated. These systems are modularized to permit removal. A partial height, shielding wall and sliding doors separate the module area from the reactor to minimize activation of materials in the system module area. The pressure boundary of the building is lined with steel to provide a leak-tight boundary and minimize tritium release in the event of a loss of primary coolant within the building.

The reactor building is 120 m long and 50 m wide (inside clear dimensions), and consists of three levels in addition to the crane maintenance balcony. A 600-tonne-capacity bridge crane spans the width of the building and runs its entire length. The reactor and system modules are located on the ground level.

The first level below ground is a pipe chase. All pipelines, conduits, and other connections between the reactor and the modules are routed through the pipe chase to provide clearance for the remote-maintenance equipment on the ground floor. The pipe chase, reactor hall, and system module areas are subject to over-pressure under accident conditions. Access to the pipe chase is by two removable hatches, one at each end of the building on the ground floor.

The second level below ground is the routinely occupied subgrade floor, which is not subject to over-pressure, significant radiation, or a tritium atmosphere. The rf system components and HVAC mechanical and electrical equipment are located on this level.

The remote-maintenance system is employed in the reactor building and also extends to the hot cell. The equipment runs on a solid monorail track, arranged to allow movement of system components into position to perform remote maintenance and repair functions on the process module equipment, the ATR systems, the HVAC equipment, as well as the reactor itself.

In addition to the floor-mounted equipment, the overhead bridge cranes and electromechanical bridge manipulators assist in the maintenance functions by providing access to the top and upper equipment areas of the reactor and various modules not accessible to other systems. The 60-tonne bridge crane at the process module end of the building is capable of lifting and removing process equipment for repair and replacement.

Preparation of blanket segments and other new reactor subsystems, equipment, and parts is carried out in the reactor service area, a ground-level high bay area between the hot cell and turbine building, and adjacent receiving end of this building, where there is also access for large vehicles and other equipment. Storage spaces for new blanket segments and other reactor and process module components are provided.

The hot cell is a carbon-steel-lined, concrete-hardened structure designed for DBE seismic loading. The building outside walls and roof are thick enough to prevent penetration by tornado missiles, to withstand a tornado-induced differential pressure, and to provide adequate shielding.

The hot cell contains three process areas. The liquid-waste processing equipment separates activated particulate matter from tritiated water that is then returned to the tritium-reprocessing building. The remote-maintenance and repair shop contains equipment for shield door seal and latch mechanism replacement and for repair of failed process module components. The blanket module and solid-waste processing cell contains the equipment required to bake out tritium from spent blanket segments and to reduce the segments into pieces conveniently sized for storage in canisters. Both wet and dry solid-waste storage facilities are provided in the blanket module and solid-waste processing cell. Wet storage for 600 m^3 of spent blanket segments and dry storage for 400 m^3 of discarded shield doors, pumps, valves, and piping is available.

The portion of the tritium-reprocessing building that contains potentially contaminated areas is hardened in accordance with applicable seismic and tornadic design criteria, and is lined with carbon steel to minimize the escape to atmosphere of potential tritium release from process equipment. Tritium-free areas of the building are occupied on an every day basis and are separated from potentially contaminated areas by an air lock. Piping to and from the reactor building and hot cell is double walled, with quick-acting isolation valves located close to the building walls. The piping is routed to the reactor building pipe chase in an underground tunnel.

7.4.3. Mechanical Systems

The power-conversion system consists of components of conventional design for use in large, central generating stations. The thermodynamic cycle and its components resemble the thermal cycle of a pressurized-water-reactor plant. The 4000 MW of thermal energy generated by the reactor is transferred to the power-conversion system by the heat-transport system.

The facility heat-rejection system consists of three hyperbolic natural draft cooling towers; eight wet pit, vertical-type centrifugal pumps; a chlorination package; an evaporation pond; and a raw water reservoir for makeup. The cooling towers are reinforced concrete, approximately 150 m in diameter at the basin, 100 m in diameter at the top, and 165 m in height. The pump house provided for the pumps and chlorination package contains a well with individual compartments for each pump and filter.

The closed-loop-cooling water system provides demineralized and deionized cooling water to reactor auxiliary components and reactor building cooling systems during normal plant operation. It consists of three half-capacity pumps, three one-third-capacity heat exchangers, a surge tank, and the necessary piping, controls, and instrumentation. The system is connected to the standby cooling-water system to provide cooling water to the atmospheric tritium cleanup systems and the solid-waste pools during normal operation. The connections include isolation valves that close automatically when the standby cooling-water system goes into operation, or when the closed-loop-cooling-water system is shut down for any reason.

The standby cooling-water system provides demineralized and deionized cooling water to the residual-heat-removal system, atmospheric-tritium-removal systems, solid-waste pools, and control building chillers during abnormal operating conditions. Operation of the system occurs, for example, during an incident when the reactor has been shut down, offsite power is unavailable, and the plant is using the onsite standby power source. The system operates whenever the reactor is shut down and maintains the blanket temperature within specified limits by removing the residual or decay heat. The system cools the blanket indirectly by providing the cooling medium to the residual-heat-removal system that in turn cools the blanket. The standby cooling-water system also provides cooling water to the three atmospheric-tritium-removal (ATR) systems located in the reactor building, the tritium-reprocessing building, and the hot cell. Components of the system include a dry cooling tower, two full-capacity pumps, a surge tank, and the necessary instrumentation and controls.

7.4.4. Building Services

The reactor hall is provided with six recirculation systems by which the atmosphere gas (CO_2) is filtered and conditioned. The recirculation systems are divided into two groups of three units each, supplied from two different power sources. A total of four units operate normally, and two units are on standby, thereby allowing flexibility of operation. In the event of a loss-of-coolant accident, the fan-coil units are capable of maintaining the temperature of the reactor building atmosphere at or below 54.4°C.

The reactor hall is at a negative pressure with respect to the environment, and two 100% capacity pressure control fans are provided. When the concentration of tritium within the reactor hall is detected to be above the normal concentration, the pressure control fans are deenergized, and automatic isolation of the building takes effect. This operation calls for automatic shift of the atmospheric air cleanup system from the normal 10% to full-capacity operation. The atmospheric cleanup system provides for removal of tritium and/or particulates within the reactor hall.

The hot cell employs 100% recirculation of filtered and tempered air to the different zones. Two 100% capacity pressure control fans are provided to maintain differing pressure requirements among confinements zones. The assignment of pressure zones ensures confinement of the potentially contaminated atmosphere and prevents its migration to other parts of the facility. Pressure assignments are such that the flow of air is from areas of lesser to areas of higher potential for radioactive contamination. The design incorporates isolation of the hot cell when it is determined that the concentration of tritium in the hot cell atmosphere is beyond the normally acceptable concentration. Upon isolation, the atmospheric-tritium-removal system located inside the hot cell runs at full capacity.

The tritium-reprocessing area, which has the highest potential for radioactive contamination within the tritium-reprocessing building, is provided with a dedicated recirculation HVAC system. Minimum outdoor purge air for ventilation is filtered and conditioned for supply to the area. As in the hot cell and the reactor hall, it is expected that the atmospheric tritium removal system will operate at 10% capacity during normal operating modes. Full capacity operation will be in effect upon the detection of an unacceptably high concentration of tritium in the area.

7.4.5. Electrical Systems

The plant is designed with adequate auxiliary electrical equipment, standby power, and protection to ensure operation of the essential station auxiliary equipment during normal operation and all emergency conditions.

Offsite power is used for startup and shutdown, that is, the plant is not designed with "black starting" capability. Power is generated continuously so that an intermediate-energy-storage system is not required. Major electrical equipment associated with power generation, auxiliary electrical power supply and distribution, onsite standby power generation, and extra-high-voltage (EHV) switchyard systems have been identified. These systems and equipment are within state-of-the-art technology or technology currently being developed.

A double-bus system is used as a minimum to supply power to equipment required for continuous operation and/or orderly shutdown. The distribution of power to two or more identical items of equipment is such that a failure of a power supply bus does not result in a complete loss of a particular mechanical or electrical function. All vital control and protection systems and equipment are powered from uninterruptible power supply (UPS) systems.

A function of the main generator and its connections is to generate 1440-MW$_e$ power (gross), to deliver power to the plant unit auxiliary transformers, and to deliver the net electrical power through the main step-up transformer and switchyard to the EHV transmission system network. The generator is provided with a breaker to allow isolation of the generator during startup, shutdown, and maintenance, while the main step-up transformers and the unit auxiliary transformers supply offsite power for the auxiliary system.

It is anticipated that a minimum of two full-capacity EHV transmission lines will be connected to the switchyard of the plant. The primary function of the switchyard is to provide an onsite EHV switching facility that can provide a power outlet for the unit and that can receive and provide the offsite power required for startup and shutdown when the main generator is off the line. The arrangement of the switchyard is based on a double-bus, breaker-and-a-half scheme that provides the flexibility and reliability required for a plant of this size.

The auxiliary electrical power supply and ac distribution system provides continuous power to the plant auxiliary equipment for startup, normal operation, and shutdown. Each auxiliary electrical power supply bus has two sources of power supply, one from the unit auxiliary transformers and the other from the reserve station service transformer is implemented when the unit auxiliary power supply fails. Electrical loads that are vital for an orderly shutdown of the plant are grouped separately. Feeders to coil and rf power supplies are duplicated or divided so that a single switchgear bus or feeder failure does not result in loss of a function.

Power at 480 V is provided by double-ended load-center-type substations located throughout the plant approximately in the center of their respective loads. Each transformer of a double-ended unit is fed from a different 13.8-kV bus for better flexibility and service continuity.

Reliable and continuous dc battery power is required for a variety of uses, such as "trip" and "close" of electrically operated circuit breakers, solenoid-operated valves, control systems, vital lighting, the station annunciators, and devices used during turbine coast-down. Through dc/ac inverters, the batteries are also used to provide power to the turbine–generator and reactor-protection system, and to instrumentation required for an orderly shutdown.

The function of the onsite standby ac power system is to generate an onsite ac source of auxiliary power if the preferred offsite source is lost. This system consists of two redundant, onsite gas-turbine generators. Each generator is connected to a pair of redundant 4160-V switchgear bus systems. The connected loads on these buses consist of equipment and systems required for an orderly shutdown of the plant.

Facilities electrical services provide adequate lighting including emergency lighting; provide the necessary grounding for systems and equipment, including a separate grounding system for "low"-signal-level instrumentation and controls; provide necessary systems for communication, including telephone, public address (PA and PAX), and sound-powered telephone systems for testing and maintenance where required; provide adequate lightning protection; and provide necessary cathodic protection.

Electrical equipment is arranged and located to minimize length of runs for interconnections and to facilitate maintenance and testing. Major electrical equipment is located in separate electrical rooms with adequate space provided for cable and raceway connections. Raceways are grouped for ease of installation and common mechanical protection. They are dedicated to the particular type of cable, and those containing different types of cable are separated. Raceways and cables serving the onsite standby ac power systems and associated with redundant equipment required for an orderly shutdown are physically separated and isolated from their redundant counterparts, and are designed to withstand DBE seismic forces.

7.5. PLANT CONSTRUCTION

The schedule for plant construction shows that 6 yr will be required between the time the first concrete is poured until initial power delivery. The pacing time in the construction schedule is the reactor building and reactor. Approximately 3 yr is required to prepare the reactor building for initial installation of the reactor components; 2 yr is required for reactor erection and 1 yr is allowed for reactor and plant tests. Reactor building construction proceeds until the roof over the reactor area is complete and the 600-tonne overhead crane is complete. One end of the reactor building is left open and temporary crane rails extended to permit direct entry of major reactor components. Modularization, with factory assembly and checkout, of reactor components is used where possible. Offsite winding of magnets and fabrication of blanket and shield sectors is planned with water shipment to the site. The overall construction schedule is discussed in Chapter 23 of Reference 1. Further reductions in overall construction time may be achievable with advances in modularization of plant components and improvements in current transportation capabilities.

7.6. OPERATION AND MAINTENANCE

A goal of the STARFIRE design has been to maximize utility compatibility not only in current practice but also with anticipated trends for future utility operations. The Utility Advisory Group has provided insight to current utility practice.

Startup power is drawn from the grid and after plasma initiation the reactor power is brought up slowly to minimize the thermal-stress effects on the blanket and steam generators. Once operating, the plant has the ability to load-follow at a rate of 5% of rated power increase or decrease per minute, although the plant is designed as a base-load unit. The plant will normally operate continuously with one scheduled shutdown per year for maintenance. It is anticipated, however, that 1–3 other major shutdowns per year will occur as a result of component failures. Once the reactor is shut down, it can be restarted to full power in approximately ½ h; however, approximately 12 h will be required for a restart if the TF coils are discharged for maintenance or if the vacuum chamber has been breached briefly. After major vacuum chamber breaches 36 h is required for restart.

Three major startup conditions are anticipated:

1. Startup after maintenance that requires opening of the vacuum chamber to the building environment. Opening the vacuum chamber requires the blanket sectors be cooled down to near ambient temperature and the TF coils to be discharged (e.g., for blanket replacement). Thirty-six hours are required to clean the vacuum walls for startup.

2. Startup after maintenance at the reactor where it is unnecessary to breach the vacuum. TF coil discharge is necessary to permit maintenance equipment manuevering about the reactor. TF coil recharging will require 12 h. In addition, if the primary coolant loop has been shut down, approximately 13 h is required for warm up.

3. Startup after shutdown not requiring reactor maintenance (turbine trip, etc.). The TF coil charge is maintained and the primary loop is kept hot.

The plant is expected to operate at full power under normal conditions; however, the fueling ratio can be adjusted to reduce the power. In event of redundant component failures the reactor power can be reduced to the appropriate level and operation continued until the failure can be corrected. Redundant systems, where continued operation at reduced power is possible, include the vacuum system where power reduction is necessary after four pump failures and the steam generators where shutdown for excessive leakage would require operation at 50% power during repairs.

Shutdown sequence of the reactor is normally the reverse of startup and requires approximately 1 h to reach a hot shutdown and approximately 20 h to reach a cold shutdown.

The STARFIRE reactor and maintenance facility designs have been developed based on the assumption that a mature fusion economy exists and that all facets of the design have been demonstrated in previous power plants. The premise that a fully understood technical basis exists for STARFIRE leads to the assumption that components can be replaced periodically to prevent unscheduled failures from dominating the maintenance needs. The plant availability goal is 75% and the system reliability requirements have been established accordingly based on the projected time for replacement of system components.

The complexities of a fusion system will reduce the achievable availability as compared with PWR if the same technology is assumed. However, advances in automated maintenance technology and simultaneous development of the maintenance system and reactor designs are believed to offer benefits that make the 75% availability goal reasonable. The maintenance system must receive as much attention from the outset of the design as any other major system in order to achieve this goal. The reactor maintenance schedule was developed to fit within the typical utility balance-of-plant maintenance scenario.

It includes an annual shutdown for 4 weeks (28 days) to perform maintenance and inspection, and a 4-month (120 days) shutdown every 10 yr for overhaul of the turbine–generator. During this period the TF coil system is also annealed. This results in an average annual scheduled outage of 37 days.

In addition, current utility experience indicates that approximately 20 days downtime per year is caused by failures in the balance-of-plant (BOP). As a result of these constraints the following downtime allocations were made as listed in Table 7.21.

The 37 days of reactor scheduled maintenance is derived from the assumption that the reactor will be maintained simultaneously with the BOP on a noninterference basis. The total of 57 days outage required for BOP maintenance leaves 34 days for reactor unscheduled maintenance if the 75% availability (91 days outage/year) is to be achieved.

A total remote-maintenance facility has been designed for all equipment located within the reactor building and hot cell. The major benefit is that this approach minimizes radiation exposure to maintenance workers. All known or foreseen maintenance operations are planned as remote; however, operational flexibility has been provided by design of the reactor shielding to permit personnel access within 24 h after reactor shutdown.

The reactor design was developed to permit removal of all components; however, the vacuum pumps and isolation valves and the blanket require scheduled, routine maintenance. Shield door seals, limiters, and rf grilles are replaced with blanket sectors. Ease of replacement of these items has therefore been emphasized. Reactor scheduled maintenance does not dominate annual plant downtime, therefore, several scheduled operations can be added without affecting reactor availability. Failure of life-of-plant items such as the magnets or shield can be permitted to result in longer outages since their failure will be infrequent.

The blanket sectors are expected to require replacement every 6 yr. An annual replacement sequence is planned whereby four sectors are replaced each year to level out the maintenance tasks and permit the reactor scheduled maintenance to fit within the normal utility practice of a 4 week/per yr outage for the balance-of-plant maintenance. Blanket sectors are removed as a unit with limiters, rf grilles, and ECRH ducts in place. When the sectors are changed, shield door seals are replaced. Total time for replacing two sectors is ~ 10 days.

Vacuum pumps and valves require replacement every 2 yrs, hence one-half are replaced at each scheduled outage. Total time for replacement of one-half the vacuum pumps and valves is 20 days annually.

TABLE 7.21. Downtime Allocations

Balance-of-Plant		Reactor	
Scheduled	Unscheduled	Scheduled	Unscheduled
37 days	20 days	37 days	34 days

The maintenance facilities consist of the reactor building, the hot cell, the reactor service area, and the remote-maintenance control room. Remote-maintenance developments in major nuclear facilities have been utilized in developing the STARFIRE maintenance approach and design.

The reactor building contains the reactor, selected support system modules, and required maintenance equipment. The reactor is located at one end of the building. The support system modules are organized within the central area. The other end of the building contains the crane bay maintenance area, the personnel airlock to the crane bay maintenance area, the equipment airlock between the hot cell and the main floor of the reactor building, laydown space for the support system modules, and parking space for the reactor building maintenance machines. The reactor and the support system are maintained with (1) equipment that is mounted on a monorail system; (2) overhead cranes; and (3) bridge-mounted electromechanical manipulators. All viewing is by means of remote CCTV. Building area lighting and special lighting in maintenance areas are required.

A shield wall isolates the reactor from the support system modules and permits maintenance of the modules while the reactor is operating. This wall serves as a neutron shield for the modules and as a missile shield for the reactor. Doors provide ingress and egress for the monorail system.

The support system modules located in the reactor building are limited to only those subsystems that are potentially contaminated with significant radioactive contamination. These are:

Tritium processing and cleanup

Reactor building HVAC

Coolant water

Steam generators

Primary-loop components

Limiter-feedwater-loop components

Vacuum pumps

The modules are designed for total remote maintenance using monorail-mounted maintenance machines and the overhead cranes and electromechanical manipulators. The modules are positioned within the reactor building to provide adequate access for the remote-maintenance machines. Maintenance of the support system modules is accomplished by removing and replacing a whole module or by removing and replacing a failed subcomponent within a module. The failed module or subcomponent is transported to the hot cell, on the monorail, for further repair or disposal.

A single monorail system provides maintenance access to the reactor and to the support system modules and allows equipment and material movement in the reactor building and the hot cell. The large monorail track provides stability for the maintenance machines as well as the required load-carrying capacity. While a maintenance machine is in position to perform a given task, additional stability is obtained by locking the machine to the monorail. Power and control for the maintenance machines is supplied by busbars on the monorail. Machines have been defined for maintenance of the reactor. The machines are preprogrammed for all planned activity. Intermediate checkpoints will be programmed into operational sequences to stop equipment for verifying position and for performing inspections.

The hot cell is located outside of the reactor building to localize contamination products and permit independent operation. An equipment air lock connects the reactor building to the hot cell. Communication between the hot cell and the reactor building is via a monorail system. The equipment airlock and hot cell are sized to handle all reactor building components.

The hot cell consists of a central corridor containing the monorail. At one end of the corridor is the airlock entrance to the reactor building. At the opposite end of the corridor is a decontamination chamber. The decontamination chamber is connected to the turbine and support building by another airlock. All processing, handling, and storage within the hot cell is on either side of the monorail corridor. Shield doors isolate the monorail corridor from the cells. Turntable switches connect the monorail with these cells.

Specific tasks performed in the hot cell are:

Blanket disposal
Solid-waste packaging
Holdup treatment of nontritiated liquid and gaseous wastes
Remote maintenance of activated components
Decontamination of nonactivated components for out-of-cell handling
Emergency tritium cleanup
Wet and dry storage of activated components

Maintenance and repair of activated components takes place in the hot cell remote-maintenance-and-repair shop. All maintenance and repair work is done using servo-manipulators, overhead bridge cranes, and bridge-mounted electromechanical manipulators. Typical maintenance operations include reactor shield door seal renewal, reactor shield door latch repair, rf duct repair, limiter replacement, blanket repair, fuel system repair, vacuum pump isolation valve repair, support system module and subcomponent repair, and maintenance equipment repair. Once repaired, testing and check-out of the refurbished components takes place. Some repair work on nonactivated components may be carried out in the hot cell. A case-by-case evaluation will be necessary to determine if decontamination is more expedient than remote maintenance.

Decontamination of nonactivated components takes place in the decontamination chamber. From the decontamination chamber the components can enter the reactor service area through an airlock. The reactor service area provides hands-on repair and testing of decontaminated, nonactivated components.

7.7. POWER FLOW

The reactor delivers 4000 MW$_{th}$ to the power-conversion system which generates 1440 MW$_e$. A circulating load of 240 MW$_e$ results in 1200 MW$_e$ of power deliverable to the grid. A power flow diagram is shown in Fig. 7.12.

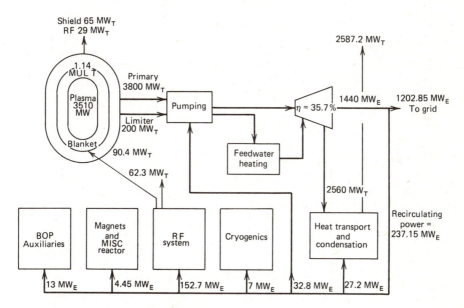

FIGURE 7.12.   Power flow diagram.

7.8. ECONOMICS

The total direct and indirect costs are summarized in Table 7.22 for both 1980 constant-year dollars and 1986 then-current dollars. Also shown is the total busbar energy cost for the fusion-generated energy along with the major contributors to the costs. These costs are higher than are currently being projected for new fission plants. However, it must be noted that the cost of energy shown here is for the initial year of operation. The levelized cost of energy (for 30-yr economic life) is actually comparable to those estimated for future LWRs and is lower than that for coal. The reason for this is that, for a given power plant, the cost of energy increases from year to year because of escalation in fuel, maintenance, and operation costs. The cost of fuel is negligible in STARFIRE while it represents ~ 25% and 40% of the cost of energy in LWR and coal, respectively.

The cost estimate of STARFIRE is specific to this system, reflecting the unique ground rules applying to this study. It should be noted that this study is a preconceptual design with some subsystems not fully developed or defined.

TABLE 7.22. STARFIRE Summary Costs

Account Number	Account Title	Costs (1980 $ $\times 10^{-6}$)
20	Land and land rights	3.30
21	Structures and site facilities	346.58
21.02	Reactor Building	157.44
21.03	Turbine Building	35.92
21.07	Hot Cell Building	53.69
21.99	Contingency Allowance	44.95
22	Reactor plant equipment	968.62
22.01	Reactor Equipment	589.26
22.01.01	Blanket and First Wall	82.36
22.01.02	Shield	186.07
22.01.03	Magnets	171.57
22.01.04	RF Heating and Current Drive	33.49
22.01.05	Primary Structure and Support	52.74
22.01.07	Power Supply, Switching, and Energy Storage	52.90
22.02	Main Heat-Transfer and -Transport Systems	69.84
22.05	Fuel-Handling and Storage Systems	38.60
22.06	Other Reactor Plant Equipment	43.75
22.06.01	Maintenance Equipment	38.30
22.06.02	Special Heating Systems	0.00
22.06.03	Coolant Receiving, Storage, and Make-Up Systems	0.24
22.06.04	Gas Systems	0.08
22.06.05	Inert Atmosphere System	0.00
22.07	Instrumentation and Control	23.41
22.07.01	Reactor IC Equipment	7.61
22.07.02	Monitoring Systems	1.76
22.07.03	Instrumentation and Transducers	14.04
22.98	Spare Parts Allowance	66.38
22.99	Contingency Allowance	117.68

Code	Item	
23	Turbine plant equipment	249.68
23.01	Turbine–Generators	77.33
23.03	Heat-Rejection Systems	44.34
24	Electric plant equipment	117.28
25	Miscellaneous plant equipment	40.77
	Total direct cost	1726.48
91	Construction facilities, equipment, and services (10%)	172.65
92	Engineering and construction management services (8%)	138.12
93	Other costs (5%)	86.32
	Subtotal	2123.57

	1980 Constant	1986 Then-Current
94 Interest during construction	246.70	671.69
95 Escalation during construction	0.00	462.63
Total capital	2400.27	3197.89
$/kW$_e$	2000	2665

Total Busbar Energy Cost

	Annual Cost ($ × 10^{-6})	
	1980 Constant	1986 Then-Current
Annualized Cost of Capital	240.43	479.68
Operations and Maintenance	19.41	26.01
Scheduled Component Replacement	17.36	23.26
Fuel	0.33	0.44
Total Annual Cost	277.13	529.39
Cost of Electricity (mill/kWh)	35.1	67.1

For these subsystems, the cost estimates were determined with implicit design allowances to account for the lack of complete definition. Fortunately, many of the balance-of-plant and heat-transport systems are similar to those of PWR systems, thus enhancing the cost credibility. The direct capital costs associated with the reactor plant equipment, the balance-of-plant equipment, land, and all the related structures and site facilities were estimated based on supplier quotes, historical data, and analogous systems. The indirect costs related to construction are assessed based on DOE recommendations with modifications specific to this design. Time-related costs account for both interest and escalation during construction. The annual costs include the annualized capital cost, the operations and maintenance costs, the fuel costs and any scheduled component replacement costs. Given these costs along with the plant capacity (net power output) and the plant availability, the busbar energy cost is determined. These costs are presented in both constant year 1980 dollars and then-current year dollars, which represent a nominal first-year facility cost.

7.9.   SAFETY

The incorporation of safety considerations into the design process, even at the conceptual stage as in STARFIRE, is done to ensure that the environmental and safety advantages inherent in fusion are fully realized. The emphasis on safety must include the concern for the safety of the general public, the personnel, and the plant itself—in that order.

Fusion power will have several significant safety advantages compared to current methods of generating electricity. The nuclear-safety aspects are decidedly improved when compared to fission reactors: the problems of accidental criticality and of prompt criticality are not applicable; the prospects and consequences of a loss-of-coolant accident are less; and the biological hazards of radioisotopes in the plant are much lower. Generally, the concerns regarding safeguarding against diversion of weapons-grade material, such as Pu or U, are eliminated. Fusion, like fission, does not involve combustion in air; thus the routine chemical releases are much lower than for fossil power plants. The dangers due to fuel mining and other associated activities, including transportation, will be greatly reduced. Radioactive-waste-storage requirements will be less complicated owing to the absence of fission products and actinides. Low-level-radioactive-waste production should be less than for fission plants. Radiation doses to the general public due to routine or accidental releases of radioactivity will also be reduced.

Most recent safety evaluations of light water reactors (LWRs) and fast breeder reactors (FBRs) have concluded that, for an adequate analysis to be conducted of public risk associated with the different concepts, the total fuel cycle must be examined. In this regard, utilizing $D-T$ for fuel should be preferable to LWRs or FBRs. The deuterium and lithium involved are not radioactive. Only the initial, start-up requirements of tritium, which is approximately 10 kg, need to be shipped to the plant. The rest of the tritium fuel cycle is contained within the plant. The breeding ratio of STARFIRE is sufficiently low that shipment of excess tritium from the site is not necessary.

It is important to note that no runaway-type accidents that would affect the public or the plant personnel have been identified by this study. No method of generating electricity is capable of completely eliminating environmental impact and risk to society; however, fusion will reduce the adverse effects and potential impacts to very low levels.

The primary emphasis in this study of deterministic methods rather than on probabilistic methods was due mainly to the timing involved. It was not possible to do a quantified probabilistic risk assessment, owing to lack of sufficient design details, statistical operating data, and physical models pertaining to hazard rates. Future efforts undoubtedly will be directed at performing a detailed probabilistic risk assessment.

In the context of a Preliminary Hazards Analysis (PHA), the following sources of hazards were identified for the STARFIRE design:

Tritium inventory

Induced activity in the first wall, blanket, magnet, shield, and structural materials

Pressurized-water primary coolant

Corrosion products in the primary coolant

Stored energy in the superconducting magnet system

Cryogenics

Kinetic and self-inductive energy associated with plasma current

RF heating

The following potential accidents were identified for STARFIRE. Safeguards were incorporated into the design to the extent possible, which prevent their occurrence or limit the damage:

Tritium release, both in a continuous and a pulsed mode

Loss of coolant flow to the first wall and/or blanket

Failure of the resistive dump for the superconductive magnet system

A superconducting magnet becoming locally normal

Gross rupture of the magnets' helium cryogenic system

Failure of first wall due to plasma disruption

Production of missiles as the result of an accident

Hydrogen detonations or explosions

REFERENCE

1. C. C. Baker et al., *STARFIRE—A Commercial Tokamak Fusion Power Plant Study*, Argonne National Laboratory Report ANL/FPP-80-1, September 1980.

8 | MARS: A TANDEM-MIRROR FUSION-POWER PLANT

8.1. INTRODUCTION

Mirror confinement of plasma is one of the oldest concepts in fusion. The basic physics of mirror confinement are described in texts such as Miyamoto [1] and in Sections 5.3 and 5.7 of this book. A review of mirror-reactor studies has been given by Moir [2]. Reactor designs, such as those carried out in the late 1970s by Livermore Laboratory scientists and engineers [3] indicated that single-cell-mirror fusion reactors are possible, but they require very high technology, such as 15-T superconducting coils, very high voltage, negative ion, neutral beams, and direct energy convertors. All single-cell-mirror reactor designs suffered from high recirculating power flow. The designs were judged workable and feasible, but the single-cell mirror is fundamentally "lossy," and the high recirculating power made the concept appear very marginal with respect to economics and eventual commercial success.

In 1976 scientists in the United States and the USSR independently proposed the tandem-mirror concept as a means to overcome the fundamental leakage of plasma from a mirror [4,5]. One way to view the tandem-mirror concept is that magnetic mirrors can be used to electrostatically "plug" the ends of a linear magnetic solenoid. A preliminary study of a reactor based on the tandem-mirror concept was published by the Livermore design team in

252

1977 [6]. That design resulted in Q values (Q = fusion power/injection power) of about 5, whereas the single-cell mirror had $Q \sim 1$. Although the solenoidal magnet (100 m long) and blanket are relatively simple, the end plugs employ very high technology. The mirror plugs needed 1.2 MeV injection energy to achieve good magnetic confinement. The plasma density in the central solenoidal cell was $n \sim 1 \times 10^{20}$ m^{-3}, and to achieve adequate potential barrier plugging, the mirror plug requires $n \sim 8 \times 10^{20}$ m^{-3}. The magnitude of the central-cell potential is determined by the required balance of ion and electron loss rates. The space potential in an open-ended mirror system is positive, since electrons are scattered into loss-cone regions significantly faster than are ions. Equation (8.1) results from electrostatic and pressure force balance for Maxwellian electrons along field lines in a simple tandem mirror. ϕ is the difference in electric potential between the higher-density plasma in the end plugs and the lower-density plasma in the central-cell solenoid. The ambipolar potential barrier at each end is

$$e\phi \approx kT_e \ln\left(\frac{n_{\text{plug}}}{n_{\text{solenoid}}}\right) \tag{8.1}$$

and the mirror magnetic fields needed are 17 T. Recirculating power was 43% of the 1000 MW$_e$.

In 1979 Baldwin and Logan introduced the thermal barrier concept into the tandem mirror configuration [7]. The main purpose of the thermal barrier is to thermally insulate the hot electrons in the end plug from those in the solenoid central cell. Their basic idea is illustrated in Figure 8.1, which shows the potentials of a basic tandem mirror and a tandem miror with a thermal barrier. In the basic tandem mirror the increased number density in the plug, relative to the central cell, causes an increase in the positive potential. Therefore, some of the ions in the central cell are confined by this positive potential barrier. By depressing the potential between the central cell and the end plug, the electrons in the central cell and the plug are thermally isolated; the potential decrease appears to the electrons as a barrier. By heating the electrons, for example with ECRF, the final potential peak needed to plug the ions in the central cell can be generated with a much lower plasma density in the end plug relative to the density in the solenoid. It is this large reduction in end plug density that is the advantage of a thermal barrier.

To create a potential depression, a mirror region is formed and its density is decreased by pumping out ions that become trapped in this region. Ion density decrease can be accomplished, for example, by using neutral-beam charge-exchange pumping or radio-frequency radial-drift pumping. The decreased ion density depresses the passing electron density, which in turn causes the potential to decrease. In addition, the potential can be further decreased by ECRF heating in the barrier to increase the mirror-trapped electron population.

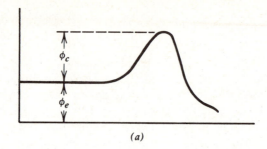

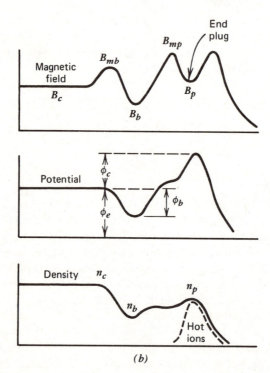

FIGURE 8.1. *(a)* Potential profile of a basic tandem mirror. *(b)* Potential, magnetic field, and density profiles of a tandem mirror with a thermal barrier (from Reference 7).

There are a number of different thermal-barrier configurations. One, called the inside barrier, has the thermal barrier and the final potential peak in separate mirror regions with the barrier between the central cell and the plug. Another version of the inside barrier has the barrier mirror region inside a yin-yang magnet pair placed at the end of the central cell. Yin-yang magnetic coil geometry evolved from the single-cell-mirror concept. This evolution is illustrated in Figure 8.2. The potential peak exists in the mirror region between the yin-yang magnet pair and a cee-shaped coil placed outside the yin-yang. The thermal barrier is created in the yin-yang coil by ECRF heating and ion

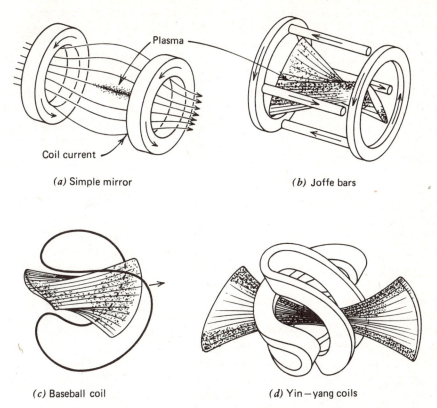

(a) Simple mirror   (b) Joffe bars

(c) Baseball coil   (d) Yin–yang coils

FIGURE 8.2. *(a)* A simple magnetic-mirror cell with the axisymmetric field concentrated at the ends (the mirrors) to reflect ions back toward the center. *(b)* Current in Joffe bars imposes transverse multipole field on the simple mirror field resulting in a magnetic minimum-*B* at the center. *(c)* The single baseball coil produces the same minimum-*B* field configuration more efficiently than the Joffe-bar system. *(d)* The two nested yin-yang coils produce the same minimum-*B* field but provide greater flexibility by permitting different currents in the two coils, and thus differing strengths of magnetic mirrors.

pumping. Neutral-beam injection and ECRF heating create the final plug. Mirror-confined hot electrons in the yin-yang help provide mhd stability for the system.

The thermal barrier reduced the technology demands on the magnets and beams needed for the tandem mirror, and increased the Q values to between 10 and 50. The Livermore reactor design team, using a thermal-barrier concept, together with tandem-mirror plugs, carried out preliminary calculations for a 500 MW$_e$ reactor with $Q \sim 11$ for a 56 m long central-cell system. With thermal barriers the central-cell-confining potential (see Reference 7) is

$$e\phi_c = kT_e\ln\left[\frac{n_p}{n_c}\left(\frac{T_{ec}}{T_{ep}}\right)^v\right] + k(T_{ep} - T_{ec})\ln\frac{n_c}{n_p} \tag{8.2}$$

where the subscript p refers to the potential peak, b to the potential minimum of the barrier, and c to the central cell. The exponent ν accounts for the non-Maxwellian ion distribution [8] and has the limiting values: $\nu = \frac{1}{2}$ for a magnetic well and $\nu = \frac{3}{2}$ for a magnetic hill. Improvements and variations of these ideas are still taking place in the tandem mirror with thermal barriers. Detailed developments of tandem mirror physics are provided in References 9 and 10. A commercial design of a tandem-mirror, thermal-barrier reactor has been developed by a design team at the University of Wisconsin. Their design is called WITAMIR-I [11]. The more recent MARS tandem-mirror reactor design is discussed in the following sections.

8.2. MARS OVERVIEW*

The MARS tandem-mirror fusion commercial power plant design was developed by the Lawrence Livermore National Laboratory group supported by teams from TRW, General Dynamics, EBASCO, SAI, Grumman, and the University of Wisconsin. Details of the design are given in Reference 12. The main objectives of the MARS project were to design an attractive fusion rector producing electricity and synfuels in a tenth-of-a-kind commercial-stage development and to exploit the potential of fusion for safety, low activation, and simple disposal of radioactive waste.

MARS is a linear magnetic-mirror fusion device using electrostatic plugs to confine a steady-state fusion plasma in a long solenoid. A 140 m long solenoid will produce about 1200 MW$_e$ net. An artist's view of the MARS powerplant is shown in Figure 8.3. The main fusion plasma in the central cell of MARS is self-sustained by alpha heating (i.e., ignition) while continuous injection of neutral beams and electron-cyclotron-resonant heating are required to maintain the plug electrostatic-confining potentials to 150 kV for the ions and 200 kV for the electrons. Plasma exhaust and vacuum pumping are accomplished in large vacuum tanks at the ends, where direct conversion of escaping ion and electron energy recovers about as much electrical power as is consumed by the plug beams and ECRH.

The MARS end plugs are the axicell type using yin-yang magnets. All reactor magnets are steady state because no plasma current is needed to provide plasma equilibrium or stability. Steady-state fueling will be accomplished in part by neutral beams in the plugs and by injection of high velocity (10^4 m/s) pellets near the ends of the central cell. Impurities and thermal alpha ash are kept to tolerable concentrations (less than 10%) by radial transport. The alpha ash and impurities diffuse into an outer radial plasma zone, or halo, which is not electrostatically plugged. This natural diverter sweeps the impurities into the large direct converter tanks, thereby minimizing plasma bombard-

---

*The name MARS comes from the first letters of mirror advanced reactor study.

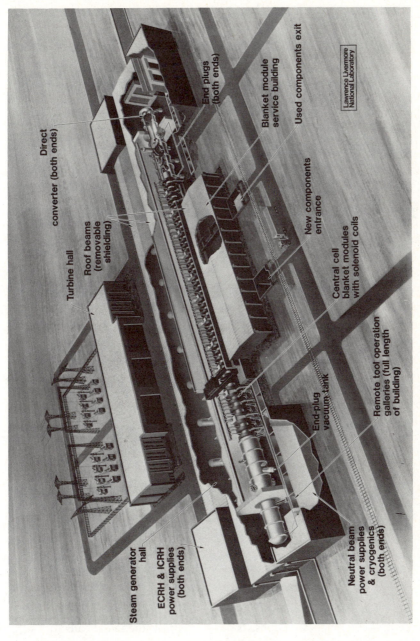

Steam generator hall

ECRH & ICRH power supplies (both ends)

Direct converter (both ends)

End plugs (both ends)

Blanket module service building

Turbine hall

Roof beams (removable shielding)

Used components exit

New components entrance

Central cell blanket modules with solenoid coils

End-plug vacuum tank

Neutral beam power supplies & cryogenics (both ends)

Remote tool operation galleries (full length of building)

Lawrence Livermore National Laboratory

FIGURE 8.3. Overview of the 1200 MW$_e$ MARS tandem-mirror fusion-power plant.

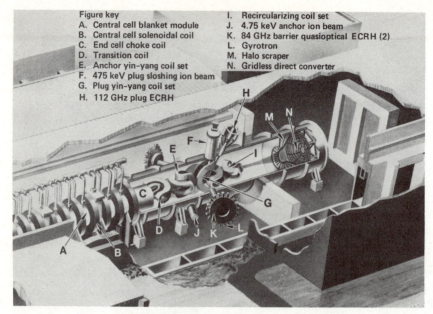

Figure key
A. Central cell blanket module
B. Central cell solenoidal coil
C. End cell choke coil
D. Transition coil
E. Anchor yin-yang coil set
F. 475 keV plug sloshing ion beam
G. Plug yin-yang coil set
H. 112 GHz plug ECRH

I. Recircularizing coil set
J. 4.75 keV anchor ion beam
K. 84 GHz barrier quasioptical ECRH (2)
L. Gyrotron
M. Halo scraper
N. Gridless direct converter

FIGURE 8.4.   MARS end-plug configuration.

ment and surface heating of the first wall in the central cell. The MARS endplug configuration is shown in Figure 8.4.

Some of the major MARS parameters are listed in Table 8.1. The plasma radius (0.49 m) is relatively small because the average central cell beta is high. This small radius permits relatively small-bore magnets: 2.4 m inner radius for the central-cell solenoid, 30 cm inner radius 24 T hybrid coils, and 3 m major radius yin-yangs.

There are plugs on each end, and both ends have direct energy convertors. The configuration of the containment building and the tritium containment walls are shown in Figures 8.3 and 8.4.

MARS employs a high-temperature, high-efficiency lithium—lead ($Li_{17}Pb_{83}$) blanket. A cross section of the central cell is shown in Figure 8.5. A high-performance double-wall heat exchanger is used to help achieve an overall steam cycle efficiency of 40%. The thin, 38 cm thick, blanket has an energy multiplication of 1.39.

8.3. MAGNETS

The central-cell magnet set consists of 44 identical solenoidal coils with axial spacing (coil midplane to midplane) of 3.16 m. Conductor bundles for each coil are 0.884 m wide and 0.932 m thick with an inner radius of 2.44 m. The total current in each central cell magnet is 12.1 MA. The nominal on axis central-cell

TABLE 8.1. MARS Parameters

Fusion power	2600 MW
Thermal power	3400 MW
Gross electric power	1500 MW
Net electric power	1200 MW
Direct convertor power	300 MW
Recirculating power	300 MW

Central-Cell Parameters

Plasma gain (Q)	26
D$-$T density	3.3×10^{20} m^{-3}
Alpha density	2.0×10^{19} m^{-3}
Ion confinement ($n\tau_i$)	5.2×10^{20} s/m^3
Electron confinement ($n\tau_e$)	5.3×10^{20} s/m^3
Alpha confinement ($n\tau_\alpha$)	0.2×10^{12} s/m^3
Central-cell magnetic field	4.7 T
Maximum magnetic field	24 T
Neutron wall loading	4.3 MW/m^2
Beta (volume average)	0.28
Plasma radius	0.49 m
First-wall radius	0.60 m
Ion temperature	28 keV
Electron temperature	24 keV
Cold-fueling current	2.1 kA
Central-cell length	131 m

End-Cell Parameters

Barrier	
Peak field	24 T
Transition/anchor	
Length	13.3 m
Passing ion density at field maximum	1.7×10^{19} m^{-3}
D$-$T particle trapping current	800 A at each end
Hot-ion density in anchor	4.2×10^{19} m^{-3}
Hot-ion energy in anchor	690 keV
Beta (volume average) in anchor	0.5
Plug	
Length	5.3 m
Passing-ion density at barrier	4.7×10^{18} m^{-3}
Trapped-ion density at barrier	8.2×10^{18} m^{-3}
Sloshing-ion density at barrier	9.4×10^{18} m^{-3}
Warm-electron temperature at potential peak	124 keV
Hot-electron energy at barrier	840 keV
Beta (volume average) in plug	0.5

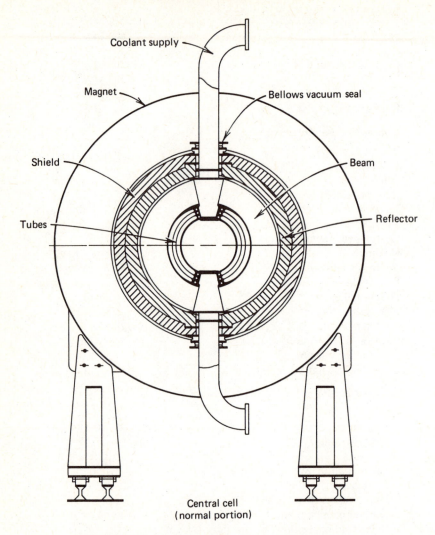

FIGURE 8.5.   A cross section of the MARS central cell.

magnet field is 4.7 T. The field ripple is about 3.8%, which causes a reduction in the mhd stability limit for beta in the central cell. Calculations indicate that this degree of field ripple degrades the stability limit by only a few percent and that, with no ripple, the central-cell beta stability limit is 70%.

The MARS end plug coils, together with their magnetic and potential profiles, are shown in Figure 8.6. The central-cell magnetic field, is followed by a barrier choke coil whose peak field is 24 T. The transition coil flares the plasma flux bundle. Two yin-yang magnet pairs provide the magnetic anchor and plug fields. The anchor provides the primary good magnetic curvature for mhd stability (hence the name anchor) and is the region for ICRH. The second

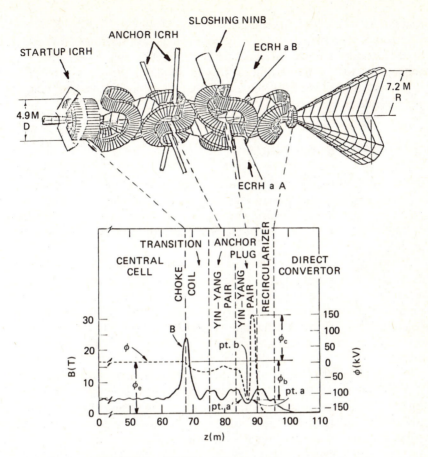

FIGURE 8.6. MARS end-plug magnetic arrangement.

yin-yang magnet pair is the plug, and it is here that a neutral beam is injected to develop the sloshing ions. Two beams of ECRH are employed to develop the thermal barrier. A cee-coil follows the plug to recircularize the plasma flux bundle. Finally, at the end, is the direct energy convertor, in which the circular flux bundle expands to about 7.2 m radius.

The potential in the central cell is slightly positive, then becomes negative in the transition and anchor regions. In the plug, the potential becomes highly negative (point b in Figure 8.6), and then very positive (about 150 kV).

The purpose of the choke coil is to limit the velocity distribution of central-cell ions that can pass to the end plug, thus reducing the requirement for trapped-particle pumping in the transition and barrier regions. The high mirror field is a magnetic reflector. The high magnetic field strengths of the choke coil are produced by a hybrid arrangement consisting of a superconducting outer coil and normal-conducting insert coils, all simple solenoids.

High-field superconducting coils using graded regions of NbTi, Nb_3Sn, and titanium-doped Nb_3Sn superconductors cooled with LHe-II at 1.8 K, are used to achieve higher fields with smaller coils. Radiation-resistant polymide and spinel insulators are employed to increase radiation tolerance of the super-conducting and resistive magnets. Design details, alternate considerations, engineering tradeoffs, stress analysis, radiation damage considerations, etc., can be found in Reference 12.

8.4. PLASMA ENGINEERING

Physics constraints together with technology tradeoffs make the design of a tandem-mirror end-plug system very complex, difficult, and interesting. Some of the constraints and engineering problems and solutions in the MARS design are briefly discussed here. The interested reader should refer to Reference 12 for further details.

The $m = 1$ ballooning criteria sets the limit for mhd stability but flute interchange is also important. A mirror-confined plasma also becomes mhd unstable if the radial electric field is too large. The plasma rotates as a result of azimuthal drift motion resulting from the radial electric field and the solenoidal magnetic field. For MARS, the on-axis central-cell potential is reduced with respect to ground, which helps overcome the possible flute interchange instability.

As central-cell ions pass through the transition region, collisions will cause some of them to scatter in a sufficient pitch angle to become trapped in this region. With a provision for removing or pumping these trapped ions, the density will increase and desired plasma conditions will be destroyed. The high density of trapped ions will cause the potential in the transition region to increase, and, because many of the trapped ions have turning points in the anchor, they will also increase the density and potential at the thermal barrier. The decreased thermal insulation between central-cell and warm electrons will cool these electrons and lower the confining potential of the center-cell ions. An additional effect is the increase in pressure in the transition region to a value that can cause mhd instability. The solution to this class of problems is to pump the trapped ions out of the transition region. This can be done using neutral-beam pumps or radio-frequency radial drift pumping at the trapped-ion frequency. Drift pumping requires an order of magnitude less power than neutral beams and helps remove impurities as well.

A sloshing-ion distribution with a peak to midplane density ratio of 3 to 1 is produced in the yin-yang anchor to help provide plasma microstability. Both ECRH and ICRH are employed in order to obtain the desired temperature profile in the plugs. Almost all the trapped injected power plus the alpha power streams out the ends of the MARS reactor in the form of energetic charged particles. By separating the positive and negative currents, electric power can be directly recovered. This is very important for the overall MARS energy efficiency. The potential profile in the MARS reactor is shown in Figure 8.6.

8.5. ASH REMOVAL

To obtain ignition and good reactor performance most of the 3.5 MeV alphas produced from D−T fusion must be magnetically trapped in the central cell until they slow down to thermalized energies of about 40 keV. After they slow down they must be disposed of so they do not dilute the reacting plasma. Computations indicate that the tolerance of thermalized alphas and impurities is only about 10% of the central-cell plasma density.

For a reactor generating 2,500 MW of fusion power, the rate of alpha production is 135 A. Of these, the fraction magnetically trapped in the central cell is the cosine of the loss-cone angle; that is, about 94%. Fokker−Planck computations indicate about 25% additional alpha particle losses will come from Coulomb scattering. The allowable lifetime of a thermalized alpha particle, in order to keep the alpha density less then 10% of the central-cell density, is found to be 3.5 s.

Thermalized alphas will be very well contained electrostatically because of their double charge and low energy compared with the confining potential. The ash and impurity removal technique chosen is radial transport and drift pumping of the alphas and impurities as they pass through and become trapped in the transition region.

8.6. PLASMA PARAMETERS

A self-consistent calculation of all the plasma parameters in MARS involves extensive computer simulation. Details of these calculations can be found in Reference 12. In Table 8.1 are shown MARS central cell and plug plasma parameters.

ECRH is required to sustain the hot-mirror-confined electrons at the barrier location and the warm-electron population at the location of the potential peak.

The largest single steady-state heating requirement in the plugs is the ECRH—82 MW for both plugs. Of this, 78 MW at 84 GHz forms hot-mirror-trapped electrons at the plug midplanes (called the thermal barrier) and 4 MW at 112 GHz heats warm electrons at the outermost plug regions, which generates and controls the potential plugging central-cell ions.

For the sloshing-plug ions to withstand the large electrostatic forces tending to expel them, they must be injected at very high energy (475 kV), which requires negative ion beams. With injection at the inboard-fan turning point, it is estimated that 56% of the incident neutral-beam current is trapped, so that a 475 kV, 10 A beam must be delivered to each plug.

The only other significant plug heating requirement is 4 MW of absorbed ion-cyclotron resonant heating to maintain a high beta (50%) in the anchor yin-yang cells, for mhd stability. A few (2.1) amperes of hydrogen gas or low-energy beams are used for fueling.

8.7. ENGINEERING AND BALANCE OF PLANT

Data on the waste management from MARS are described in Reference 12. They are similar to that described for other fusion reactor designs such as STARFIRE.

The lithium–lead blanket design was chosen to produce electricity in the most cost-effective manner. A high-temperature blanket (350/500°C) is used to help achieve cost-effective power.

Shielding is required to protect the central-cell solenoids from neutron and gamma radiation. The blanket is a poor attenuator of low-energy neutrons because of the large volume fraction of lead, and a separate shield is needed. For an estimated 5 FPY (full power years) before the copper magnet insert suffers its first anneal, the damage rate in copper should not exceed 2.2×10^{-5} dpa/FPY. Polymides used for irradiated magnet insulators are 5–10 times more resistant to damage than comparable epoxies, and are therefore used in the central-cell coils.

Tritium is bred in the $Li_{17} Pb_{83}$ eutectic at the rate of 0.42 g/day and is extracted at this rate by vacuum pumping. Secondary and tertiary containment systems are designed to minimize personnel exposure to tritium and environmental effects of tritium by preventing its escape by permeation, leakage, or system malfunction.

The blanket and shield reflector are illustrated in Figure 8.7. The blanket for the central cell develops 2081 MW_{th}, and uses HT-9 structural steel. The neutron wall loading is 4.3 MW/m^2 and the maximum dpa/FPY is 60.7. The blanket/reflector multiplication is 1.36 and the tritium breeding ratio is 1.15.

The basic power cycle, thermal hydraulics, and power components for MARS are similar to other modern thermal power plants. The circulating

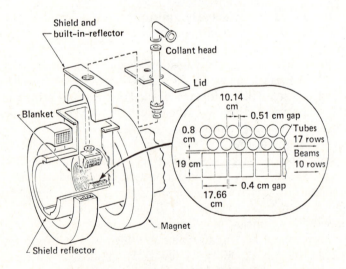

FIGURE 8.7. MARS Li–Pb power blanket.

power is 20%. Maintenance of MARS is somewhat easier than that of tokamaks because of the linear geometry. The interested reader is referred to details described in Reference 12.

REFERENCES

1. K. Miyamoto, *Plasma Physics for Nuclear Fusion*. MIT Press, Cambridge, MA, 1980.
2. R. W. Moir "Mirror-Reactor Studies," *Proc. IEEE* **69**, 958 (1981).
3. R. W. Moir et al., *Standard Mirror Fusion Reactor Design Study*, Lawrence Livermore National Laboratory, UCID-17644, January 1978.
4. G. I. Dimov, V. V. Zakaidakov, and M. E. Kishincvshy. *Fiz. Plazmy*, **2**, 597 (July/Aug. 1976); see also *Proc. 6th IAEA International Conference* Berchtesgaden, Germany, 1976.
5. T. K. Fowler and B. G. Logan, "The Tandem Mirror Reactor," *Comments Plasma Physics Controlled Fus. Res.* **II**, 167 (1977).
6. R. W. Moir et al, *Preliminary Design Study of the Tandem Mirror Reactor (TMR)*, UCRL-52302, July 1977.
7. D. E. Baldwin and B. G. Logan, *Phys. Rev. Lett.* **43**, 1318 (1979).
8. R. Cohen *et al.,* UCRL-84147 (1980).
9. *Physics Basis for MFTF-B*, D. E. Baldwin, B. G. Logan, and T. C. Simonen (eds.), UCID-18496, Parts 1 and 2, January 16, 1980.
10. *Physics Basis for an Axicell Design for the Eng Plugs of MFTF-B*, D. E. Baldwin and B. G. Logan, (eds.), UCID-19359, April 21, 1982.
11. B. Badger et al., *WITAMIR-I*, University of Wisconsin Report UWRDM-400, September 1980.
12. *Mirror Advanced Reactor Study, Final Report*, UCRL-53333-83, C. D. Henning et al., 1983.

9 | OTHER FUSION REACTOR CONCEPTS

9.1. INTRODUCTION

Fusion is a field of research and development that is rich in ideas. There are many different concepts for achieving fusion energy, and in this chapter we examine several that are actively being explored. None of them, as yet, has been experimentally tested to the extent of the tokamak concept. None of them has the number of reactor design studies nor the design detail that exists for the tokamak concept. Nonetheless, each concept has its advocates, and reasons why it may, in the long run, be superior to a tokamak power reactor. No one knows which, if any, of the present concepts will be successful in achieving a truly commercial energy resource. It is expected that, if research and development funding is available in sufficient quantity, these alternate concepts will be explored further and for those that survive the tests of experiment, analysis, and marketplace criteria, reactors will be developed.

Plasma physics relevant to an alternate concept may be substantially different than that of tokamak confinement. The basic objectives, however, are the same—namely, to heat a light element plasma (usually deuterium and tritium) to ignition and confine it sufficiently long to permit a substantial net energy gain from fusion reactions. Fusion concepts differ in geometry, time and length scales, input power requirements, and the technology that is employed. Some alternate magnetic fusion concepts have been summarized and reviewed by Baker et al. [1] and inertial-confinement reactor concepts have been surveyed by Monsler et al. [2]. A number of different fusion reactor designs have been presented in a 1981 special issue of *Nuclear Engineering and Design* [3]. Every hopeful magnetic-confinement fusion approach, in the opinion of Teller, is discussed in the two volumes he edited, and these provide advanced treatment of many of the topics described more briefly in this chapter [4].

9.2. STELLARATORS, HELIOTRONS, AND TORSATRONS

In the history of controlled fusion one of the earliest and most beautiful confinement concepts is the stellarator. It is a toroidal configuration that achieves rotational transform, closed magnetic surfaces, and equilibrium by means of external currents carried in toroidal and helical coils. The earliest stellarators used separate toroidal and helical coil sets in either a figure 8 or racetrack configuration. Current in the helical windings of a stellarator flows in opposite directions in alternate windings. A heliotron is similar to a stellarator but has the alternate helical windings removed. The torsatron concept eliminates the toroidal coils, achieving its magnetic trap from a single set of helical windings. Schematics of stellarator and torsatron windings are illustrated in Figure 9.1. The number of helical windings carrying current in a given direction passing through the poloidal plane is exactly equal to the l number. The internal magnetic topology of stellarators, heliotrons, and torsatrons are, for the same l number, very similar and, therefore, these three confinement concepts are treated together. Stellarator magnetic surfaces for $l = 1, 2, 3$, and 4 are shown in Figure 9.2a, and torsatron $l = 3$ surfaces are illustrated in 9.2b. Stellarator research has been reviewed by Miyamoto [5] and Shafranov [6]. A comprehensive and informative scientific and technical status report on stellarators and recommendations for future work has been recently published by a joint US−Euratom steering committee [7].

The mechanical stresses in the windings of a large stellarator device are very substantial, making its design very difficult. The torsatron, on the other hand, can be created by using nearly force-free helical coils. Recently, it was suggested that the helical coils be replaced by distorted (twisted) toroidal coils [7]. This allows the generation of torsatron magnetic surfaces from a highly modular system, with lower stresses than in comparable stellarator configuration. A modular torsatron constructed from twisted toroidal coils has important maintenance and manufacturing advantages. Illustrations of a modular stellarator coil and a torsatron torus are shown in Figure 9.3.

The potential advantages claimed for the stellarator−torsatron as a fusion reactor are:

1. Steady-state operation.
2. Capable of ignition and low recirculating power.

3. Ash and impurity removal by means of the helical poloidal divertor that naturally occurs in this magnetic geometry.

4. Free of magnetic disruptions.

5. High-aspect-ratio design affords good maintenance access.

The stellarator-torsatron concept can operate without any toroidal plasma current. Plasma heating may be achieved by neutral beams and/or RF heating. The beta stability limit of these helical devices is uncertain but has been estimated to be similar to that of the tokamaks. Plasma-energy-confinement time is also thought to scale in a way similar to that found for tokamaks. Some parameters of recent preliminary stellarator/torsatron reactor designs are summarized in Reference 1.

9.2.1.   The UWTOR—M Stellarator Power Reactor*

A modular stellarator reactor study was carried out by a Wisconsin design group to evaluate its feasibility and competitiveness [8]. A beta of 6% was chosen arbitrarily. The design objective was to maximize rotational transform and shear, avoid magnetic island formation, and provide an adequate magnetic volume within a practical coil system.

A 5000-MW_{th} size unit was selected. The reactor has 18 modular, $l = 3$ twisted coils, as shown in Figure 9.4. Note that there are only two different coil geometries in this coil set. Ignition is calculated to be achieved with ICRF heating of 80−100 MW for about 5 s. Table 9.1 lists the primary parameters of this reactor. The reactor is housed in a toroidal building, which is maintained at a reduced pressure of 10^{-6} torr. The cross section of the stellerator reactor in its evacuated building is shown in Figure 9.5. The building is steel lined and reinforced to withstand 1.5 atm of over-pressure.

The modular $l = 3$ magnet coils were chosen to maximize rotational transform, and they constitute a difficult design challenge. They are large, having an 11 m outside diameter, and heavy, each weighing about 960 tonnes. The forces on each coil come from two sources: the self-force due to their own field and the mutual attractive force between adjacent coils. At the twist extremities where the coils are close, the mutual attractive forces are substantial and result in a large toroidal force. The poloidal forces, when integrated around the coil, produce a net centering force. The self-field produces radial forces that attempt to expand the coil. The designers found that the best way to sustain the radial force was to use a radial ring surrounding the coil at its midplane. This ring is welded to the coil frame at the points of intersection and is enclosed within its own Dewar. The coil operates at 1.8 K with He-II and has NbTiTa in the 11.6-T region. Calculations show that the coil stresses are below two-thirds of the yield limit of 304 LN stainless steel. The centering force on the coils is taken up

---

*This section is taken from Reference 8.

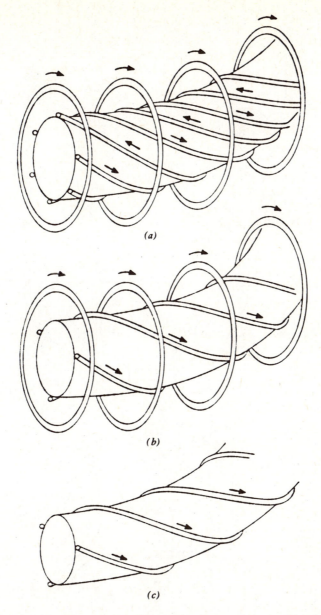

FIGURE 9.1. Windings for *(a)* stellarators, *(b)* heliotrons, and *(c)* torsatrons; heliotrons are similar to stellarator coils but with alternate helical windings removed. l refers to the number of helical windings carrying current in a given direction; in the above drawings, all are $l = 3$.

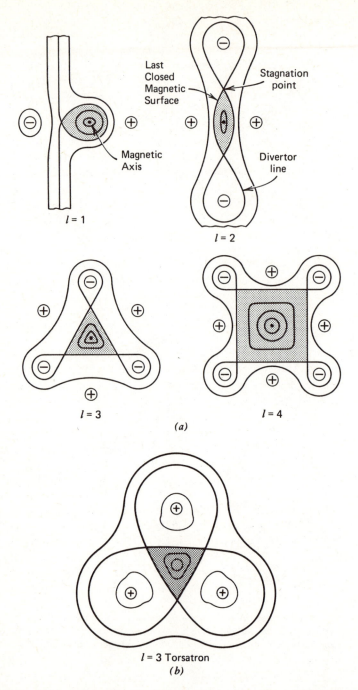

FIGURE 9.2. (a) Magnetic surfaces showing separatrix points, and separatricies for $l = 1, 2, 3,$ and 4 stellarators. (b) Magnetic surface for $l = 3$ torsatron; from Reference 7.

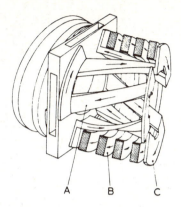

A   B   C

FIGURE 9.3. *(a)* Modularized classical stellarator coil for an $l = 3$ system. *A* is the helical segment, *B* is the toroidal-field coil, and *C* is the connection of helical segment. *(b)* Modular $l = 3$ twisted-coil system approximating opposing torsatron windings, from Reference 7.

by the structural ring bearing against a central support structure. Toroidal forces are handled by allowing the coils to make contact with each other at the twist extremities.

The UWTOR-M reactor blanket is made of ferretic steel HT-9, and the breeding material is $Li_{17} Pb_{83}$ with 35% enriched ^6Li. The breeding material is static, and has steam tubes immersed in it. Steam enters at 330 °C and exits at 500 °C. Tritium diffuses into the steam and then is removed in the same way as heavy water is extracted. The resulting tritium inventory in the blanket is 180 g.

Impurity control is achieved with a magnetic divertor. A natural divertor occurs in a stellarator as a consequence of a magnetic separatrix that bounds regions of closed nested flux surfaces. Inside the separatrix magnetic flux links all the coils. Outside the separate flux links only some of the coils; and some flux therefore emerges between magnets. The divertor is modular in a modular coil stellarator. Divertor targets are used to recover the energy at a high temperature and to prevent neutron streaming through the divertor slots. This is achieved using cylindrical actively cooled shields surrounded with a rotating (100 rpm) cylindrical graphite surface. The divertor targets can be seen in Figure 9.5. The hot graphite radiates to cooled surfaces in the divertor target housings and is then converted for use in the overall power cycle.

The blanket, reflector, and shield were designed to give adequate tritium breeding of 1.08 and relatively high energy multiplication of 1.15. The superconducting coils are sufficiently shielded so that they will not have to be annealed during the reactor lifetime.

The reactor is fueled with $D-T$ pellets. The He ash, which goes through the divertor system, is exhausted into the reactor building and is subsequently pumped out by cryopumps. The impurities are separated by a fuel clean-up unit, and the hydrogen species then go to an isotope-separation system. The total tritium inventory in the plant, including a day's supply of fuel (14.9 kg), is 16.7 kg.

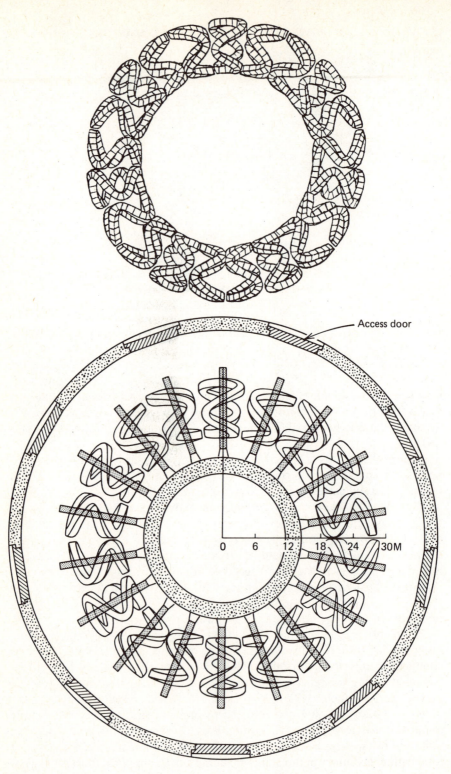

Access door

TABLE 9.1. UWTOR-M Reactor Parameters

D−T fusion power, MW_{th}	4300
Major radius, m	24.09
Minor coil radius, m	4.77
Plasma aspect ratio	14
Assumed average β,%	6
Rotational transform on edge	1.13
Number of field periods	6
Number of coils	18
Field on axis, T	4.5
Maximum field on the conductor, T	11.6
Average ion density, m^{-3}	1.46×10^{20}
Average ion temperature, keV	9.8
Ion-energy-confinement time, s	3.7
$n\tau$ average through plasma, s/m^3	5.4×10^{20}
Centerline Z_{eff}	1.28
Average neutron wall flux, MW/m^2	1.41
Global breeding ratio	1.08
Blanket energy multiplication	1.15
Total thermal power, MW_{th}	4820
Gross electric power, MW_e	1898
Net electric power, MW_e	1836

The power cycle employs a steam−steam heat exchanger to confine tritium to the primary (blanket) side. The secondary steam is 454 °C at 13.8 MPa pressure. At 40% thermal efficiency, the system has a gross electric output of 1898 MW_e. Steam pumping power and other household power require about 62 MW_e, leaving a net electric power of 1836 MW_e.

The UWTOR-M blanket is estimated to have a life of three full power years which, at 75% availability, result in a 4-year actual life. It is expected that one-third of the blanket segments will be changed every 16 months. The blanket change requires radial extraction of every other coil from the reactor building. A reactor module, as viewed from its back, is shown in Figure 9.6. The reactor building has an access door spaced every other coil, as seen in Figure 9.4. A downtime of 4 weeks is estimated for blanket exchange.

Using DOE guidelines, an economic analysis of UWTOR-M shows that the magnets account for 52.4% of the reactor plant equipment cost and the reactor plant equipment comprises 67.4% of the total direct costs. The unit plant cost is \$2,034/kW_e in 1982 dollars. Busbar costs are calculated to be 36 mills/kWh.

For a more complete description and discussion of physics and engineering details, the reader is referred to Reference 8.

FIGURE 9.4.   Top view of the UWTOR-M modular superconducting magnets inside the reactor building.

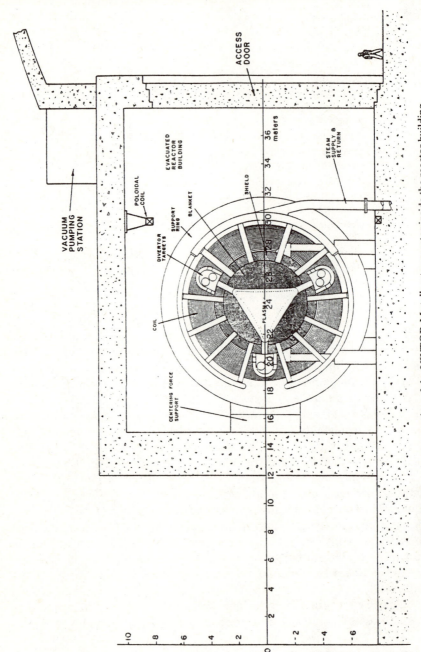

FIGURE 9.5. A cross section of the UWTOR-M stellarator reactor inside the reactor building, which is maintained at evacuated conditions.

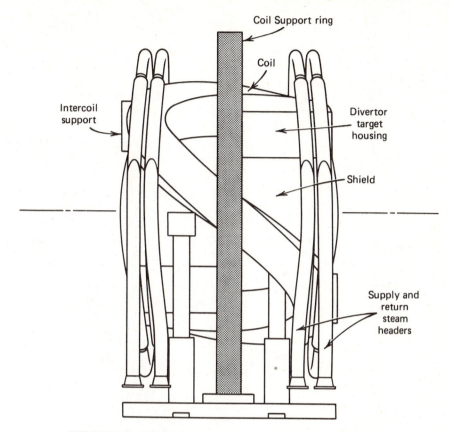

FIGURE 9.6.   UWTOR-M reactor module as viewed from its back.

9.3.   ELMO* BUMPY TORUS (EBT)

The basic concept of a bumpy torus is to join a number of mirror sectors end to end into a high-aspect-ratio torus, which can operate continuously. An important discovery by Dandl [9] is the fact that, when ECH is applied, a stable hot electron ring or annulus forms in the midplane of a mirror. This high-beta ring produces a local magnetic well that stabilizes the torodially confined plasma, and particle confinement is greatly improved by the resulting ambipolar potential. The physics of ECH bumpy toroids have been studied at Oak Ridge National Laboratory in EBT-1 and EBT-S, and at Nagoya University Institute of Plasma Physics in NBT. The state of understanding of plasma confinement in ECH bumpy toroids has been reviewed by Fujiwara et al. [10], and EBT reactor design projections have been summarized by Uckan [11].

*ELMO = electrons with large magnetic orbits.

In an electron-cyclotron-heated bumpy torus there are two principal plasma components. A relatively low density, mirror-trapped, very-hot and high-beta electron annulus or ring forms at the location of the second harmonic ($\omega \sim 2\omega_{ce}$) resonance. It has a radial width of a few hot-electron gyro-radii, and an axial length less than the plasma radius. A higher-density, warm-toroidal-core plasma threads the hot-electron rings. A stable, well-confined mode of operation is called the T mode, and under these operating conditions the rim of the potential well is located in the region of the rings and this defines the radius of the stable core plasma. Core plasma temperature and density profiles are nearly flat inside the rings and decrease sharply to where the edge plasma is cold and dilute. The electric field is radially inward inside the ring and outward outside the ring. EBTs are closed-field-line devices except for relatively small field-error asymmetries that cause some field lines to spiral outward. There is thus natural divertor action for wall-generated impurities at the plasma edge (outside the hot-electron rings) where most drift orbits intersect the wall, and the plasma potential effectively screens the toroidal plasma from incident impurity ions.

The toroidal core plasma is stable only if the electron ring beta exceeds a critical value, β_{crit}, required to produce a distinct local minimum in B. The mhd analyses suggest that

$$\beta_{crit} \sim \frac{\text{ring radial thickness}}{\text{mean radius of curvature}} \sim 10 - 40\%$$

The core beta is estimated to be stable to $\beta_{core} \sim 20-30\%$ [11]. Plasma confinement in bumpy toroids depends on the modification of the toroidal drift by the poloidal ∇B and $\mathbf{E} \times \mathbf{B}$ drifts produced by the bumpiness in the toroidal field and by the ambipolar electric field. When operated in the stable T mode, radial plasma losses are expected to be dominated by neoclassical transport. Collisionless scaling, that is, $\nu/\omega < 1$, where ν is the collision frequency and ω is the poloidal precessional drift frequency, leads to "EBT scaling,"

$$n\tau_E \sim A^2(kT_e)^{3/2} \left(1 + \frac{R_B}{R_E} \frac{e\phi}{kT_e}\right) \tag{9.1}$$

where A is the magnetic aspect ratio R/R_B, and R, R_B, and R_E are the major radius of the torus, the magnetic radius of curvature, and the electric field scale length, respectively. The potential $e\phi$ is proportional to the ion temperature, where, from experimental data, $e\phi \sim (1.5-2)kT_i$.

In the present bumpy toroid experiments, the plasma is produced, heated, and stabilized by electron-cyclotron heating (ECH). Heating power is continuous at frequencies of 8.5 GHz (NBT), 10.6 GHz (EBT-0), 18 GHz (EBT-1), and 28 GHz (EBT-S). For a fusion reactor, ECH at $\sim$ 110 GHz is needed at a power level much higher than the present state of the art. Electron ring loss processes are drag, scattering, and radiation. For ring temperatures less than 1 MeV, the drag loss is the dominant one. In the present steady-state experiments, fueling is accomplished by neutral gas influx at the plasma boundary.

9.3.1. ELMO Bumpy Torus Reactor* (EBTR)

The EBT principle was first examined for its potential as a power reactor in a design study published in 1976 [13]. A detailed reassessment of the EBTR has recently been completed by a design team from Los Alamos National Laboratory, the McDonnel−Douglas Astronautics Co., and other industrial firms [12]. The size of the power plant has been reduced to about 1200 MW_e, in which the toroidal radius is 35 m. The reactor operates in a steady-state ignited mode. This new design achieves a high magnetic aspect ratio in a system of lower physical aspect ratios through the use of "aspect-ratio-enhancement" (ARE) coils. A summary of the EBTR design point parameters is given in Table 9.2. The overall reactor layout is shown in Figure 9.7. A sector of the torus is shown in Figures 9.8 and 9.9 illustrating the location of the ARE coils, the vacuum duct, blanket, shield, etc. A major advantage of an EBT reactor is the high aspect ratio, which permits a design with good accessibility and relatively easy maintenance.

The mechanical forces generated by the toroidal field and ARE coils are transmitted to the building walls through a support structure. The torus is located in a concrete structure containing an inert gas. The shield forms the primary vacuum boundary. The limiter is used as a vacuum-divertor system similar to that used in the STARFIRE design. The toroidal and ARE superconducting coils are designed for life-of-plant operation. Major scheduled maintenance is for blanket and shield replacement. Each sector has a life expectancy of 15−20 yr. Cryo-pumps for each sector are attached directly to the outside of the shield. The ECRH duct, coolant lines, ICRH coaxial lines, instrumentation and control cables for each midplane section enter from the inboard side (an advantage for large-aspect-ratio devices) and need not be retracted for removal of the midplane section. The coolant lines for the coil-throat section are separate and exit on the outboard side next to the TF coils. All sectors are sealed by flexible, welded joints to accommodate vacuum requirements and differential thermal expansion.

Blanket and shield sections are moved by a transporter that runs along a circular track outboard of the torus between the reactor torus and hot cells. Sections are removed and installed by a single unit maintenance machine. Complete replacement of a midplane and throat sector is expected to require 7 days. Balance-of-plant maintenance operations are expected to require 28 days per year. Unscheduled maintenance is estimated at 42 days per year.

The Elmo bumpy torus thus appears to offer an attractive reactor concept. It is the only fusion-plasma-confinement concept that has as yet experimentally

---

*This section is taken from Reference 12.

TABLE 9.2. Summary of EBTR Design Parameters[a]

Net electric power, P_E (MW$_e$)	1214.
Engineering power density (MW$_{th}$/m^3)	0.23
Major radius, R_T (m)	35
Average minor plasma radius, r_p (m)	1.0
Average bulk-plasma beta, β	0.17
Plasma temperature, T_i/T_e (keV)	27.9/29.0
Average ion density, n_i (10^{20}/m^3)	0.95
Fusion neutron wall loading,	
$\quad I_w$ (MW/m^2)	1.4
Average mirror ratio, M	2.24
ARE-coil/TR-coil current ratio,	
$\quad I_{ARE}/I_{TF}$	−0.22
Average magnetic field, B (T)	3.64
Blanket multiplier/breeder	Be/LiAlO$_2$ (natural)
Primary coolant	H$_2$O (ℓ)
Shielding	B-H$_2$O/SS/Pb-W/TiB$_2$/TiH$_2$
Plasma volume, V_p (m^3)	691.
Plasma chamber volume (m^3)	838.
Blanket volume (m^3)	1108.
First-wall area (m^2)	1537.
Number of sectors, N	36.
Machine circumference (m)	220.

Plasma Parameters

Transport	Neoclassical/plateau
Major radius, R_T (m)	35.
Average minor radius, r_p (m)	1.0
Coil-plane plasma radius, r_{CP}	0.76
Midplane plasma radius, r_{MP}	1.24
Average ion density, n_i (10^{20}/m^3)	0.95
Average plasma temperature,	
$\quad T_i/T_e$ (keV)	27.9/29.0
Electron/ion particle-confinement time,	
$\quad \tau_p$ (s)	4.21
Energy confinement time, τ_E (s)	1.80
Core-plasma Lawson parameter, $n_i\tau_E$	
$\quad (10^{20}$ s/m^3)	1.71
Average mirror ratio, M	2.24
Average magnetic field, B (T)	3.64
Average core-plasma beta, β	0.17
Maximum (midplane) core-plasma beta,	
$\quad \beta_{MP}$	0.46
Electron collisionality, $\xi_e = v_e/\Omega_{\Delta B}$	0.26
Plasma fusion power density at	
$\quad$ 17.6 MeV/fusion, P_F (MW/m^3)	4.13
Plasma Q-value, Q_p	$\gtrsim$ 100. (ignited)

TABLE 9.2. *(Continued)*

Electron-Ring Parameters

Heating	First-harmonic ECRH
Frequency (GHz)	52.
Temperature, T_R (keV)	$\gtrsim$ 1500.
Density, n_R ($10^{20}/m^3$)	0.05
Length (m)	2.4[b]
Average minor radius (m)	1.1
Thickness (m)	0.024[c]
Total volume, V_R (m^3)	14.1(V_R/V_p = 0.02)

Output/Input Power Flow

Plasma fusion power at 17.6 MeV/ fusion, P_F (MW$_{th}$)	2857.
Thermal power from wall/limiter/ blanket (MW$_{th}$)	644./300./3048.
Total recoverable thermal power, P_{TH} (MW$_{th}$)	4028.
Power to electron rings, P_R (MW)	42.
Total electrical power (η_{TH} = 0.355), P_{ET} (MW$_e$)	1430.
Net electrical power, P_E (MW$_e$)	1214.
First-wall fusion neutron loading, L_w (MW/m^2)	1.4
Blanket thermal power density (MW$_{th}$/m^3)	3.33
Engineering power density (MW$_{th}$/m^3)	0.24

Magnet Coil Parameters

Major radius of TF-coil set, R_{TF} (m)	35.08
Average, midplane, and coil-plane fields, $B/B_{MP}/B_{CP}$ (T)	3.64/2.25/5.03
Number of coils, TF/ARE	36/72
Mean coil radius, TF/ARE (m)	2.90/4.44
Peak field at conductor, TF/ARE (T)	9.7/6.6
TF-coil length/width, l_{TF}/w_{TF} (m)	1.98/0.99
Coil currents, I_{ARE}/I_{TF} (MA)	$-6.97/31.35(-0.22)$
TF/ARE-coil mass/volume (tonnes/m^3 per coil set)	726/164
Total stored energy (GJ)	131.
Superconductor current density (MA/m^2)	16.

Blanket/Shield/Coil Neutronics Parameters

Theoretical tritium breeding ration, T	1.06
Energy multiplication, M_N	1.50

(continued)

TABLE 9.2.   *(Continued)*

Blanket/Shield/Coil Neutronics Parameters

First-wall radiation damage [(dpa/yr)/ (ppm He/yr)]	17.4/164
Peak coil nuclear-heating rate (W/m^3)	14.0
Peak coil radiation damage and doses [(dpa/yr)/(Gy/yr)]	$9.6(10)^{-6}/18.1(10)^4$

Reactor Plant Operational Characteristics/Parameters

Impurity control	pumped-limiter
Refueling	pellet injection/edge refueling
Vacuum volume (m^3)/pressure (Pa)	$1506/10^{-3}$
Plasma startup power [type/power (MW)/frequency (GHz)]	LHH/69./0.55−1.4
Electron-ring power [type/power (MW/frequency (GHz)]	ECRH/42./50.
Blanket structure/multiplier/breeder/ coolant	PCASS/Be/LiAlO$_2$/H$_2$O(ℓ)
Blanket and shield thickness at coil- plane/midplane locations (m)	1.21/1.44
Basic radiation shield materials	B-H$_2$O/SS/W-Pb
Blanket coolant outlet temperature (K)/ pressure (Pa)/flow rate (kg/s)	593/15.1/2420.
Number of sectors/modules	36/72
Weight of blanket coil-plane/midplane modules (tonnes)	44.4/69.9
Weight of coil-plane/midplane shield modules (tonnes)	160.9/203.4
Reactor building height/diameter/ volume (m/m/m^3)	36./107./167,000.
Thermal-conversion efficiency, η_{TH}	0.355
Recirculating power fraction, ε	0.15
Net plant efficiency, $\eta_p = \eta_{TH} (1-\varepsilon)$	0.302
Availability of reactor/plant	0.85/0.77
Plant life (yr)	40.

[a]From Reference 12.
[b]Distance between tips of limiter.
[c]Ten electron Larmor radii.

demonstrated steady-state operation. The physics, particularly stability and plasma transport, require further experimental evaluation. The electron-cyclotron heating technology (i.e., high-power dc gyrotrons at $\sim 50-100$ GHz) remain to be developed. Further work on these and other physics and engineering aspects of EBTRs need to be done before this concept can be evaluated relative to other fusion reactor concepts.

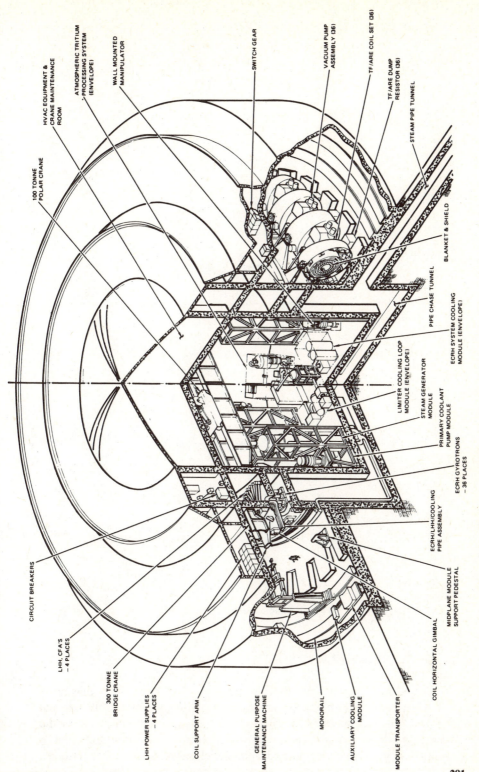

FIGURE 9.7. EBT reactor showing the key components from Reference 12.

CIRCUIT BREAKERS

100 TONNE POLAR CRANE

HVAC EQUIPMENT & CRANE MAINTENANCE ROOM

ATMOSPHERIC TRITIUM PROCESSING SYSTEM (ENVELOPE)

WALL MOUNTED MANIPULATOR

SWITCH GEAR

VACUUM PUMP ASSEMBLY (36)

TF/ARE COIL SET (36)

TF/ARE DUMP RESISTOR (36)

STEAM PIPE TUNNEL

BLANKET & SHIELD

LHH, CFA'S – 4 PLACES

300 TONNE BRIDGE CRANE

LHH POWER SUPPLIES – 4 PLACES

COIL SUPPORT ARM

GENERAL PURPOSE MAINTENANCE MACHINE

MONORAIL

AUXILIARY COOLING MODULE

MODULE TRANSPORTER

COIL HORIZONTAL GIMBAL

MIDPLANE MODULE SUPPORT PEDESTAL

ECRH/LHH/COOLING PIPE ASSEMBLY

ECRH GYROTRONS – 36 PLACES

PRIMARY COOLANT PUMP MODULE

STEAM GENERATOR MODULE

LIMITER COOLING LOOP MODULE (ENVELOPE)

ECRH SYSTEM COOLING MODULE (ENVELOPE)

PIPE CHASE TUNNEL

281

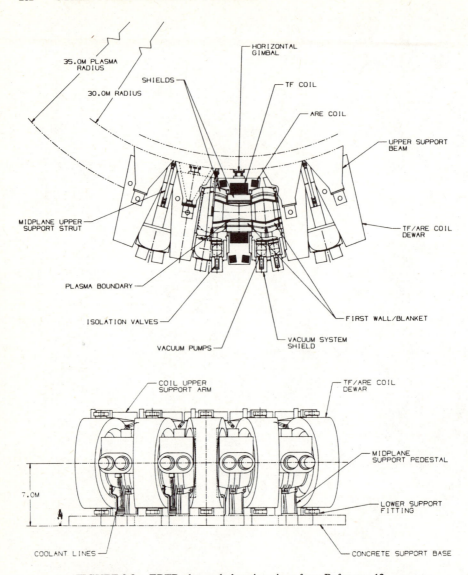

FIGURE 9.8.   EBTR plan and elevation views from Reference 12.

9.4.   REVERSED-FIELD PINCHES

The reversed-field pinch concept is a toroidal system in which plasma is confined by a combination of a toroidal magnetic field generated from external coils and a poloidal magnetic field due to a toroidal current flowing in the plasma. In contrast with a tokamak, however, the reversed-field pinch has a safety factor of $q < 1$ and toroidal and poloidal fields of comparable magnitudes.

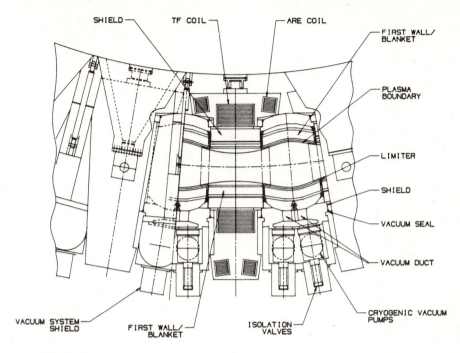

SHIELD   TF COIL   ARE COIL   FIRST WALL/ BLANKET   PLASMA BOUNDARY   LIMITER   SHIELD   VACUUM SEAL   VACUUM DUCT   CRYOGENIC VACUUM PUMPS   VACUUM SYSTEM SHIELD   FIRST WALL/ BLANKET   ISOLATION VALVES

FIGURE 9.9.   Details of two of the 36 reactor sectors of an EBTR from Reference 12.

A comparison of tokamak and reversed-field pinch magnetic field configurations is illustrated in Figure 9.10. The reverse-field pinch configuration is stabilized by high magnetic shear ($-d\ln q/d\ln r$) and the presence of an electrically conducting wall close to the plasma boundary. It is important for plasma stability that dq/dr not change sign. Since $q < 1$ in the center, q reverses sign near the plasma edge; that is, for the reversed-field pinch, the magnetic shear should not exhibit a minimum in the region enclosed by a conducting shell. The reversed-field-pinch configuration can be created by fast field programming* or, in slow-mode operation, it will form naturally. A comprehensive review of reversed-field pinch research has been given by Bodin and Newton [14], and reversed-field pinch reactor designs can be found in Reference 15.

The claimed advantages of a reversed-field-pinch configuration as a fusion reactor are:

1. High-beta ($\sim 30\%$) stable operation.
2. Can achieve fusion ignition from purely ohmic heating.
3. Large aspect ratio permits designs with good maintenance access.

*Fast field reversal has not been very successful in creating long-lasting toroidal plasmas, and it requires high-voltage technology. The slow-mode operation appears preferable for RFP reactors.

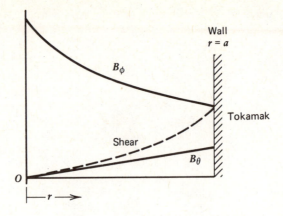

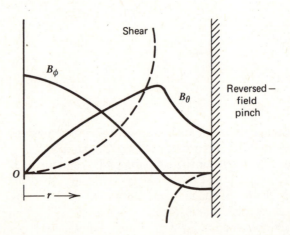

FIGURE 9.10. Magnetic field configurations of the reversed-field pinch and the tokamak.

On the other hand, the need for a conducting shell close to the plasma boundary to enable self-reversal of the toroidal field and to ensure mhd stability is a handicap for the reactor designer. By its very nature, the reversed-field pinch is a pulsed system with a burn time of the order of 20 s. The recirculated power fraction in a recent 750-MW$_e$ design (see Reference 1, p. 57) is 17%.

An example of a Culham reversed-field-pinch reactor module is shown in Figure 9.11 along with a Los Alamos design of a similar reversed-field reactor cut-away.

9.5. COMPACT FUSION REACTOR CONCEPTS

Leading magnetic-fusion-confinement concepts, such as tokamaks and tandem mirrors, are envisaged as having reactors that produce about 1000 MW_e and have a relatively low power density. They are physically large, employ complex technology, and are correspondingly expensive. In competition with these large fusion reactors is a class of fusion concepts whose proponents seek to develop compact power reactors. Their development, if successful, will lead to qualitatively different types of commercial products. A compact reactor may be defined as one that has a higher power density and, for the same total power, is significantly smaller than a conventional magnetic fusion reactor. A survey of reactor aspects of compact fusion concepts has been written by Gross [16].

For example, conventional magnetic fusion systems operate with a relatively low engineering power density of the order of 0.3−0.5 MW_{th}/m^3. This power density is the ratio of the total thermal power to the volume enclosed by and including the magnet coils. Conventional magnetic fusion systems use large superconducting coils and are usually designed to have a total power output below about 4000 MW_{th}. They have large plasma volumes, typically 500−1000 m^3 and even a larger volume first-wall container so as to operate with a fusion neutron first-wall flux of 1−3 MW/m^2.

A mass utilization factor is defined as the weight of the first wall, blanket, shield, and coil divided by the total thermal power. For conventional fusion systems the mass utilization factor has values between 5 and 10 tonnes/MW_{th}. Large values of mass utilization suggest a relatively high cost for the power.

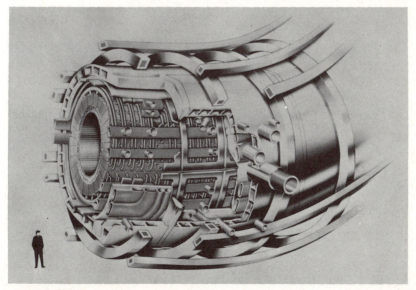

FIGURE 9.11. *(a)* Isometric view of a reversed-field pinch reactor module (Culham design from Reference 15).

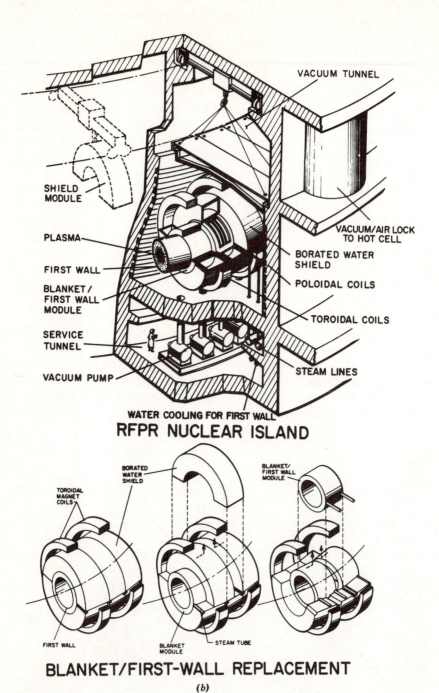

VACUUM TUNNEL

SHIELD
MODULE

PLASMA

FIRST WALL

BLANKET/
FIRST WALL
MODULE

SERVICE
TUNNEL

VACUUM PUMP

VACUUM/AIR LOCK
TO HOT CELL

BORATED WATER
SHIELD

POLOIDAL COILS

TOROIDAL COILS

STEAM LINES

WATER COOLING FOR FIRST WALL

RFPR NUCLEAR ISLAND

BORATED
WATER
SHIELD

BLANKET/
FIRST WALL
MODULE

TOROIDAL
MAGNET
COILS

FIRST WALL

BLANKET
MODULE

STEAM TUBE

BLANKET/FIRST-WALL REPLACEMENT

(b)

FIGURE 9.11. *(b)* Cut-away view of a reversed-field pinch reactor (Los Alamos design from Reference 15).

Conventional fusion system parameters can be compared with the corresponding values for light-water fission reactors; they are $\sim 5-10$ MW$_{th}$/m^3 and $0.3-0.4$ tonne/MW$_{th}$, based on pressure vessel mass or volume.

9.5.1. Compact Reversed-Field Pinch Reactor

The Los Alamos National Laboratory fusion group has engaged in studies of high-power-density approaches for fusion and in particular they studied a compact reversed-field pinch reactor (CRFPR) [17]. Believing that the choice of fusion systems will ultimately be made on economic grounds, they focused on a high-power-density RFP using thin blanket/shields and resistive coils. The major goal of their approach is to reduce the percentage of total direct cost attributable to the reactor plant equipment from $60\%-80\%$, as projected for conventional magnetic fusion systems, to 25% to 30%, which is typical of commercial fission plants.

The Los Alamos design group searched, using computer codes, for a cost-optimized design point, and they found one that met their objectives. They studied the effects of changes in plasma transport, beta, blanket thickness, and normal vs superconducting coils. Their high-power-density reactor design is found to be surprisingly resilient to changes in key, but relatively unknown physics and system parameters. A summary of key parameters for the cost optimized CRFPR is given in Table 9.3, and a cross section of this reactor plus shield is shown in Figure 9.12. There are unresolved physics and technology issues associated with this preliminary design. It has a high heat flux to the first wall (average of $4-5$ MW/m^2) and high power density (100 MW/m^3) in the breeder blanket. There are many material-lifetime problems, and for long pulse lengths, it may need some form of divertor. A more-detailed discussion of the physics and technology issues for this compact reactor design can be found in References [16] and [17].

9.5.2. OHTE Reactor

The OHTE* concept has a magnetic configuration similar to a reversed-field pinch, but with the addition of helical stellarator windings. Like the reversed-field pinch its poloidal and toroidal fields are of comparable magnitude. The plasma is confined primarily by the poloidal field, which is produced the toroidal plasma current. A close-in copper shell provides wall stabilization. The magnetic field has both rotational transform and shear provided by the external helical coils. These helical coils can be in configurations $l = 1, 2,$ or 3, where l indicates the number of coil pairs. OHTE requires an ohmic-heating transformer winding as well as equilibrium (or vertical field) coils. Like the

*OHTE is formed from the first letters of ohmically heated toroidal experiment. It is also a play on words, for in Japanese, it can mean "checkmate" and/or "gateway to the palace." The inventor of the OHTE concept is Tihiro Ohkawa.

**TABLE 9.3. Summary of Parameters for the D−T
Cost-Optimized 1000-MW$_e$ Compact Reversed-Field
Pinch Reactor (from Reference 17)**

Design Parameter	D−T
First-wall radius, r_w (m)	0.75
Major radius, R_r (m)	4.3
Minor system radius, r_s (m)	1.6
Toroidal-coil mass (tonne)	159
Poloidal-coil mass (tonne)	729
First-wall/blanket mass (tonne)	356
Mass utilization, M/P_{th}	
[tonne/MW (thermal)]	0.37
Plasma temperature, T (keV)	20.0(10.0)
Plasma density, n ($10^{20}/m^3$)	3.4(6.7)
Energy-confinement time, τ_E (s)	0.23
Alcator coefficient, τ_E/nr_p^2 (10^{-21} s · m)	1.37(0.76)
Toroidal plasma current, $I\phi$ (MA)	18.5
PF at plasma, B_θ (T)	5.2
Poloidal-coil field, $B_{\theta c}$ (T)	2.6
Initial toroidal-coil field, $B_{\phi 0}$ (T)	3.3
Poloidal-coil energy, $W_{B\theta}$ (GJ)	1.11
Toroidal-coil energy, $W_{B\phi}$ (GJ)	0.54
Magnetic energy recovery time, τ^* (s)	0.49
Total thermal power, P_{th} [MW(thermal)]	3350
Engineering power density,	
P_{th}/V_c (MW/m^3)	15.0
Recirculating power fraction, $\varepsilon = 1/Q_E$	0.147
Ohmic Q value, $Q_T = P_{th}/(P_{ohm} + P_{th})$	37.1
Neutron wall loading, l_w (MW/m^2)	19.5
Unit total cost [\$/kW (electric)]	1490
COE [mill/kW(electric) · h]	40.7

reversed-field pinch, OHTE operates at low q (less than 1) and the current density can be sufficiently large that pure ohmic heating to the ignition state is believed possible. The OHTE concept is being studied by G. A. Technologies and Phillips Petroleum Company in a private venture.

The OHTE concept, like the reversed-field pinch, can operate at relatively high beta. It appears to be limited in power density—not by plasma stability—but rather by the technology associated with neutron flux on the first wall and on first-wall heat removal rate limitations. The OHTE reactor concept has been studied by Bourque [18]. He calculates that a few millimeters of aluminum alloy or copper can absorb about 5 MW/m^2 surface loading under cyclic conditions. He accepts this specific heat power flux as the average alpha power heating of the OHTE first wall. This corresponds to a neutron wall loading of 20 MW/m^2. Bourque developed computer codes that simulate different operat-

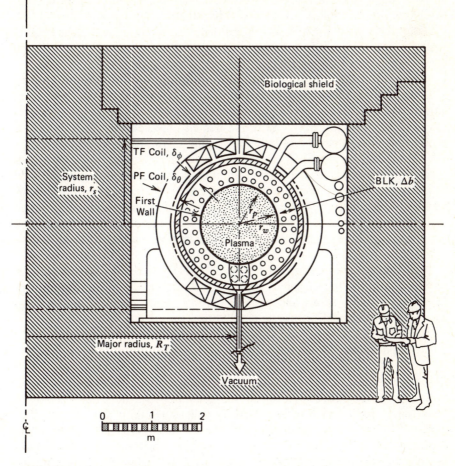

FIGURE 9.12. A scaled layout of the 1000-MW$_e$ compact reversed-field pinch reactor (from Reference 17).

ing modes, power levels, l numbers, etc., subject to a variety of engineering constraints. Ion confinement is taken to be classical conduction plus ambipolar diffusion. The actual values of plasma parameters must await the analysis of experiments in progress now.

The basic OHTE reactor component arrangement, moving from the plasma outward, is (1) first wall and copper shell, (2) helical coils, (3) blanket and (4) OH and equilibrium coils. Power considerations require that the helical coils be close-in and the blanket behind them. The coils are sufficiently thin that most of the neutron flux passes through them.

Because of the substantial number of ohmic-heating coils needed for OHTE, the neutron-absorbing blanket must be positioned within the ohmic heating and equilibrium coils. This results in a smaller, lower cost blanket provided that liquid breeding material is used that also serves as a coolant. A separate coolant would require technologically very difficult ducting at these high heat fluxes. Liquid $Li_{17}Pb_{83}$ eutectic with 70% ^6Li is the primary candidate for the cooling and breeding media. It is estimated that the copper shell and helical coils will require replacement after about 4 MW-yr/m^2 exposure.

An OHTE ignition test reactor (ITR) is suggested by G. A. that will have a 30 s burn and recirculating power of about 200 MW, which can be drawn from an external electric grid system. A commercial electricity producer (DEMO) has been sized at 900 MW$_e$. Bourque has also studied OHTE reactors for process heat and fission fuel breeding. Some of the basic specifications for OHTE reactors are given in Table 9.4. Although the reactor is small, the power supplies and the power-conversion equipment for recirculating power add significantly to the reactor cost. The cost for an ignition device is estimated to be about $300 million. The fusion reactor can produce large amounts of fissile fuel quickly and when combined with the production and sale of electricity, the fissile fuel cost is negligible. One OHTE fuel factor is calculated to be able to support 21 1000-MW$_e$ fissile high-temperature gas-cooled reactors.

Major uncertainties associated with the OHTE concept concern plasma-physics questions. In particular, the energy-confinement scaling is unknown. High-beta stability limits and general plasma performance have yet to be established at reactorlike plasma conditions.

9.5.3. Spheromak Reactor

A spheromak is a toroidal-plasma-confinement configuration where the confinement is accomplished by internal (self-generated) magnetic fields and by an equilibrium field, which is produced by coils not linking the plasma. The spheromak confinement geometry is illustrated in Figure 9.13. Experiments exploring the creation and the stability of the spheromak plasma configuration are in progress in several laboratories in the United States (Princeton and Los Alamos National Laboratories). There is as yet no complete design for a spheromak reactor but some preliminary studies have been made by Yamada and colleagues at Princeton (see Reference 16). If the spheromak-confinement concept should prove to have desirable plasma stability and confinement properties, then it could lead to a compact fusion reactor.

9.5.4. A Compact Tokamak Reactor—Riggatron[TM]

Tokamak confinement will permit ignition to be achieved from purely ohmic heating if a very large toroidal magnetic field is employed and if the plasma-confinement physics, observed at lower magnetic fields and lower beta, continue to be valid at high-field strengths. This leads to the Riggatron[TM] concept being developed by the INESCO company. Some details of the Riggatron concept have been published by Baker et al. [1] and by Krakowski, et al. [19]. They propose a high magnetic field of 16 T using water cooled copper coils

TABLE 9.4. OHTE Reactor Specifications and Estimated Costs.[a]

	ITE	DEMO	Electricity	Fuel	Heat
Reactor Specifications					
Major radius (m)	2.43	3.97	5.91	4.86	5.81
Plasma radius (m)	0.27	0.45	0.67	0.55	0.66
Wall radius (m)	0.30	0.50	0.74	0.61	0.73
Burn time (s)	36.7	127	242	241	242
Cycle time (s)	56	141	262	253	258
Average fusion power (MW_{th})	253	1595	3795	841	3727
Useful blanket power		1060	2740	2700	2700
Peak wall load (MW_n m^2)	11.4	18.9	19.5	6.1	19.5
I_p at burn (MA)	6.6	9.2	12.4	8.1	12.3
Net electric (MW_e)	−186	69	904	915	267
Gross electric (MW_e)	0	394	1393	1258	759
Helical coil power (MW_e)	107	150	306	204	308
OH coil power (MW_e)	64	138	123	89	135
VF coil power (MW_e)	15	30	32	25	28
Circular power (MW_e)	0	8	28	25	21
^{233}U production, kg/yr				2100	
Number of 1000-MW_e HTGRs supported				21	
Heat production (10^{12} Btu/yr)					20
Helical coil thickness (cm)	4.9	7.6	5.5	4.0	5.5
Helical coil wall coverage (%)	75	73	71	66	70
Helical coil heating (W/cm^3)	192	153	195	145	201
Approximate helical coil field (T)	4	4	4	4	4
OH peak field (T)	12.3	15.1	11.2	11.0	11.6
OH coil average stress (ksi)	24	24	23	23	23
VF (T)	1.5	1.3	1.1	0.9	1.1
Estimated Costs *(Capital costs in millions)* *(Costs of electricity in mill/kWh)*					
First wall/copper shell	0.6	1.6	3.4	2.3	3.3
Blanket		23	40	77	41
Helical coil	0.9	3.7	5.4	2.5	5.1
OH/VF coil	11	81	163	89	165
Helical coil power supply	6.4	7.7	20	17	20
OH/VF power supply	7.4	26	32	30	40
OH/VF MG sets	17	78	136	76	139
Recirculating power conversion		114	124	88	149
Reactor total	43	334	524	382	561
Balance of plant	178	380	843	697	718
Indirects	86	278	533	421	499
Total plant cost	307	992	1900	1500	1778

(continued)

TABLE 9.4. *(Continued)*

Estimated Costs
(Capital costs in millions)
(Costs of electricity in mill/kWh)

$/kW(electric) net		2100	1639	
Wall/helical coil changeout/yr	3.1	3.3	1.1	3.4
COE for capital		34.2		
COE for changeouts		5.3		
COE for operations and maintenance		3.3		
Total COE (FCR = 0.10)		42.7		
COE with ^{233}U sold at $10/g			25.6	
$/10^6 Btu (electricity sold at 35 mill/ kWh)				3.69

*a*ITE is an ignition test experiment; DEMO refers to a small demonstration power plant. Electricity, fuel, and heat refer to the purpose of OHTE.

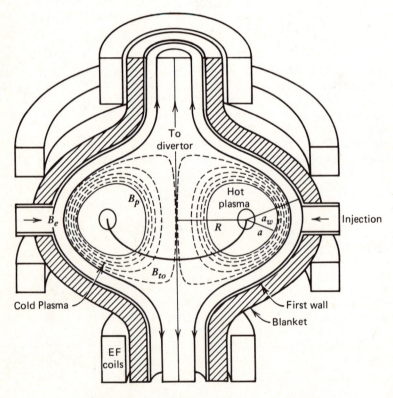

FIGURE 9.13. Spheromak confinement geometry.

positioned near the first wall. A high current density of 8 MA/m^2 and a high beta leads to a very compact tokamak reactor.

The severe thermal and mechanical problems associated with this high-energy-density tokamak power core suggest a short lifetime for the core elements of the power plant. The INESCO concept is to mass produce the Riggatron fusion power core and throw it away after its useful lifetime has transpired. The ~ 8 tonne power core will generate about 1 GW$_{th}$ and they predict that, when mass produced, each will cost about \$250,000. A Riggatron power plant would consist of a cluster of four to six power core modules, with two additional stand-by modules. The modules are designed for a rapid "plug-in" capability giving high availability without the need for *in situ* remote maintenance. A sketch of a Riggatron power core plugged into its surrounding blanket and shield is shown in Figure 9.14.

The safety factor q for a Riggatron was set equal to 2 at the plasma edge. Ignition is predicted for $B = 16$ T, $A = 2.5$, and $0.57 \leq R_0 \leq 0.95$ m. The first wall has a heat flux between 20 and 40 MW/m^2. A summary of one Riggatron design is given in Table 9.5. The Riggatron is started at low plasma density and high toroidal field strength. After ignition (some rf additional heating may be necessary to ignite) the density of the plasma is increased by gas puffing, the magnetic fields are reduced (i.e., the beta is increased), and q is

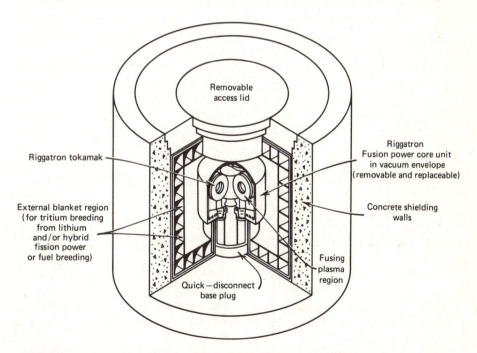

FIGURE 9.14. The Riggatron™ high-field tokamak reactor core plugged into its blanket and shield.

**TABLE 9.5.   A Riggatron High-Field Tokamak
Performance Parameters**

Minor radius (m)	0.34
Major radius (m)	0.85
Plasma volume (m^3)	2.0
Density (10^{20}/m^3)	20–30
Temperature (keV)	12–20
Averaged beta	0.2
Plasma power density (MW/m^3)	460
Ignited/driven burn	Ignited
Magnetic field (T)	16.0
Pulsed energy (MJ)	200
Burn time (s)	36
Off time (s)	3
Neutron current (MW/m^2)	68
Heat flux (MW/m^2)	17
Thermal power (MW$_{th}$)	1325
Net power (MW$_e$)	355
System power density (MW$_{th}$/m^3)	14
Recirculating power fraction	0.33
Net plant efficiency ($\eta_{th} = 0.40$)	0.27

reduced to the operating burn conditions. The burn continues for about 30 s
until impurity build-up forces termination.

The Riggatron concept represents a different approach and philosophy to
develop commercial fusion power. The plug-in, and later throw-away, reactor
concept is distinct from all other fusion concepts. There are many physics and
engineering questions that must be answered for the Riggatron compact-
tokamak reactor. Among these are the plasma-reactor state needed to achieve
ohmic ignition, the maximum stable beta for tokamaks, material radiation
lifetime limits, and the rate of impurity build-up at burn conditions. The
physics, engineering, and commercial desirability of Riggatrons await future
testing.

9.6.   INERTIAL-CONFINEMENT FUSION (ICF)

The inertial-confinement concept has its origins in the development of nuclear
explosives. In the 1940s, the idea of compressing solids by detonation of
chemical explosives was developed for nuclear weapons. For thermonuclear
devices, intense compression can heat the fuel, increase the rate at which the
reaction takes place (recall that fusion reaction rate $\sim n^2 \langle \sigma v \rangle$), and assist in
retention of alpha products within the reacting media. The invention of high-
power lasers opened the possibility of using their radiation to compress and
heat very small spherical D–T pellets to thermonuclear burn conditions. The

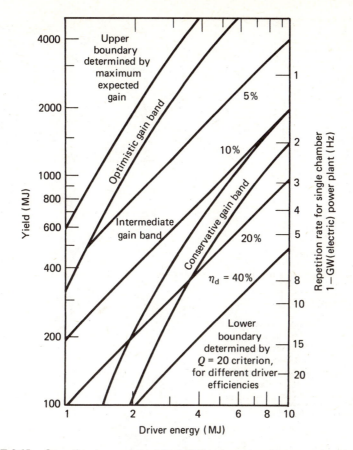

FIGURE 9.15. Operational parameters of a 1-GW$_e$ power plant illustrates the driver energy, fusion yield, and repetition rate required for a single reaction chamber. Conservative and optimistic refer to assumptions in the implosion physics and η_d is the driver (photon or ion-beam) efficiency. Data from Reference 27.

respectively. Light- and heavy-ion beams are also being considered as drivers. Their energy efficiencies are between 15 and 30%, but very high voltages ($\sim 2-8$ MV) are required. Switches, dielectrics, pulse forming networks, etc., all will require extensive technology development to reliably achieve the $\sim 10^9$ shots required for reactors.

There is as yet no consensus on the construction details of an inertial-fusion reactor chamber. There are several proposed configurations. The cavity must be able to withstand the pulsed irradiation and stress cycling resulting from repetitive pellet explosions and it also must be constructed so as to provide efficient driver-beam transport through the chamber to the pellet. Pellet yields being considered for reactors range from 100 to 3000 MJ. The pellet explosion releases energetic x-rays, pellet debris, and high-energy neutrons. The reactor wall and reactor components must be protected from excessive damage from

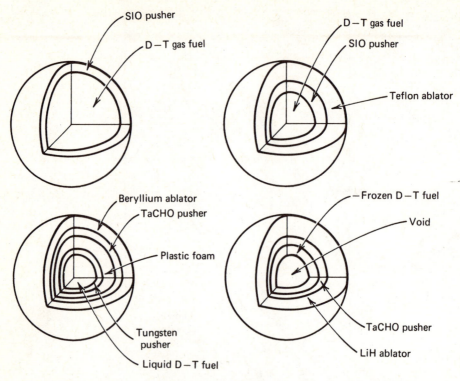

FIGURE 9.16. Different designs of inertial fusion pellets. From Reference 28.

these explosion products. To help achieve this protection, reactor designers have proposed wetted-wall chambers, dry-wall gas-filled chambers, magnetically protected walls, and liquid curtains. ICF reactor designs, dating from the early 1970s, are reviewed by Monsler et al. and these designs helped identify some of the more critical ICF design issues.

An example of a laser-driven liquid lithium "waterfall" reactor concept proposed by a Livermore Laboratory design team [29] is shown in Figure 9.17. It features a thick continuous fall of an annular curtain of lithium. The pellet is injected from the top of the chamber and the driver beam is transported through the liquid curtain by means of two beam transport tubes that penetrate the lithium-fall. This curtain is intentionally maintained away from the chamber wall to minimize impact on the wall from the blast pressure waves. This system is really a self-pumping vacuum system, and the deposition of neutron energy in the liquid lithium will break it into droplets. These droplets then act as condensation sites for the high-temperature lithium vapor produced by the short-ranged fusion products. This particular design, however, has unacceptable wall stresses.

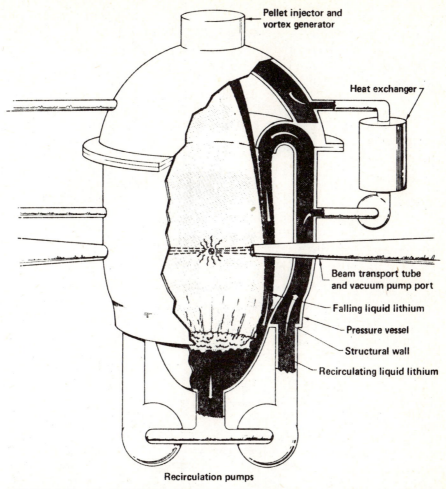

Pellet injector and
vortex generator

Heat exchanger

Beam transport tube
and vacuum pump port

Falling liquid lithium

Pressure vessel

Structural wall

Recirculating liquid lithium

Recirculation pumps

FIGURE 9.17. The liquid-lithium-fall, laser-driven reactor concept from Reference 29. The thick annular lithium curtain forms both the first wall and the blanket.

9.6.1. SOLASE* (A Laser-Driven Inertial Fusion Reactor)

The SOLASE power plant is designed as a 1000 MW_e unit for central station elective power generation. The power plant is illustrated in Figure 9.18 and the reactor is shown in Figure 9.19. The net plant efficiency is 30% and energy on target is 1 MJ and 20 pellets per second are exploded, each with a gain of 150. The gas-phase laser relies on multipassing to achieve an efficiency of 6.7%. A

---

*This section is based on Reference 30. The reactor name is inspired by James Joyce's *Finnegans Wake*, Viking Press, New York, 1939, p. 470.

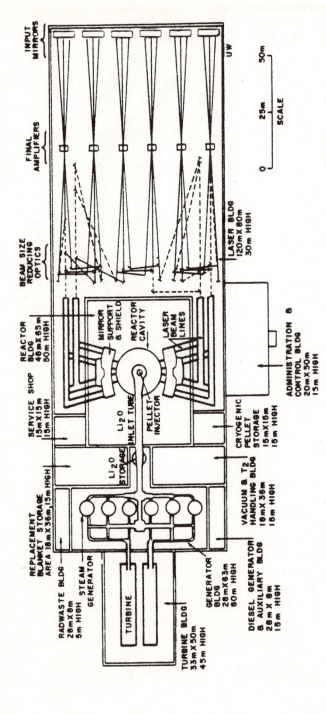

FIGURE 9.18. Top view of SOLASE, a 1000-MW$_e$ laser-driven, inertially confined power plant [29].

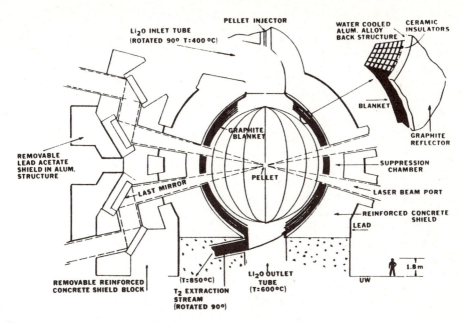

FIGURE 9.19.   SOLASE reactor system.

specific laser wavelength is not chosen by the Wisconsin designers. The reactor cavity is spherical with a 6 m radius and it is constructed of graphite designed to guide the flow of ceramic microspheres (100 μm radius) of lithium oxide (Li_2O), which serves as both a tritium-breeding and heat-transport medium. A buffer gas of neon (high breakdown threshold) or xenon (good x-ray absorption) is proposed to prevent pellet debris and x-rays from directly impacting the cavity first wall.

The pellet has a fuel mass of 1 mg, is multilayered and cryogenic, and is calculated to have a fractional burnup of 45% of its fuel. The neutron wall loading is 5 MW/m^2 and the tritium-breeding ratio is 1.33. The pellet target is illuminated by a 1-ns pulse from two sides, by six beams on each side. The last mirrors are copper, diamond turned on an aluminum back structure. The mirrors are cooled to 50°C to minimize neutron-radiation effects. The average Li_2O flow velocity is 0.7 m/s. The reactor is accessible and a simple procedure is proposed for blanket sector replacement. The 16 blanket segments are sufficiently inexpensive that replacement is suggested, rather than maintenance and repair. The overall radioactivity levels of the blanket are low and the activity, 50 yr after shutdown, is just 3 Ci. SOLASE power plant parameters are summarized in Table 9.6. The power flow diagram for SOLASE is shown in Figure 9.20. High-gain targets, some with classified design features, have been proposed for inertial confinement, but their predicted performance has not yet been verified by experiment. The product of laser efficiency η_L and pellet gain $G(G$ = energy yield/energy absorbed) should be such that $\eta_L G \geq 10$. Target

TABLE 9.6. Solase Laser-Driven Inertial Reactor Power Plant Parameters, from Reference 29

Cavity shape	Spherical
Cavity radius	6 m
14-MeV Neutron wall loading	5 MW/m^2
Thermal power	3340 MW
Gross electrical power	1334 MW
Net electrical power	1000 MW
Recirculating power fraction	28%
Net plant thermal efficiency	30%
Laser type	Gas phase
Laser energy on target	1 MJ
Laser efficiency (with multipassing)	6.7%
Number of final amplifiers	6
Number of final beams	12
Energy output/amplifier pass	45.8 kJ
Pulse Width	1 ns
Pulse repetition rate	20 Hz
Pellet yield and gain	150 MJ
Fractional burnup of fuel	45%
Initial fuel mass	1 mg
Generic target design	Multilayered-cryogenic
Target illumination	Two sided
Number of final mirrors	12
F/No. of final mirror	7.5
Distance from last mirror to pellet	15 m
Diameter of last mirror	3.5 m
Composition of last mirror	Cu on Al
Manufacturing procedure	Diamond turning
First-wall protection method	Ne or Xe Buffer gas
Blanket structure	Graphite composite
Blanket-breeding and heat-transport media	Lithium oxide (Li_2O)
Tritium-breeding ratio	1.33
Total energy per fusion event	18.6 MeV
Total Li_2O flow velocity	3.12×10^7 kg/h
Average Li_2O flow velocity	0.7 m/s
Li_2O Inlet temperature	400°C
Li_2O Outlet temperature	600°C
Tritium inventory	
Glass encapsulation of target	24.7 kg
Polymer encapsulation of target	10.9 kg
Total reactor radioactivity level	
50-yr after shutdown	3 Ci

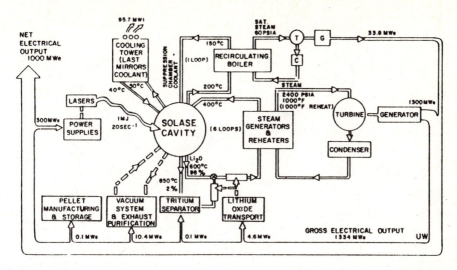

FIGURE 9.20.   Power flow for the SOLASE laser driven inertial fusion power plant [30].

delivery can be by pneumatic guns, but accuracy of trajectory and very precise observation and coordination of laser firing are required. The SOLASE inventory of tritium (1 week's supply of pellets) is between 10 and 25 kg, depending on the type of target and the time required to permeate and fill the pellet with D−T. The last mirrors present no insuperable problems unless dielectric coatings (required for $\lambda \leqslant 8000$ Å) are used. It is calculated that the reradiated heat flux from the buffer gas to the 6 m radius wall will have two peaks of about 130 and 140 kW/cm^2 at about 1 and 2 s after the explosion. The maximum pressure at the wall is estimated to be about 0.018 atm.

9.6.2. Light-Ion-Driven Inertial Fusion

The use of light ions, like He^{++} to drive and implode pellets has the advantages of higher driver efficiency and better coupling to the target than lasers. Particle-beam inertial confinement is discussed in References 26−28. The technology requirements of the pulsed power accelerators for light-ion-driven inertial confinement has been described by Martin et al. of the Sandia Laboratory [31].

A cutaway view of a light-ion fusion reactor is shown in Figure 9.21. The modular-oil-insulated prime energy stores are charged in less than 0.1 s and the pulse rate is 10 Hz. Storage is by either Marx generators or capacitors. These are discharged into water-insulated storage lines, which are discharged in turn by synchronized laser-triggered, gas-insulated switches. The pulse length is about 300 ns. These pulses are further compressed and transformed to 20-ns positive pulses. In the diodes, the pulses (10 TW/diode) are used to extract ions (3 kA/cm^2). These ions are focused onto the entrances of laser-initiated

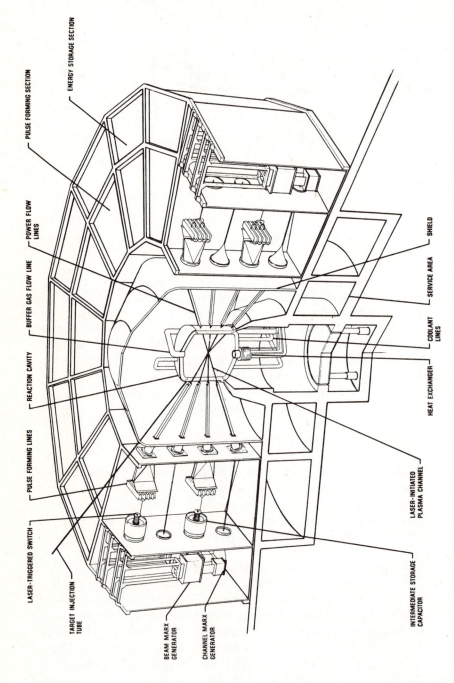

LASER-TRIGGERED SWITCH

TARGET INJECTION TUBE

BEAM MARX GENERATOR

CHANNEL MARX GENERATOR

PULSE FORMING LINES

REACTION CAVITY

BUFFER GAS FLOW LINE

POWER FLOW LINES

PULSE FORMING SECTION

ENERGY STORAGE SECTION

SERVICE AREA

SHIELD

HEAT EXCHANGER

COOLANT LINES

INTERMEDIATE STORAGE CAPACITOR

LASER-INITIATED PLASMA CHANNEL

FIGURE 9.21. A preliminary design of a light-ion-driven fusion reactor by the Sandia design team. From Reference 27.

TABLE 9.7. Light-Ion-Driven ICFR

Number of modules	40
Module diode voltage	8 MV
Module current (He^{++})	1.68 MA/channel
Total stored energy	15 MJ
Stored to pellet efficiency	0.27
Total energy on pellet	4.0 MJ
Pulse width on pellet	10 ns
Total power on pellet	400 TW
Pellet radius	0.8 cm
Power density on pellet	50 TW/cm^2
First-wall radius	4 m
Repetition rate	10 Hz
Radiant heat to first wall	1.6 MJ in 2.5 ms
Maximum heat flux to wall	470 W/cm^2
Shock overpressure at wall	0.257 MPa
Wall material	HT-9
Cell-wall thickness	0.5 cm
Buffer-gas density	1.8×10^{18} cm^{-3} Ar + 0.2% Na
Maximum wall thermal stress	36.9 MPa
Maximum wall thermal rise	15 K

plasma channels. The ion beams propagate along these plasma channels through a buffer gas to the target pellet. The plasma channels, formed by the laser, conduct currents that produce magnetic fields strong enough to confine the ion-beam transverse pressure during its propagation to the target.

Some parameters for a light-ion-driven fusion reactor as envisaged by a Sandia Lab design group are given in Table 9.7 taken from Reference 27.

9.6.3. Some Remarks about ICF

There are many variations possible in inertial-confinement fusion, such as different types of drivers and different pellet and reactor cavity designs. All of them have the advantage over magnetic fusion concepts that there is no need for complex magnet systems. This advantage is, however, offset by a complex beam-transport and delivery system.

A serious problem associated with inertial confinement involves the fact that the design of pellets with the highest gain involve classified concepts that originated from nuclear weapons development. It seems unrealistic and perhaps unwise to expect that classified pellets would be used in commercial applications such as electric power plants. Whether small pellets using unclassified designs can succeed in obtaining the needed gain remains to be seen. Certainly the technology of mass production of precision pellets, the development of efficient drivers, and the design of blast-resistant reactor chambers presents a very high challenge to the engineering professions.

9.7. CONCLUDING REMARKS

The international research effort on plasma physics and magnetic confinement made during the past 30 yr has produced a very significant increase in the understanding of fusion physics. There are now one or two magnetic-confinement concepts that are well enough understood that they could be developed into fusion power sources.

However, fusion must be more than a scientific feat. To be successful, it must eventually evolve into commercial products that find acceptance in the energy marketplace. This implies that fusion reactors produce energy at an economically competitive price. Most fusion scientists believe scientific success will be demonstrated within a few years. It is the great challange to fusion scientists and engineers to transform this great scientific accomplishment into a reliable, economical energy resource.

REFERENCES

1. C. C. Baker, G. A. Carlson, and R. A. Krakowski, "Trends and Developments in Magnetic Confinement Fusion Reactor Concepts," *Nucl. Tech./Fusion* **1**, 5 (1981).
2. M. J. Monsler, J. Hovingh, D. L. Cook, T. G. Frank, and G. A. Moses, "An Overview of Inertial Fusion Reactor Design," *Nucl. Tech./Fusion* **1**, 302 (1981).
3. *Nuclear Engineering and Design*, Special Issue, R. A. Conn (ed.), **63** (1981). North Holland Publishing, Amsterdam, Netherlands.
4. *Fusion*, E. Teller (ed.), Academic Press, New York, 1981, Vol. I, Parts A and B.
5. K. Miyamoto, "Recent Stellarator Research," *Nuclear Fusion* **18**, 243 (1978).
6. V. D. Shafranov, "Stellarators," *Nuclear Fusion* **20**, 1075 (1980).
7. *Stellarators, Status and Future Directions*, Joint U.S.–Euratom Report, IPP-2/254 (DE 81026572), Max Planck Institute für Plasmaphysik, Garching, Germany, July 1981.
8. B. Badger, et al, *UWTOR-M, A Conceptual Modular Stellarator Power Reactor*, University of Wisconsin, UWFDM-550, October 1982.
9. R. A. Dandl et al., "Electron Cyclotron Heated 'Target' Plasma Experiments," *IAEA 3rd International Conference on Plasma Physics and Controlled Nuclear Fusion Research*, Novosibirsk, Vol. II, p. 435, 1968.
10. M. Fujiwara et al., *Plasma Confinement in ECH Bumpy Torus*, IPPJ-522, June 1981, Nagoya, University of Japan.
11. N. A. Uckan, "Overview of EBT Reactor Projections," *Nucl. Eng. Design* **63**, 271 (1981).
12. C. G. Bathke, et al., *ELMO Bumpy Torus Reactor and Power Plant*, Los Alamos National Laboratory Report LA-8882, August 1981.
13. D. G. McAlees et al., *The ELMO Bumpy Torus Reactor (EBTR) Reference Design*, ORNL/TM-5669, Nov. 1976.
14. H. A. B. Bodin and A. A. Newton, "Reversed-Field-Pinch Research," *Nuclear Fusion* **20**, 1255 (1980).
15. R. Hancox, R. A. Krakowski, and W. R. Spears, "The Reversed Field Pinch Reactor," *Nucl. Eng. Design* **63**, 251 (1981).
16. R. A. Gross, "Survey of Reactor Aspects of Compact Fusion Concepts," *Nucl. Tech. Fusion* **4**, 305 (1983).

17.   R. L. Hagenson and R. A. Krakowski, *Compact Reversed-Field Pinch Reactors (CRFPR). Sensitivity Study and Design Point Determination*, LA-9389-MS, Los Alamos National Laboratory, July 1982.

18.   R. F. Bourque, OHTE *Reactor Concepts*. GA-A-16458. General Atomic Co. Sept. 1981; also paper 67−25 presented at the *9th Symposium on Engineering Problems of Fusion Research*, Chicago, Ill., October 26−29, 1981.

19.   R. A. Kradowski et al., "Reactor System Studies of Alternate Fusion Concepts," *Proceedings of the 8th International Conference on Plasma Physics and Controlled Nuclear Fusion Research*, Brussels, IAEA-CN-38/V-4, July 1980.

20.   J. Nuckolls, J. Emmet, and L. Wood, "Laser Induced Thermonuclear Fusion," *Physics Today*, 46−53 (Aug. 1953).

21.   H. Motz, *The Physics of Laser Fusion*. Academic Press, New York, 1979.

22.   R. E. Kidder, "Laser Compression of Matter: Optical Power and Energy Requirements," *Nuclear Fusion* **14**, 797 (1974).

23.   G. S. Fraley, E. J. Linnebur, R. J. Mason, and R. L. Morse. "Thermonuclear Burn Characteristics of Compressed DT Microspheres," *Phys. Fluids* **17**, 474 (1974).

24.   R. E. Kidder, "Laser-Driven Compression of Hollow Shells: Power Requirements and Stability Limitations," *Nuclear Fusion* **16**, 1 (1976).

25.   J. Maniscalco, J. Blink, R. Buntzen, J. Hovingh, W. Meirer, M. Monsler and P. Walker, *Civilian Applications of Laser Fusion*, UCRL-52349, Nov. 1977.

26.   S. G. Varnado, J. L. Mitchiner, and G. Yonas, *Civilian Applications of Particle-Beam-Initiated Inertial Confinement Fusion Technology*, Sand. 77-0516, May 1977.

27.   M. J. Monsler, J. Hovingh, D. L. Cook, T. G. Frank, and G. A. Moses, "An Overview of Inertial Fusion Reactor Design," *Nucl. Tech./Fusion* **1**, 302 (1981).

28.   T. G. Frank and C. E. Rossi, "Technology Requirements for Commerical Applications of Inertail Confinement Fusion," *Nucl. Tech./Fusion* **1**, 359 (1981)

29.   W. R. Meier and J. A. Maniscalco, *Reactor Concepts for Laser Fusion*, UCRL-79694 (1977).

30.   R. W. Conn et al., *SOLASE, A Conceptual Laser Fusion Reactor Design*, University of Wisconsin, UWFDM-220, December 1977.

31.   T. H. Martin, G. W. Barr, J. P. Van Devender, R. A. White, and D. L. Johnson, "Pulsed Power Accelerators for Particle Beam Fusion," *Conference Record of the 14th Pulse Power Modulator Symposium*, Orlando, Florida, June 1980.

APPENDIX A

UNITS AND CONVERSION FACTORS

The units employed in this book are the International System of Units (SI), which is the metric system adopted as the common language of the world for expressing scientific and technical data. The SI system is described in the U.S. National Bureau of Standards Publication 330, 1970. The seven base units are the meter (m) for length, kilogram (kg) for mass, second (s) for time, kelvin (K) for thermodynamic temperature, ampere (A) for electric current, mole (mol) for amount of substance, and candela (cd) for luminous intensity. Two supplementary units are the radian (rad) for plane angle and steradian (sr) for solid angle. Many derived units are defined from the base and supplementary units. These include, for example, newton (force), joule (energy), pascal (pressure), hertz (frequency), watt (power), coulomb (electric charge), weber (magnetic flux), tesla (magnetic flux density), etc. Some useful conversion factors between SI and other units are shown in Table A.1.

Energy

$1 \text{ J} = 10^7 \text{ ergs} = 0.74 \text{ ft-lb} = 0.2388 \text{ cal} = 9.488 \times 10^{-4} \text{ Btu}$
$1 \text{ Btu} = 1055 \text{ J} = 778 \text{ ft-lb} = 252 \text{ cal} = 1.054 \times 10^{10} \text{ ergs}$
$1 \text{ cal} = 4.184 \text{ J}$
$1 \text{ food cal} = 1 \text{ kcal} = 4184 \text{ J}$
$1 \text{ eV} = 1.60219 \times 10^{-19} \text{ J} = 1.60219 \times 10^{-12} \text{ erg}$
$1 \text{ amu} = 1.66 \times 10^{-24} \text{ g} = 1.9 \times 10^{-10} \text{ J} = 931 \text{ MeV}$
$1 \text{ ft-lb} = 1.35 \text{ J}$
$1 \text{ kWh} = 3.596 \times 10^6 \text{ J} = 3412 \text{ Btu}$
$1 \text{ MeV} = 1.60 \times 10^{-13} \text{ J}$
$1 \text{ W-yr} = 3.154 \times 10^7 \text{ J}$
$1 \text{ ft-lb} = 1.3549 \text{ J}$
$1 \text{ erg} = 1 \times 10^7 \text{ J}$
$1 \text{ eV} = 11,605 \text{ K}$
$1 \text{ quad} = 1 \times 10^{15} \text{ Btu} = 1.054 \times 10^{18} \text{ J}$

$1 \text{ ton of coal (2000 lb)} \approx 2.9 \times 10^{10} \text{ J}$
$1 \text{ barrel crude oil} \approx 6.3 \times 10^9 \text{ J}$
$1 \text{ cu ft nat. gas} \approx 1.0 \times 10^6 \text{ J}$
$1 \text{ g}^{235} \text{ U} \approx 8.3 \times 10^{10} \text{ J}$
$1 \text{ tonne of TNT (1000 kg)} \approx 4.3 \times 10^9 \text{ J}$

Power

$1 \text{ w} = \text{J/s}$
$1 \text{ hp} = 746 \text{ W} = 2545 \text{ Btu/h}$
$1 \text{ kW} = 1.34 \text{ hp}$
$1 \text{ Btu/s} = 1055 \text{ W}$
$1 \text{ cal/s} = 4.187 \text{ W}$

Pressure

$1 \text{ kPa} = 0.1450 \text{ psi} = 0.7501 \text{ torr}$
$1 \text{ MPa} = 10 \text{ bar} = 9.869 \text{ atm}$
$1 \text{ atm} = 1.01325 \times 10^5 \text{ Pa}$
$1 \text{ torr} = 1.333 \times 10^2 \text{ Pa}$
$1 \text{ lb/in.}^2 = 6.895 \times 10^3 \text{ Pa}$
$1 \text{ bar} = 1.00 \times 10^5 \text{ Pa}$
$1 \text{ in. Hg} = 3.386 \times 10^3 \text{ Pa}$

Radiation Units

$1 \text{ Bq} = 1 \text{ disintegrations/s} = 2.703 \times 10^{-11} \text{ Ci}$
$1 \text{ Gy} = 1 \text{ gray} \equiv 1 \text{ J/kg} = 2.388 \times 10^{-4} \text{ cal/g} = 10^2 \text{ rad}$
$1 \text{ Ci} = 3.70 \times 10^{10} \text{ Bq}$
$1 \text{ R} = 2.58 \times 10^{-4} \text{ c/kg}$
$1 \text{ rad} = 1 \times 10^{-2} \text{ Gy}$

TABLE A.1 *(Continued)*

Mass

1 kg = 2.205 lbm
1 Mg = 1 tonne = 1.102 tons (2000 lbm)
1 u = unified atomic mass unit = $\frac{1}{12}$ mass ^{12}C = 1.66057 × 10^{-27} kg
1 tonne = metric ton ≡ 1 × 10^3 kg
1 lbm = 4.536 × 10^{-1} kg
1 ton (long) ≡ 2240 lbm = 1.016 × 10^3 kg
1 ton (short) ≡ 2000 lbm = 9.072 × 10^2 kg

Force

1 N = 0.2248 lbf

Time

1 yr = 3.1536 × 10^7 s = 5.256 × 10^5 min = 8760 h

Density

1 kg/m^3 = 6.243 × 10^{-2} lbm/ft^3

Length

1 m = 3.281 ft = 39.37 in.
1 nm = 10 Å (1 Å = 10^{-8} cm)
1 km = 0.6214 miles
1 ft = 0.3048 m
1 statute mile = 1.609 × 10^3 m
1 nautical mile = 1.852 × 10^3 m
1 mil = 10^{-3} in. = 2.54 × 10^{-5} m

Area

1 hectare = 1.00 × 10^4 m^2
1 barn = 1.00 × 10^{-28} m^2
1 sq mi. = 2.590 km^2
1 acre = 4.047 × 10^3 m^2
1 sq yd = 8.361 × 10^{-1} m^2
1 sq ft = 9.290 × 10^{-2} m^2
1 sq in. = 6.452 × 10^2 mm^2

Volume

1 liter = 1 × 10^{-3} m^3
1 cu yd = 7.646 × 10^{-1} m^3
1 cu ft = 2.832 × 10^{-2} m^3
1 cu inch = 1.639 × 10^4 mm^3
1 gallon (imperial) = 4.546 × 10^{-3} m^3
1 gallon(U.S. liq.) = 3.785 × 10^{-3} m^3
1 barrel = 42 U.S. gal. = 35 British gal. = 1.585 × 10^2 liters
        = 0.1591 m^3

(continued)

Some useful physical quantities

Charge on electron $e = 1.6019 \times 10^{-19}$ C
Boltzmann constant $k = 1.38047 \times 10^{-23}$ J/K
Electron mass $m_e = 9.1096 \times 10^{-31}$ kg $= 0.5110$ MeV
Proton mass $m_p = 1.6726 \times 10^{-27}$ kg $= 938.26$ MeV
Planck's constant $h = 6.625 \times 10^{-34}$ J-s
Speed of light $c = 2.99793 \times 10^8$ m/s
1 amu $= 1.6605 \times 10^{-27}$ kg $= 931.48$ MeV
Neutron mass $m_n = 1.6749 \times 10^{-27}$ kg $= 939.553$ MeV
Particle mass $m_i = 1.67 \times 10^{-27} A_i$ kg

Prefixes that can be combined with the names of SI units to indicate multiples are

$$
\begin{aligned}
\text{E (exa)} &= 10^{18} \\
\text{P (peta)} &= 10^{15} \\
\text{T (tera)} &= 10^{12} \\
\text{G (giga)} &= 10^{9} \\
\text{M (mega)} &= 10^{6} \\
\text{k (kilo)} &= 10^{3} \\
\text{m (milli)} &= 10^{-3} \\
\mu \text{ (micro)} &= 10^{-6} \\
\text{n (nano)} &= 10^{-9} \\
\text{p (pico)} &= 10^{-12} \\
\text{f (femto)} &= 10^{-15} \\
\text{a (atto)} &= 10^{-18}
\end{aligned}
$$

The use of c (centi) $= 10^{-2}$ is used sometimes.

APPENDIX

B | THE LIGHT ELEMENT ISOTOPES

Isotope	Natural Abundance (%)	Atomic Mass (amu)	Half−Life
$_0^1 n$	—	1.0086652	12 m
$_1 H$	—	1.00797	
$^1 H$	99.985	1.007825	
$^2 H$ or $^2 D$	0.015	2.014102	
$^3 H$ or $^3 T$	—	3.016046	12.36 yr
$_2 He$	—	4.0026	
$^3 He$	0.00013	3.01603	
$^4 He$	100.	4.00260	
$^5 He$	—	5.0123	
$^6 He$	—	6.01888	0.81 s
$^8 He$	—	8.0375	0.1225 s
$_3 Li$	—	6.939	
$^5 Li$	—	5.0125	
$^6 Li$	7.42	6.01512	
$^7 Li$	92.58	7.01600	
$^8 Li$			0.855 s
$^9 Li$			

(continued)

Isotope	Natural Abundance (%)	Atomic Mass (amu)	Half−Life
$_4$Be	—	9.0122	
^{6}Be	—	6.0197	
^{7}Be	—	7.0169	53.37 d
^{8}Be	—	8.0053	2×10^{-16} s
^{9}Be	100	9.01218	
^{10}Be	—	10.0135	2.5×10^6 yr
^{11}Be	—	11.0216	13.6 s
^{12}Be	—	—	0.01 s
$_5$B	—	10.811	
^{8}B	—	8.0246	0.77 s
^{9}B	—	9.0133	8×10^{-19} s
^{10}B	19.78	10.0129	
^{11}B	80.22	11.00931	
^{12}B	—	12.0143	0.02 s
^{13}B	—	13.0178	0.019 s
$_6$C	—	12.01115	
^{10}C	—	—	19.45 s
^{11}C	—	—	20.3 m
^{12}C	98.89	12.00000	
^{13}C	1.11	13.00335	
^{14}C			5.730 yr
^{15}C			2.4 s
^{16}C			0.74 s
$_7$N	—	14.0067	
^{12}N	—	—	0.011 s
^{13}N	—	—	10.1 min
^{14}N	99.63	14.00307	
^{15}N	0.37	15.00011	
^{16}N	—	—	7.2 s
^{17}N	—	—	4.16 s
^{18}N	—	—	0.63 s
$_8$O	—	15.9994	
^{13}O	—	—	0.0087 s
^{14}O	—	—	71.0 s
^{15}O	—	—	124 s
^{16}O	99.759	15.99491	

Isotope	Natural Abundance (%)	Atomic Mass (amu)	Half−Life
^{17}O	0.037	—	
^{18}O	0.204	—	
^{19}O	—	—	29 s
^{20}O	—	—	14 s

Data taken from *Handbook of Chemistry & Physics*, 55th ed. Robert C. Weast (ed.), CRC Press, Cleveland, OH, (1974–75); and from *Table of Isotopes*, 7th ed. C. Michael Lederer and V.S. Shirley (eds.), Wiley, New York, 1978.

Column 1 lists the isotopes with the atomic number Z as subscript and atomic mass number A as superscript. The first listing of each species (without mass number) lists the average mass of the naturally occurring isotopes.

Column 2 lists the natural abundance in percent. Only the stable isotopes have values in this column.

Column 3 lists the atomic mass (sometimes called the atomic weight). All masses are relative to carbon 12, which is, by international agreement, arbitrarily assigned 12.000000. All data are for neutral atoms.

Column 4 lists the half-life of the radioactive isotopes.

1 amu = 1.660531×10^{-27} kg = 931.4812 MeV.
Electron mass: $m_e = 9.1091 \times 10^{-31}$ kg = 5.4857×10^{-4} amu = 0.51100 MeV.
Proton mass: $m_p = 1.67252 \times 10^{-27}$ kg = 1.007220 amu = 938.259 MeV.
Neutron mass: $m_n = 1.67482 \times 10^{-27}$ kg = 1.008605 amu = 939.5527 MeV.
For many plasma physics purposes the mass of a particle can be approximated by $m = 1.67 \times 10^{-27}A$ kg.

APPENDIX
C

SOME FUSION REACTIONS

The following nuclear fusion reactions have been compiled primarily from References 1–8 of Chapter 2. Note that some of the particles are meta stable but with half–lives $\geq$ seconds.

Number	Reaction	Q (MeV)
1	$^1\text{H} + {}^1\text{H} \rightarrow {}^2\text{D} + e^+$	1.4
2	$^1\text{H} + {}^2\text{D} \rightarrow {}^3\text{He}$	5.5
3	$^1\text{H} + {}^3\text{T} \rightarrow {}^4\text{He}$	19.6
4	$^1\text{H} + {}^3\text{T} \rightarrow {}^3\text{He} + {}^1n$	−0.8
5	$^1\text{H} + {}^6\text{Li} \rightarrow {}^3\text{He} + {}^4\text{He}$	4.0
6	$^1\text{H} + {}^7\text{Li} \rightarrow 2{}^4\text{He}$	17.5
7	$^1\text{H} + {}^9\text{Be} \rightarrow {}^4\text{He} + {}^6\text{Li}$	2.2
8	$^1\text{H} + {}^9\text{Be} \rightarrow {}^2\text{D} + 2{}^4\text{He}$	0.7
9	$^1\text{H} + {}^{10}\text{B} \rightarrow {}^{11}\text{C}$	8.6
10	$^1\text{H} + {}^{11}\text{B} \rightarrow 3{}^4\text{He}$	8.7
11	$4{}^1\text{H} \rightarrow {}^4\text{He} + 2e^{+a}$	26.7
12	$^2\text{D} + {}^2\text{D} \rightarrow {}^3\text{He} + {}^1_0n$	3.27
13	$^2\text{D} + {}^2\text{D} \rightarrow {}^3\text{T} + {}^1\text{H}$	4.03
14	$^2\text{D} + \text{T} \rightarrow {}^4\text{He} + {}^1_0n$	17.59
15	$^2\text{D} + {}^3\text{He} \rightarrow {}^1\text{H} + {}^4\text{He}$	18.3
16	$^2\text{D} + {}^4\text{He} \rightarrow {}^6\text{Li}$	1.7
17	$^2\text{D} + {}^6\text{Li} \rightarrow {}^3\text{He} + {}^4\text{He} + {}^1_0n$	2.56

Number	Reaction	Q (MeV)
18	$^2D + {}^6Li \rightarrow {}^3He + {}^4He + {}_0^1n$	1.8
19	$^2D + {}^6Li \rightarrow 2{}^4He$	22.4
20	$^2D + {}^6Li \rightarrow {}^1H + {}^7Li$	5.0
21	$^2D + {}^6Li \rightarrow {}^7Be + {}_0^1n$	3.4
22	$^2D + {}^7Li \rightarrow {}^8Be + {}_0^1n$	15.0
23	$^2D + {}^7Li \rightarrow 2{}^4He + {}_0^1n$	15.1
24	$^2D + {}^7Be \rightarrow {}^1H + 2{}^4He$	16.5
25	$^3T + {}^3T \rightarrow {}^4He + 2{}_0^1n$	11.3
26	$^3T + {}^3He \rightarrow {}^2D + {}^4He$	14.3
27	$^3T + {}^3He \rightarrow {}^5Li + {}_0^1n \rightarrow 2{}^1H + {}^4He$	12.1
28	$^3T + {}^3He \rightarrow {}^5He + {}^1H \rightarrow {}^4He + 2{}_0^1n$	12.1
29	$^3T + {}^6Li \rightarrow {}^7Li + D$	0.9
30	$^3T + {}^6Li \rightarrow {}^7Li + {}^1H + {}_0^1n$	−1.2
31	$^3T + {}^6Li \rightarrow 2{}^4He + {}_0^1n$	15.8
32	$^3T + {}^3He \rightarrow {}^4He + {}^1H + {}_0^1n$	12.1
33	$^3T + {}^7Li \rightarrow 2{}^4He + 2{}_0^1n$	8.9
34	$^3T + {}^7Be \rightarrow 2{}^4He + {}^1H + {}_0^1n$	10.5
35	$^3He + {}^3He \rightarrow {}^4He + 2{}^1H$	12.9
36	$^3He + {}^4He \rightarrow {}^7Be$	1.5
37	$^3He + {}^6Li \rightarrow {}^1H + 2{}^4He$	16.8
38	$^3He + {}^6Li \rightarrow {}^7Be + {}^2D$	0.1
39	$^3He + {}^7Li \rightarrow {}^1H + 2{}^4He + {}_0^1n$	9.6
40	$^3He + {}^7Be \rightarrow 2{}^1H + 2{}^4He$	11.3
41	$^3He + {}^7Be \rightarrow {}^{10}C$	15.1
42	$^3He + {}^9Be \rightarrow 3{}^4He$	18.7
43	$^4He + {}^7Li \rightarrow {}^{11}B$	8.5
44	$^4He + {}^7Be \rightarrow {}^{11}C$	7.5
45	$^4He + {}^9Be \rightarrow {}^{12}C + {}_0^1n$	5.7
46	$^4He + {}^9Be \rightarrow 3{}^4He + {}_0^1n$	−1.6
47	$^4He + {}^{11}B \rightarrow {}^{14}C + {}^1H$	0.8
48	$^4He + {}^{11}B \rightarrow {}^{14}N + {}_0^1n$	0.2
49	$^4He + {}^{12}C \rightarrow {}^{16}O$	7.3
50	$^6Li + {}^6Li \rightarrow 3{}^4He$	20.5
51	$^6Li + {}^7Be \rightarrow 3{}^4He + {}^1H$	15.0

[a]The reaction $4{}^1H \rightarrow {}^4He + 2e^+$ is the overall reaction of the carbon–nitrogen catalyzed cycle in the sun (see Reference 1 of Chapter 2).

INDEX

320   INDEX